Vectors in three-dimensional space

TO MY DEAREST MONTY

# *Vectors in three-dimensional space*

J. S. R. CHISHOLM

M.A., Ph.D. (Cantab.), M.R.I.A., F.I.M.A.
Professor of Applied Mathematics, University of Kent at Canterbury

CAMBRIDGE UNIVERSITY PRESS
CAMBRIDGE
LONDON · NEW YORK · MELBOURNE

CAMBRIDGE UNIVERSITY PRESS
Cambridge, New York, Melbourne, Madrid, Cape Town, Singapore, São Paulo, Delhi

Cambridge University Press
The Edinburgh Building, Cambridge CB2 8RU, UK

Published in the United States of America by Cambridge University Press, New York

www.cambridge.org
Information on this title: www.cambridge.org/9780521218320



First published 1978
Re-issued in this digitally printed version 2008

*A catalogue record for this publication is available from the British Library*

*Library of Congress Cataloguing in Publication data*

Chisholm, John Stephen Roy.
Vectors in three-dimensional space.

Includes bibliographical references and index.
1. Vector algebra. 2. Vector analysis. I. Title.
QA200.C47 515'.63 77-82492

ISBN 978-0-521-21832-0 hardback
ISBN 978-0-521-29289-4 paperback

# *CONTENTS*

# PREFACE

It is a fundamental fact of nature that the space we live in is three-dimensional. Consequently, many branches of applied mathematics and theoretical physics are concerned with physical quantities defined in 3-space, as I shall call it; these subjects include Newtonian mechanics, fluid mechanics, theories of elasticity and plasticity, non-relativistic quantum mechanics, and many parts of solid state physics. The Greek geometers made the first systematic investigation of the properties of 'ordinary' 3-space, and their work is known to us mainly through the books of Euclid; our basic geometrical ideas about the physical world have their origins in Euclidean geometry. A major advantage of Euclid's work was its presentation as a deductive system derived from a small number of definitions and axioms (or 'basic assumptions'); although Euclid's axioms have turned out to be inadequate in a number of ways, he nevertheless provided us with a model of what a proper mathematical system should be [Reference P.1].

Through the introduction of coordinate systems, Descartes linked geometry with algebra [Reference P.2]; geometrical structures in 3-space such as lines, planes, circles, ellipses and spheres, were associated with algebraic equations involving three Cartesian coordinates $(x, y, z)$. Then in the nineteenth century, Hamilton [Reference P.3] and Gibbs [Reference P.4] introduced two similar types of algebraic objects, 'quaternions' and 'vectors', which treated the three coordinates simultaneously; the rules of operation of these new sets of objects were different from those of real or complex numbers, giving rise to new types of 'algebra'; a more general algebra of $N$-dimensional space $(N = 3, 4, 5, \ldots)$ was introduced by Grassmann [Reference P.5]. Over several decades, the vector concept developed in two different ways: in a wide variety of physical applications, vector notation and techniques became, by the middle of this century, almost universal; on the other hand, pure mathematicians reduced vector algebra to an axiomatic system, and introduced wide generalisations of the concept of a three-dimensional 'vector space', not only to $N$-dimensional spaces, but also to Hilbert space and other infinite-dimensional metric spaces, and to topological spaces. These two developments proceeded largely independently, and many books dealing with the applications of vectors have approached the fundamentals of the subject intuitively rather than axiomatically, assuming some prior knowledge of Euclidean and Cartesian geometry. In recent decades, however, hard-and-fast distinctions between 'pure' and 'applied' mathematics have been disappearing; in particular, the concept of an abstract 'space', especially Hilbert space, has become familiar in many applications of mathematics, including quantum mechanics, numerical analysis and statistics, and in the study of differential and integral equations. Also, the concept of 'basic assumptions' or 'structure', in dealing with number systems, has taken its place in school mathematics [Reference P.6]; while these basic assumptions are not presented as a complete logical scheme in the way that Euclid intended, they nevertheless familiarise students with the concept of an axiomatic scheme. For these reasons, it seems appropriate to take account of both pure and applied mathematical points of view when treating the subject of 'vectors', which is now a fundamental part of both these modes of thought.

This book deals with vector algebra and analysis, and with their application to three-dimensional geometry and to the analysis of fields in 3-space. In order to bring out both the 'pure' and 'applied' aspects of the subject, my main objectives have been:

(i) to base the work on sound algebraic and analytic foundations;
(ii) to develop those intuitive relations between algebraic equations and geometrical concepts which are of fundamental importance in physical applications;
(iii) to establish standard vector techniques and theorems, giving numerous examples of their use.

In the first three chapters, the algebra of vectors is developed, based upon the axioms of vector space algebra; as the axioms are introduced, their geometrical interpretation is given, so that they can be understood intuitively. The axiomatic scheme is extended to provide a definition of Euclidean space, consisting of 'points' and 'displacements'; this provides an axiomatic basis for Euclidean geometry, linking it directly with the algebra of linear vector spaces. This linkage has the reciprocal advantage (not apparent in this book) that it enables geometrical intuitions to be developed in dealing with more general types of linear space, in particular with finite-dimensional spaces and Hilbert space. In the process of interpreting the algebraic axioms geometrically, algebraic definitions of elementary geometrical concepts such as 'length' and 'angle' have to be given and justified; we also define Cartesian or 'rectangular' coordinates and establish their fundamental properties, such as Pythagoras' theorem. By this means, Cartesian geometry and trigonometry, as well as Euclidean geometry, are seen to arise out of a single set of axioms. The first three chapters also develop the techniques of vector algebra, and apply them to problems in geometry, in particular the geometry of lines and planes.

The fourth chapter deals with transformations of the components of a vector in two or three dimensions, in particular with transformations representing rotations and reflections. A clear distinction is made between 'active transformations', due to a change of the vector itself, and 'passive transformations', due to change of the frame of reference. The idea of groups of transformations is introduced, and the study of rotations in two dimensions is linked with the intuitively familiar concept of 'addition of angles'. Transformations in 3-space are represented by $3 \times 3$ matrices. Although it has been assumed that the reader has some familiarity with matrices, the necessary theory of $3 \times 3$ matrices and their determinants has been developed in the first two sections of Chapter 4, using the properties of vectors established earlier. This emphasises the fact that vectors and matrix algebra are simply two different aspects of the algebra of vector spaces. It is of interest to note that this vectorial approach to matrix algebra can be made quite general, and is not restricted to $3 \times 3$ matrices.

The study of functions $f(x)$, where $x$ is a variable lying in a continuous range, depends to a great extent upon the differential and integral calculus. When we study functions defined in 3-space, it is necessary to develop an extension of calculus appropriate to regions of this space. There are several difficult problems to solve before this extended calculus can be defined. First, we have to study how points in 3-space are specified by systems of coordinates; second, we have to give definitions of curves, surfaces and volume regions in 3-space; third, if we consider a specific surface or volume region, we need to define the 'boundary' of that region. These problems are dealt with in Chapter 5. Since points in 3-space are described by three coordinates, this work necessarily involves using analytic properties of functions of up to three variables, and of their derivatives and integrals. This raises a problem of presentation: establishing the necessary analytic properties of functions of one, two and three variables requires a substantial amount of work, whose incorporation in Chapter 5 would break the continuity of ideas developed there. Elementary analysis of functions of one variable is normally dealt with early in university mathematics courses, and is the subject of a large number of textbooks; so when I use an analytic property of one-variable functions, I simply quote a reference in one of the most readable elementary books on the subject, J. C. Burkill's *A First Course in Mathematical Analysis*. The analysis of functions of several variables is appreciably more complicated, and it is arguable that in an elementary textbook, we should not trouble about proofs of properties of partial derivatives and multiple integrals. In a book for students of mathematics, however, it is unsatisfactory to omit explanations simply because they are complicated. I have met this difficulty by establishing the essential properties of functions of *two* variables in Appendix A, to which reference is made when these properties are used in the main text; the necessary properties of functions of *three* variables are simple generalisations of those of two variables, and when they are used, I again refer to the analogous property of two-variable functions. A reader can therefore either accept the analytic properties assumed in the main text, or refer to Appendix A for a justification of these assumptions. By omitting this analytic detail from Chapter 5, it is possible to give a fairly detailed account of surfaces, volume regions, and especially of curves.

Scalars and vectors whose value depends upon their position in space are called scalar and vector 'fields', provided that they satisfy suitable analytic conditions. Since these fields in general depend upon three coordinates, variations in a field throughout 3-space depend upon the derivatives of the field with respect to three coordinates; certain combinations of derivatives, 'divergence', 'gradient' and 'curl', known as vector operators, are closely associated with physical concepts such as flux and vorticity. In the final chapter, these operators are defined and studied, and their physical significance is emphasised. As in Chapter 5, it is necessary to be careful over analytic details; the vector operators are defined in a mathematically sound way, but in the discussion of their physical significance, I have thought it best to omit some analytic details. One theorem (Stokes' theorem) is difficult to prove in full generality: its significance is brought out by proving it under special conditions in the main text; the general proof is given in Appendix B. The discussion of physical examples leads naturally to the introduction of the 'Laplacian' operator; this completes the definition and discussion of the principal differential operators used in a variety of branches of mathematical physics, and provides a natural point at which to end.

The first three chapters of this book arose out of a course of lectures given to first-year mathematics students at the University of Kent. Although the book is written primarily for students of mathematics in the early part of their University course, those interested in the more mathematical aspects of physics and engineering may prefer this treatment of vectors based on linear space algebra, since linear spaces have a rapidly widening relevance in these disciplines. A number of my former students have chosen to follow this approach in sixth-form mathematics teaching, and those studying advanced school mathematics may find that the first four chapters of the book provide a coherent picture of a number of sixth-form topics which are often treated separately.

While writing this book, I have had many helpful discussions with other members of staff of the School of Mathematics in the University of Kent. I am particularly indebted to Dr R. Hughes Jones for many exchanges of ideas, not only while the book was being written, but also when I was formulating the approach to Chapters 1–3. I am very grateful to Mrs Sandra Bateman and Miss Diane Mayes for their careful preparation of the manuscript, and for their patience in coping with a long series of additions and amendments. The Cambridge University Press have been most helpful and thorough in checking and tidying up the manuscript; I wish to thank them for their help, and also Miss Ruth Farwell for checking the examples and problems. I have been pleased to have the student's-eye comments of my daughter Carol, and I very much appreciate the interest that my whole family have shown in the book, despite the nuisance value of books and papers strewn all over the house.

Roy Chisholm

Mathematical Institute
University of Kent
February 1977

# 1

# Linear spaces and displacements

## 1.1 *Introduction*

Our understanding of the physical world depends to a great extent on making more or less exact measurements of a variety of physical quantities. All single measurements on a physical system consist of observing a single real number, and very often this single real number is, by itself, the value of an important physical quantity; examples are the measurement of a mass, a length, an interval of time, an electrical potential, the frequency or wavelength of an electromagnetic wave, a quantity of electrical charge, and the electric current in a wire. Physical quantities of this kind are called **scalar quantities**, or, more frequently, **scalars**. We shall make a distinction between these two expressions: 'scalar' will be used as a *mathematical* expression; scalars, for our purposes, are real algebraic variables $\lambda$, $\mu$, . . . , which can, in general, take values in the whole range $(-\infty, \infty)$; they possess other properties which will be defined in Chapter 4, but for the present we shall regard them simply as real numbers. The expression 'scalar quantity' will refer to any specific physically measurable quantity, such as a mass or a charge, which is found experimentally to have the mathematical properties of a scalar. One important property of scalar quantities is that they are intrinsic properties of a physical system, and do not change if the whole physical system is translated to a different position in three-dimensional space, or is rotated in space. For example, if a metallic conductor is at a certain potential in an electric field produced by certain electric charges, this potential is unchanged if the conductor *and* the charges are translated or rotated as a whole, their relative positions remaining unchanged. Similarly, the mass of a body is independent of the position and orientation of the body in three-dimensional space.

Not all measured quantities are best understood as a single number. A change in position in three-dimensional space from a point $P$ to a point $Q$, known as a **displacement**, depends upon a number (the distance from $P$ to $Q$), but also depends upon the direction from $P$ to $Q$. There are various ways of defining a displacement; the most familiar is to define a set of Cartesian axes, with the origin at $P$, as in Fig. 1.1. Then the displacement $PQ$ is defined by giving the projections $(x, y, z)$ of the line $PQ$ on the three axes. Other examples of physical quantities with which we intuitively associate both a real number (the magnitude) and a direction in space are force, velocity, the electric, magnetic or gravitational field at a point in space, and the direction normal (that is, perpendicular) to a given plane in space. Physical quantities of this type are known as **vector quantities**; the corresponding abstract mathematical entities, whose properties we now start to define, are called **vectors.**

We shall define vectors by assuming that they obey certain basic algebraic equations, the **axioms** of vector algebra. From these axioms we shall be able to deduce the usual geometric properties of displacements in three-dimensional space; for example, we can show that the lengths $PQ$, $x$, $y$ and $z$ in Fig. 1.1 obey the Pythagorean relation

$$PQ^2 = x^2 + y^2 + z^2. \tag{1.1}$$

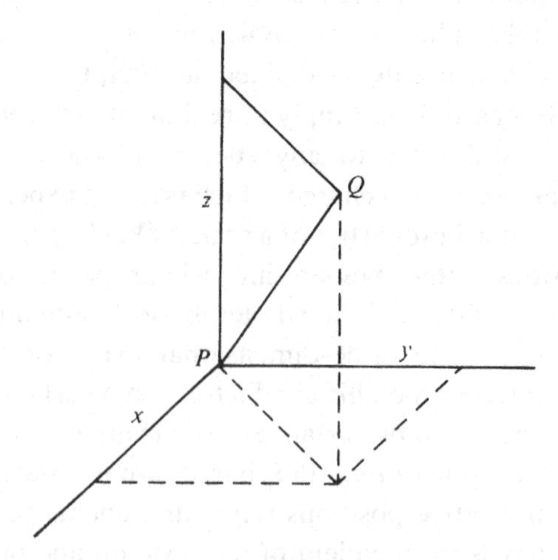


Fig. 1.1

The algebraic axioms are not necessarily associated with any geometrical interpretation; however, the interpretation of vectors as displacements is such a natural and familiar one that we inevitably think in geometrical terms when we discuss vectors; moreover, the geometric picture is a great aid to our intuition about vector quantities. So, on the one hand, we shall derive vector algebra from axioms written in algebraic form, and shall eventually deduce three-dimensional Euclidean geometry from these axioms; on the other hand, we shall, from the beginning, interpret the axioms and other equations intuitively in terms of three-dimensional geometry, with the vectors represented by displacements.

We denote vectors by symbols such as **a**, **b**, **r**, **n** and **u**. A set of vectors satisfying certain conditions is denoted by {**a**}, for example. Geometrically, a vector **a** is represented by a 'directed line' in space, as in Fig. 1.2. With any vector **a** we associate a unique non-negative

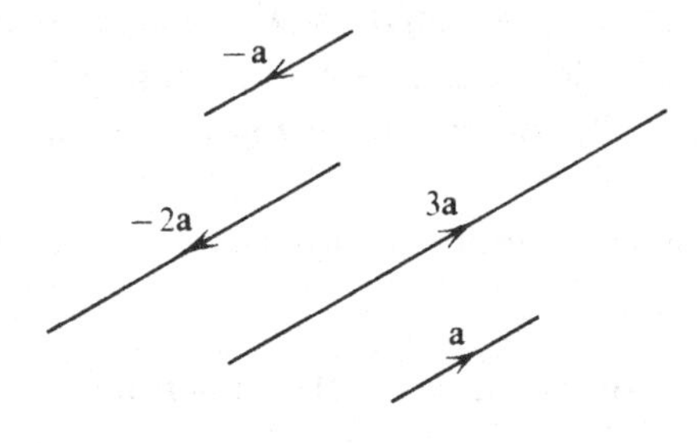


Fig. 1.2

real number $a$, called the **modulus** or **magnitude** of the vector. We frequently say that $a$ is the 'length' of the vector; in saying this, we are using the geometrical interpretation of a vector as a spatial displacement. Although we do not give a definition of modulus or length $a$ until Chapter 2, we shall use the concept in talking about the geometrical interpretation of vectors.

## *1.2 Scalar multiplication of vectors*

A vector **a** can be multiplied by any real number $\lambda$ to give another vector $\lambda\mathbf{a}$. If $\lambda > 0$ and if **a** represents a displacement, $\lambda\mathbf{a}$ is a displacement in the same direction as **a**, but with magnitude $\lambda a$; so $1\mathbf{a}$

is simply **a** itself. The displacement 3**a** is indicated in Fig. 1.2. We have drawn the displacements representing 3**a** and **a** in different positions. As we shall discuss fully in §1.4, a displacement has a definite 'initial point' in space, and the displacement is 'from' this point; because displacements have a definite position in space, they are often referred to as 'fixed vectors'. The abstract vectors **a** and 3**a**, however, have no initial points in space – it is, in fact, rather misleading to represent them by directed lines in a diagram. In order to remind ourselves that vectors are *not* associated with points in space, we represent them (as in Fig. 1.2) by directed lines at arbitrarily chosen points; abstract vectors are for this reason sometimes called 'free vectors'.

Multiplication of a vector **a** by $-1$ gives a vector denoted by $-\mathbf{a}$; this vector is represented by a displacement of the same length as **a**, and in exactly the opposite direction, as indicated in Fig. 1.2. When $\lambda < 0$, the vector $\lambda\mathbf{a}$ is again represented by a displacement in the opposite direction to **a**; its length is $|\lambda|a$, where $|\lambda|$ is the absolute value of $\lambda$. For example, a displacement representing the vector $-2\mathbf{a}$ is as shown in Fig. 1.2; note that the arrows on $-\mathbf{a}$ and $-2\mathbf{a}$ are in the opposite sense to those on **a** and 3**a**.

The formal axioms governing multiplication by finite real scalars $\lambda$ and $\mu$ are:

(1A) If **a** is a vector, and $\lambda$ any real number, then $\lambda\mathbf{a}$ is a vector,

(1B) $$1\mathbf{a} = \mathbf{a}, \tag{1.2}$$

(1C) $$\lambda(\mu\mathbf{a}) = (\lambda\mu)\mathbf{a}.$$

The Axiom (1B) tells us that multiplication by unity does not change a vector **a**. Since $\lambda\mu = \mu\lambda$ on the right of Axiom (1C), we can extend the axiom to give

$$\lambda(\mu\mathbf{a}) = \mu(\lambda\mathbf{a}) = (\mu\lambda)\mathbf{a}. \tag{1.3}$$

So Axiom (1C) tells us that the order of multiplication by two scalars ($\lambda$ and $\mu$) does not matter, since the result is equivalent to multiplication by $\lambda\mu$. In formal language, (1.3) tells us that scalar multiplication of vectors is **associative** and **commutative**. If $\lambda \neq 0$ and $\mu = \lambda^{-1}$, (1.2) and (1.3) give

$$\lambda^{-1}(\lambda\mathbf{a}) = 1\mathbf{a} = \mathbf{a}.$$

This means that **a** is a scalar multiple of all vectors $\lambda\mathbf{a}$ ($\lambda \neq 0$). Geometrically, displacements corresponding to **a** and $\lambda\mathbf{a}$ are said to be 'parallel'; this is explained more fully in §1.4. Although we have referred to the modulus of a vector in discussion, we note that this has not been defined by the Axioms (1A)–(1C).

When $\lambda = 0$, the Axiom (1A) implies the existence of a vector $0\mathbf{a}$. A displacement corresponding to $0\mathbf{a}$ is of zero length, and so is no displacement at all; so for all vectors $\mathbf{a}$ we write

$$0\mathbf{a} = \mathbf{0} \tag{1.4}$$

defining the **zero vector 0**. The essential point of equation (1.4) is that $\mathbf{0}$ is the same vector, whatever $\mathbf{a}$ is. We formalize this into the axiom:

(1D) There is a unique vector $\mathbf{0}$, called the zero vector, which satisfies

$$0\mathbf{a} = \mathbf{0},$$

for all vectors $\mathbf{a}$.

The uniqueness of the zero vector is an important property of three-dimensional space, and is also a property of many more complicated 'spaces' occurring in mathematics and mathematical physics. Equation (1.4) is 'intuitively obvious', but this is only because of our everyday experience of displacements; in formulating an abstract mathematical theory of vectors, the obvious needs to be stated explicitly.

## *1.3 Addition and subtraction of vectors*

The second set of axioms for vectors $\{\mathbf{a}\}$ define the laws of **addition of vectors**. They embody many familiar properties of displacements in space, and after stating the axioms, we shall discuss their geometric meaning. The operation of addition is denoted by the symbol '+'. It may appear confusing to use the same symbol for addition of numbers (scalars) and for addition of vectors; there are two reasons why confusion does not arise:

(i) the sum of two scalars $\lambda + \mu$ contains scalars ($\lambda$ and $\mu$), while the sum of two vectors $\mathbf{a} + \mathbf{b}$ contains vectors ($\mathbf{a}$ and $\mathbf{b}$);

(ii) the axioms of addition and scalar multiplication of vectors are very similar to axioms of addition and multiplication of scalars.

The axioms of vector addition are:

(2A) $\mathbf{a} + \mathbf{b}$ is a vector, for any two vectors $\mathbf{a}$, $\mathbf{b}$,

(2B) $$\mathbf{a} + \mathbf{b} = \mathbf{b} + \mathbf{a}, \tag{1.5}$$

(2C) $$\mathbf{a}+(\mathbf{b}+\mathbf{c})=(\mathbf{a}+\mathbf{b})+\mathbf{c}, \qquad (1.6)$$

(2D) $$(\lambda+\mu)\mathbf{a}=\lambda\mathbf{a}+\mu\mathbf{a}, \qquad (1.7)$$

(2E) $$\lambda(\mathbf{a}+\mathbf{b})=\lambda\mathbf{a}+\lambda\mathbf{b}. \qquad (1.8)$$

The addition of **a** and **b** to give the **vector sum a+b** is represented in Fig. 1.3. The two vectors are represented by displacements **PQ** and **PS** from the point *P*. The point *R* is chosen so that *PQRS* is a parallelogram; then **PR** represents the vector sum **a+b**. Geometrically, this rule of combination, known as the **parallelogram law**, is obviously symmetrical between **a** and **b**; this symmetry is built in to vector algebra in Axiom (2B); algebraically, this axiom is known as the **commutative law of addition** of vectors. It has been already pointed out that representing vectors by displacements can be misleading; this shows up in Fig. 1.3, where it is more natural to think

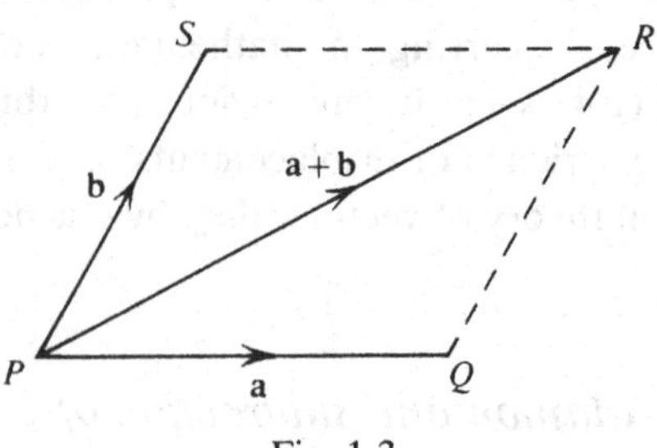


Fig. 1.3

of combining displacements **PQ** and **QR** to give the displacement **PR**. We shall see in §1.4, however, that this is not an accurate way of representing vector addition. A closer physical analogy to vector addition is the experimental law of combination of two forces acting at a point *P*: if they are represented by the vectors **a** and **b**, then they are equivalent to a force represented by **a+b**, also acting at *P*; Fig. 1.3 is then interpreted as the 'parallelogram of forces'.

Axiom (2C) is the **associative law of addition** of vectors; **a+(b+c)** is the vector formed by first adding **b** and **c** to give **(b+c)** and then adding this to **a**; this process is represented in Fig. 1.4, with **a+(b+c)** represented by **PT**. Likewise, **(a+b)+c** is represented by **PT** in Fig. 1.5. Axiom (2C) has the interpretation that the same displacement **PT** is defined by the two processes.

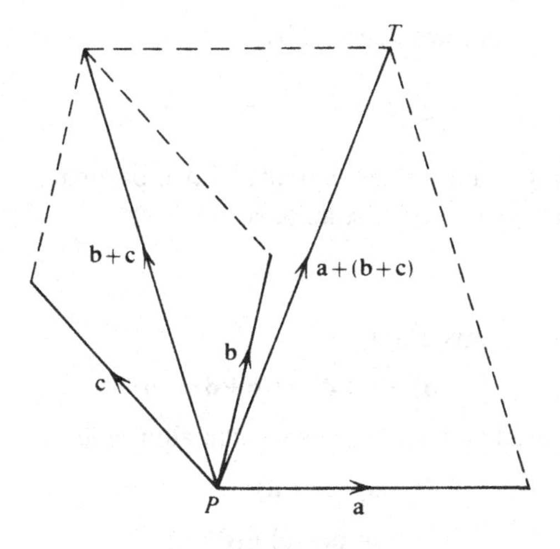


Fig. 1.4

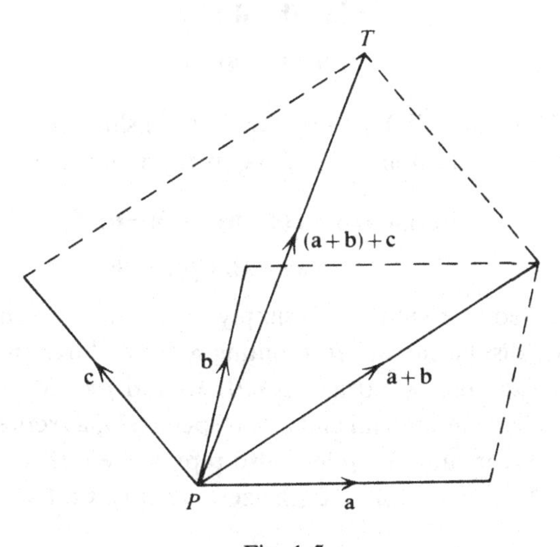


Fig. 1.5

Axioms (2B) and (2C) together tell us that all possible sums of three vectors **a**, **b** and **c**, for example $(\mathbf{a}+\mathbf{b})+\mathbf{c}$ and $\mathbf{b}+(\mathbf{a}+\mathbf{c})$, are equal; the sum can therefore be written as $\mathbf{a}+\mathbf{b}+\mathbf{c}$, without brackets. This property can be readily extended to sums of more then three vectors; for example, all possible sums of four vectors **a**, **b**, **c** and **d** are equal, and are denoted by $\mathbf{a}+\mathbf{b}+\mathbf{c}+\mathbf{d}$. In general, all sums of $n$

vectors $\mathbf{a}_1, \mathbf{a}_2, \ldots, \mathbf{a}_n$ are denoted by

$$\sum_{r=1}^{n} \mathbf{a}_r \equiv \mathbf{a}_1 + \mathbf{a}_2 + \ldots + \mathbf{a}_n.$$

In the following example, the equality of two particular sums of four vectors is established from the axioms.

**Example 1.1**

Using only the axioms, show that

$$(\mathbf{a}+\mathbf{b})+(\mathbf{c}+\mathbf{d}) = [(\mathbf{b}+\mathbf{d})+\mathbf{a}]+\mathbf{c}.$$

The axiom used at each stage is written alongside:

$$\begin{aligned}
&(\mathbf{a}+\mathbf{b})+(\mathbf{c}+\mathbf{d}) \\
&\quad = (\mathbf{a}+\mathbf{b})+(\mathbf{d}+\mathbf{c}) && \text{(2B)} \\
&\quad = [(\mathbf{a}+\mathbf{b})+\mathbf{d}]+\mathbf{c} && \text{(2C)} \\
&\quad = [\mathbf{a}+(\mathbf{b}+\mathbf{d})]+\mathbf{c} && \text{(2C)} \\
&\quad = [(\mathbf{b}+\mathbf{d})+\mathbf{a}]+\mathbf{c}. && \text{(2B)}
\end{aligned}$$

Axioms (2D) and (2E) define the relationship between scalar multiplication and addition; they allow us to 'multiply out' products so that

$$\begin{aligned}
(\lambda+\mu)(\mathbf{a}+\mathbf{b}) &= \lambda(\mathbf{a}+\mathbf{b})+\mu(\mathbf{a}+\mathbf{b}) \\
&= \lambda\mathbf{a}+\lambda\mathbf{b}+\mu\mathbf{a}+\mu\mathbf{b},
\end{aligned}$$

for example. Geometrically, (1.7) simply means that distances along the line of the displacement representing $\mathbf{a}$ are additive in the usual sense. If, for example, $\lambda > 0$ and $\mu < 0$, $\lambda\mathbf{a}$ and $\mu\mathbf{a}$ are in opposite directions and have moduli $\lambda a$ and $|\mu|a$ respectively, as represented in Fig. 1.6; the vector sum $(\lambda+\mu)\mathbf{a}$ is also represented. If we put $\lambda = 1$ and $\mu = 0$ in (1.7), so that $\mu\mathbf{a} = \mathbf{0}$ (the zero vector), we find

$$\mathbf{a} = \mathbf{a} + \mathbf{0}. \tag{1.9}$$

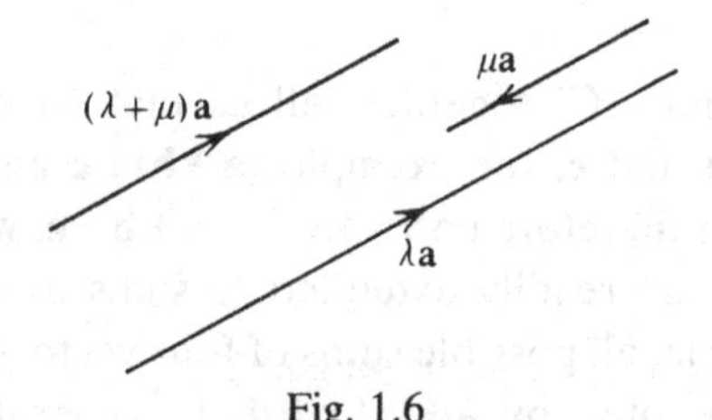


Fig. 1.6

So addition of the zero vector to any vector **a** leaves **a** unaltered, in accord with our intuitions about zero displacements.

The final axiom (2E) has a very important geometrical meaning, and can be called the 'similarity postulate'. In Fig. 1.3 we have already pictured the vectors **a**, **b** and **a** + **b** as displacements connected by the parallelogram law. Axiom (2E) means that if each of these displacements is multiplied by a number $\lambda$, the resulting three displacements representing $\lambda \mathbf{a}$, $\lambda \mathbf{b}$ and $\lambda(\mathbf{a}+\mathbf{b})$, still obey the parallelogram law. This property is represented in Fig. 1.7. The similarity

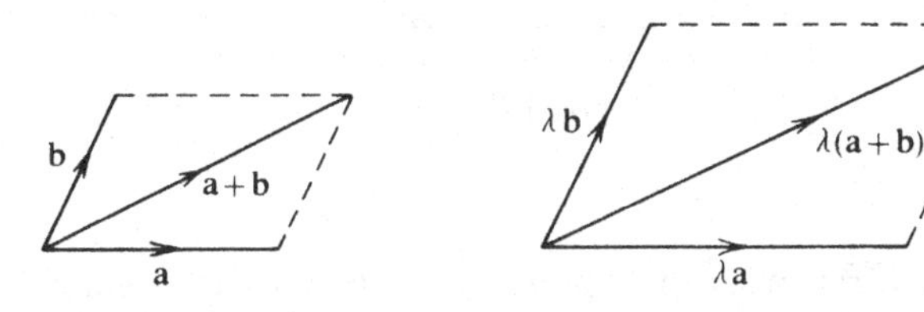


Fig. 1.7

postulate is characteristic of Euclidean space, and is one of the axioms closely associated with the 'flatness' of the space. If, for example, we tried to represent displacements along great circles on a sphere by vectors $\{\mathbf{a}\}$, it is clearly hard to satisfy the commutative law of addition (1.5). It is also clear that the similarity postulate (1.8) is going to fail: one cannot expand or contract a figure drawn on the surface of a sphere without changing some angles or the ratios of some of the lengths; preservation of angles and of the ratios of all lengths are the essential properties of 'similarity'.

We have defined the vector $-\mathbf{a} \equiv (-1)\mathbf{a}$. From (1.7),

$$\mathbf{a}+(-\mathbf{a})=(1-1)\mathbf{a}=\mathbf{0}, \tag{1.10}$$

in accord with our intuitive concept of the vector $-\mathbf{a}$, represented in Fig. 1.2. Equation (1.10) is often written

$$\mathbf{a}-\mathbf{a}=\mathbf{0}. \tag{1.10}$$

More generally, **subtraction of a vector b** from a vector **a** is defined by

$$\mathbf{a}-\mathbf{b} \equiv \mathbf{a}+(-\mathbf{b}), \tag{1.11}$$

this difference of displacements is represented by **PT** in Fig. 1.8. If we replace **b** by $-\mathbf{b}$ in Axiom (2A) and use (1.11), we see that $\mathbf{a}-\mathbf{b}$ is a vector, for all vectors **a** and **b**. Making similar replacements in

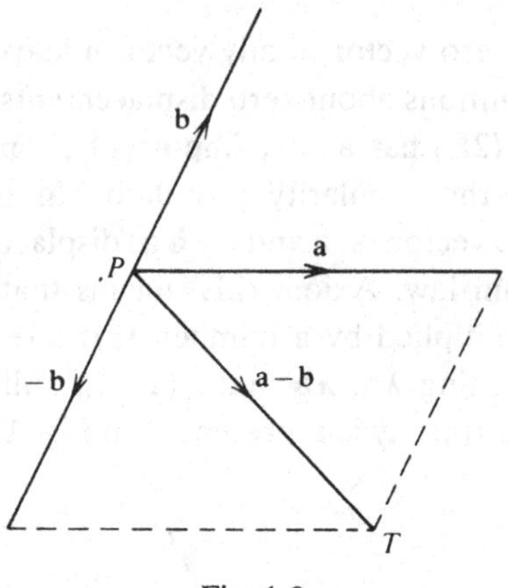


Fig. 1.8

Axioms (2B)–(2E), we can express these axioms in a number of different forms, in which both plus and minus signs may occur. Some examples of these different forms are:

$$\mathbf{a}-\mathbf{b}=-(\mathbf{b}-\mathbf{a}), \tag{1.12}$$

$$\mathbf{a}-(\mathbf{b}+\mathbf{c})=(\mathbf{a}-\mathbf{b})-\mathbf{c}, \tag{1.13}$$

$$(\lambda-\mu)\mathbf{a}=\lambda\mathbf{a}-\mu\mathbf{a}, \tag{1.14}$$

$$\lambda(\mathbf{a}-\mathbf{b})=\lambda\mathbf{a}-\lambda\mathbf{b}. \tag{1.15}$$

**Example 1.2**

Establish (1.13).

$$\begin{aligned}\mathbf{a}-(\mathbf{b}+\mathbf{c}) &= \mathbf{a}+[-1(\mathbf{b}+\mathbf{c})] && \text{by (1.11)}\\ &= \mathbf{a}+[(-\mathbf{b})+(-\mathbf{c})] && \text{by (1.8)}\\ &= [\mathbf{a}+(-\mathbf{b})]+(-\mathbf{c}) && \text{by (1.6)}\\ &= (\mathbf{a}-\mathbf{b})-\mathbf{c}. && \text{by (1.11)}\end{aligned}$$

Summarising, the axioms of scalar multiplication and addition, together with the definition (1.11) of $\mathbf{a}-\mathbf{b}$, ensure that

(i) any linear combination of vectors, for example $\lambda\mathbf{a}+\mu\mathbf{b}+\nu\mathbf{c}$, is a vector;

(ii) products can be 'multiplied out', and terms rearranged and cancelled, as in the algebra of real or complex numbers.

**Example 1.3**

Simplify

$$(\lambda+\mu)(\mathbf{a}-\mathbf{b})-\lambda(\mathbf{a}+\mathbf{c})+(\mu-\nu)\mathbf{b}.$$

This expression is a vector, being a linear combination of vectors **a**, **b** and **c**. Multiplying out, cancelling, and 'collecting terms', we obtain

$$\begin{aligned}&\lambda\mathbf{a}+\mu\mathbf{a}-\lambda\mathbf{b}-\mu\mathbf{b}-\lambda\mathbf{a}-\lambda\mathbf{c}+\mu\mathbf{b}-\nu\mathbf{b}\\&=\mu\mathbf{a}-\lambda\mathbf{b}-\lambda\mathbf{c}-\nu\mathbf{b}\\&=\mu\mathbf{a}-(\lambda+\nu)\mathbf{b}-\lambda\mathbf{c}.\end{aligned}$$

The Axioms (1A)–(1D) and (2A)–(2E) define a set $\{\mathbf{a}\}$ which is called a **linear space** or a **vector space**. Many sets of mathematical objects satisfy these axioms and form a linear space. We therefore need to impose on the set $\{\mathbf{a}\}$ further restrictions or axioms which define the structure of three-dimensional Euclidean space. One important concept which needs to be properly defined is 'modulus' or 'length'; we also need to introduce the concept of 'angle'. Both of these elementary geometric concepts will be introduced in Chapter 2.

## ■ *Problems 1.1*

1 If **a**, **b**, **c** are three vectors, simplify the following expressions, using only axioms and definitions given in the text:

$$\lambda(\mathbf{a}+\mathbf{b})-\lambda(\mathbf{b}+\mathbf{c}),$$
$$3(\mathbf{a}+\mathbf{b})-[\mathbf{a}+3(\mathbf{b}+\mathbf{c})].$$

2 If vectors **u**, **v** and **w** are defined in terms of two vectors **a** and **b** by

$$\mathbf{u}=\mathbf{a}-3\mathbf{b},$$
$$\mathbf{v}=-2\mathbf{a}+\mathbf{b},$$
$$\mathbf{w}=3\mathbf{a}+2\mathbf{b},$$

express the following vectors in terms of **a** and **b**:

$$2\mathbf{u}+\mathbf{v} \qquad \mathbf{u}+\mathbf{v}+\mathbf{w}$$
$$3\mathbf{u}-2\mathbf{v}+2\mathbf{w} \qquad 2\mathbf{u}+4\mathbf{v}+\mathbf{w}.$$

3 Establish Equation (1.15), using only (1.11) and the axioms.

## 1.4 *Displacements in Euclidean space*

Although we have not yet stated all the axioms of vector algebra, some familiar properties of the geometry of points and lines can be derived from the axioms already given. In previous sections, we have used displacements in geometrical or Euclidean space to exemplify properties of vectors, but we have not yet stated clearly the relationship between abstract vectors and displacements. Further, the concept of 'a point in space' has not been introduced in the axioms. So before we derive any geometrical results, we must discuss the relationship between an abstract vector space $\{\mathbf{a}\}$ and three-dimensional Euclidean space.

We shall describe Euclidean space in terms of **points**, which we denote by $O, P, Q, R, \ldots$, and **displacements**. A displacement is determined if we are given a point $P$, called the **initial point**, and an abstract vector $\mathbf{a}$; the displacement is then denoted by $\mathbf{a}_P$, and we say that $\mathbf{a}_P$ is a displacement 'from the point $P$'. In practice, one normally omits the suffix '$P$' on $\mathbf{a}_P$; for the present, we shall retain the suffix, since it is essential for a clear understanding of displacements.

We shall set out the properties of displacements as a third set of axioms, commenting on them and explaining them as we go along. The first axiom assumes the existence of one 'point' $O$; other points are introduced later:

(3A) The set of displacements $\{\mathbf{a}_O\}$ from the given point $O$ is in one-to-one correspondence with the abstract vectors $\{\mathbf{a}\}$, and obeys the same Axioms (1A)–(1D) and (2A)–(2E).

The two sets $\{\mathbf{a}_O\}$ and $\{\mathbf{a}\}$ therefore have identical mathematical properties; we say that they are **isomorphic**. Since $\{\mathbf{a}\}$ is a linear space, so is $\{\mathbf{a}_O\}$. Other sets of physical quantities are also isomorphic to the vector space $\{\mathbf{a}\}$; for example, **forces** $\mathbf{F}_O$ acting at the given point $O$ are found experimentally to have vector properties; so the set $\{\mathbf{F}_O\}$ also forms a vector space isomorphic to $\{\mathbf{a}\}$. It is important to realise that the spaces $\{\mathbf{a}\}$, $\{\mathbf{a}_O\}$ and $\{\mathbf{F}_O\}$ are essentially different, and contain different types of element: each $\mathbf{a}_O$ is a *displacement*, with the dimension of length; each $\mathbf{F}_O$ is a *force*, with the dimension of force; each $\mathbf{a}$ is an abstract algebraic object, and has no dimension associated with it.

Points other than the given point $O$ are introduced by the second axiom:

(3B) The set of all points $\{P\}$ is in one-to-one correspondence with the displacements $\{\mathbf{a}_O\}$ from $O$, and hence in one-to-one correspondence with the vectors $\{\mathbf{a}\}$. The point $O$ itself corresponds to the zero vector $\mathbf{0}$.

The point $A$ associated with a particular displacement $\mathbf{a}_O$ from $O$ is called the **end-point** of $\mathbf{a}_O$, and the displacement is written

$$\mathbf{a}_O \equiv \mathbf{OA}; \tag{1.16}$$

the abstract vector $\mathbf{a}$ corresponding to $\mathbf{OA}$ is called the **position vector** of $A$ relative to origin $O$. Displacements $\mathbf{a}_O$, $\mathbf{b}_O$ and $\mathbf{c}_O$ from $O$ are represented in Fig. 1.9($a$) by directed lines from the point $O$. The corresponding vectors $\mathbf{a}$, $\mathbf{b}$, $\mathbf{c}$ are represented in Fig. 1.9($b$) by directed lines in arbitrary positions, since they are not associated with any point in space. The displacements $\{\mathbf{a}_O\}$, with a given initial point $O$, are often called **fixed vectors**; a force $\mathbf{F}_O$ acting at $O$ is also a 'fixed vector'. The abstract vectors $\{\mathbf{a}\}$ are then called **free vectors**, since they are not associated with any initial point. The fact that the sets $\{\mathbf{a}\}$, $\{\mathbf{a}_O\}$ and $\{A\}$ are in one-to-one correspondence means that if we are given a member of one of the sets, there is exactly one member of each of the other two sets corresponding to it. For example, any given displacement $\mathbf{a}_O$ from $O$ corresponds both to a unique end-point $A$ and to a unique vector $\mathbf{a}$.

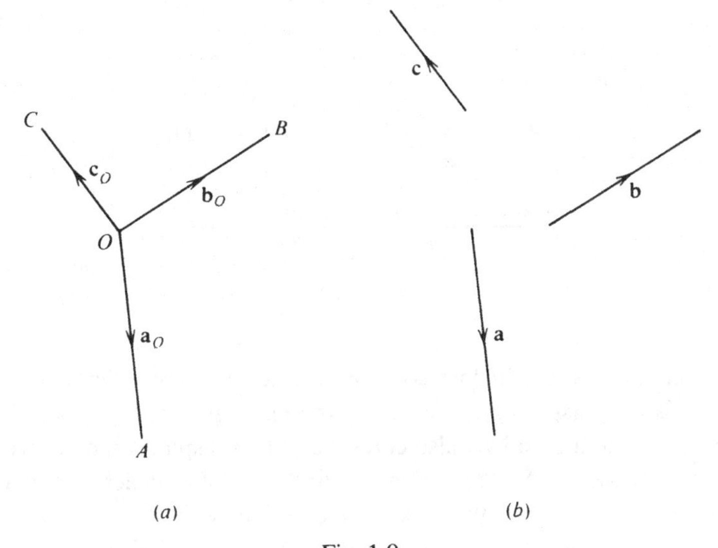


Fig. 1.9

Axiom (3A) only defines displacements $\{\mathbf{a}_O\}$ from the single point $O$. Now that other points $\{A\}$ have been introduced, we define displacements from an arbitrary point $P$ by further axioms, which are represented in Fig. 1.10:

(3C) If points $A$ and $P$ correspond to displacements $\mathbf{a}_O$ and $\mathbf{p}_O$, and hence to vectors $\mathbf{a}$ and $\mathbf{p}$, a unique displacement

$$\mathbf{a}'_P \equiv \mathbf{PA}$$

from $P$ to $A$ exists.

(3D) For every two points $A$ and $P$, and corresponding vectors $\mathbf{a}$ and $\mathbf{p}$, the displacement $\mathbf{a}'_P$ from $P$ to $A$ corresponds to the abstract vector $\mathbf{a}-\mathbf{p}$. For this reason, we use the notation

$$\mathbf{a}'_P \equiv (\mathbf{a}-\mathbf{p})_P. \tag{1.17}$$

(3E) For each point $P$, the displacements $\{\mathbf{a}_P\}$ obey the same algebraic rules as the abstract vectors to which they correspond.

In Fig. 1.10, the given displacements $\mathbf{a}_O$ and $\mathbf{p}_O$, defining $A$ and $P$, are drawn as solid lines. The displacement $\mathbf{a}'_P \equiv \mathbf{PA}$, defined by the axioms, is represented by a broken line $PA$. At first sight it seems that Fig. 1.10 merely represents the fact that two vectors $\mathbf{p}$ and $\mathbf{a}$ obey the equation

$$\mathbf{p}+(\mathbf{a}-\mathbf{p})=\mathbf{a}. \tag{1.18}$$

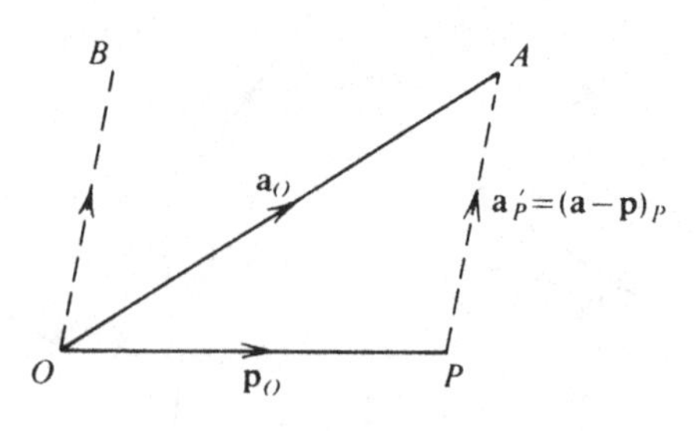


Fig. 1.10

The situation is not in fact so simple. Each abstract vector $\mathbf{a}$ corresponds to a displacement $\mathbf{a}_O$ from the given point $O$. Axiom (3C) now tells us that $\mathbf{a}$ and $\mathbf{a}_O$ also correspond to a displacement $\mathbf{a}_P$ from $P$, *for every point P*. So corresponding to the set of abstract vectors $\{\mathbf{a}\}$ represented in Fig. 1.9($a$), we have a whole collection of sets $\{\mathbf{a}_O\}, \{\mathbf{a}_P\}, \ldots$, one for each of the points $O, P, \ldots$. Axiom (3D) tells us how displacements from $O$ are related to displacements from $P$, and is in fact the rule governing **change of origin** from $O$ to $P$; the set of displacements $\{\mathbf{a}_P\}$, for example, are said to 'have $P$ as origin'.

The axioms defining displacements appear to single out one point, denoted by $O$. We shall now show that all points $O, P, \ldots$, are in fact equivalent. First note that from any point $P$, there is a unique displacement $\mathbf{a}'_P$ related to $\mathbf{a}-\mathbf{p}$, defined by (1.17); further, the displacements $\{\mathbf{a}'_P\}$ obey the same axioms as the displacements $\{\mathbf{a}_O\}$ from the given point $O$, each forming a linear space. Let us consider how the axioms for the sets $\{\mathbf{a}\}$, $\{\mathbf{a}_O\}$ and $\{\mathbf{a}'_P\}$ correspond; take for instance the Axiom (2B),

$$\mathbf{a}+\mathbf{b}=\mathbf{b}+\mathbf{a}. \tag{1.5}$$

If the vectors $\mathbf{a}$, $\mathbf{b}$ correspond to displacements $\mathbf{a}_O$, $\mathbf{b}_O$ from $O$, the corresponding axiom for these displacements is

$$\mathbf{a}_O+\mathbf{b}_O=\mathbf{b}_O+\mathbf{a}_O. \tag{1.19}$$

Using Axiom (1.6), the Axiom (1.5) can be written in the form

$$(\mathbf{a}-\mathbf{p})+(\mathbf{b}-\mathbf{p})=(\mathbf{b}-\mathbf{p})+(\mathbf{a}-\mathbf{p}), \tag{1.20}$$

where $\mathbf{p}$ is any vector; we take $\mathbf{p}$ to correspond to the displacement $\mathbf{p}_O \equiv \mathbf{OP}$. Then, since $(\mathbf{a}-\mathbf{p})$ and $(\mathbf{b}-\mathbf{p})$ correspond to displacements $\mathbf{a}'_P$ and $\mathbf{b}'_P$ by Axiom (3D), we see that (1.20), or (1.5), corresponds to

$$\mathbf{a}'_P+\mathbf{b}'_P=\mathbf{b}'_P+\mathbf{a}'_P, \tag{1.21}$$

by virtue of Axiom (3E). So the vector Axiom (2B) corresponds directly to an equivalent property (1.21) of displacements from *any* point $P$. The axiom for displacements $\{\mathbf{a}_O\}$ from the point $O$ is recovered from (1.21) by putting $\mathbf{p}=\mathbf{0}$ and using Axiom (3B).

It is not hard to check that, in a similar way, all the axioms (1A)–(1D) and (2A)–(2E) give rise to the same set of properties of displacements from an arbitrary point $P$; so displacements from $O$ are in no way specially related to the abstract vector space. The fact that $\mathbf{a}_O$ corresponds to the vector $\mathbf{a}$, while $\mathbf{a}'_P$ corresponds to $\mathbf{a}-\mathbf{p}$, does appear to single out the point $O$; for this reason, the point $O$ is often called the **origin**; we emphasise, however, that it is in no basic sense a special point, and that any point may be chosen as origin.

We have already noted that the relation (1.17) between vectors and displacements from $P$, indicated in Fig. 1.10, is related to the vector equation (1.18). Because we choose to relate displacements from different points in this way, it is possible to use (1.18) as an equation relating displacements between different points; the equation could in fact be written in the form

$$\mathbf{a}_O = \mathbf{p}_O + (\mathbf{a}-\mathbf{p})_P; \tag{1.22}$$

this equation can be interpreted directly as 'a displacement from $O$ to $P$ followed by a displacement from $P$ to $A$ is equivalent to a displacement from $O$ to $A$'. Whether (1.22) is regarded as an equation defining 'change of origin', or as a rule governing physical displacements in space, it is quite different from the *mathematical* relation (1.18) governing abstract vectors. It is customary to omit the suffixes '$O$' and '$P$' in Equation (1.22) and to write it in the form (1.18); in discussing geometric applications, we shall frequently do this; the initial points of the various displacements are usually fairly obvious in any application.

It is important to realise that, in some applications, Equation (1.18) can be interpreted in terms of vector quantities at a single point. We noted earlier the experimental fact that forces $\{\mathbf{F}_P\}$ at a single point $P$ obey the vector axioms. So if $\mathbf{F}_P$ and $\mathbf{G}_P$ are two such vectors, they obey the relation

$$\mathbf{G}_P + (\mathbf{F}_P - \mathbf{G}_P) = \mathbf{F}_P,$$

corresponding directly to the vector relation (1.18); this relation, unlike (1.22), involves only vector quantities at a single point $P$. We must therefore bear in mind that a simple vector relation like (1.18) can have several different interpretations which should be clearly distinguished.

Equation (1.17) relates the abstract vector $\mathbf{a}-\mathbf{p}$ to the displacement $\mathbf{PA} \equiv \mathbf{a}'_P$. The vector $\mathbf{a}-\mathbf{p}$ is called the position vector of $A$ relative to $P$. The vector $\mathbf{a}-\mathbf{p}$ also corresponds to the displacement

$$\mathbf{OB} \equiv (\mathbf{a}-\mathbf{p})_O$$

from $O$, shown in Fig. 1.10. Displacements such as **OB** and **PA** which correspond to the same position vector ($\mathbf{a}-\mathbf{p}$ in this case), are said to be **equal and parallel**. More generally, a non-zero displacement $\mathbf{a}_P$ from a point $P$ is said to be **parallel** to any displacement $\lambda\mathbf{a}_M$ from another point $M$, provided $\lambda \neq 0$; $\mathbf{a}_P$ and $\mathbf{a}_M$ correspond, of course, to the same non-zero vector $\mathbf{a}$.

If we are given a displacement $\mathbf{a}_P$ from a point $P$ to a point $Q$, as in Fig. 1.11, the set of displacements of the form $\lambda\mathbf{a}_P$ are said to form a **ray**, and the end-point $R$ of a displacement $\lambda\mathbf{a}_P$ 'lies on the line $PQ$';

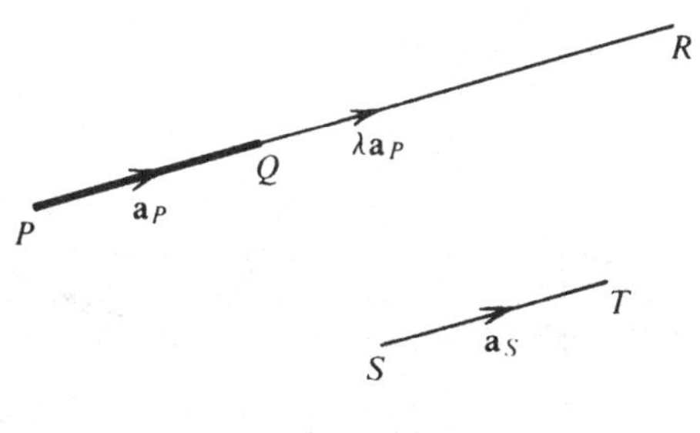


Fig. 1.11

the line $PQ$ is the set of all such points. We say that $PR$ and $PQ$ are 'in the ratio $\lambda : 1$', or that

$$PR : PQ = \lambda : 1.$$

If $\mathbf{ST} \equiv \mathbf{a}_S$ is equal and parallel to $\mathbf{a}_P$, as in Fig. 1.11, we also say that

$$PR : ST = \lambda : 1.$$

## *1.5 Geometrical applications*

Now that we have defined displacements, points and lines in Euclidean space, we can establish some simple geometric results. Normally, we shall adopt the standard convention of labelling displacements $\mathbf{a}_O, \mathbf{b}_P, \ldots$ by the corresponding position vectors $\mathbf{a}, \mathbf{b}, \ldots$; the initial point of each displacement is usually clear, and is often represented in a diagram. When displacements are written as $\mathbf{OR}, \mathbf{PQ}, \ldots$, the initial point is the first of the two points named. Again following normal practice, we shall use equations such as

$$\mathbf{OB} = \mathbf{PA} = \mathbf{a} - \mathbf{p}, \tag{1.23}$$

relating different displacements in Fig. 1.10 to the position vector $\mathbf{a} - \mathbf{p}$. Equation (1.23) tells us which displacements of the two sets $\{\mathbf{a}_O\}$, $\{\mathbf{a}_P\}$ correspond to the vector $\mathbf{a} - \mathbf{p}$; geometrically, it tells us that $\mathbf{OB}$ and $\mathbf{PA}$ are equal and parallel, and it is a different kind of equation from (1.21), say, which relates displacements from a single point only.

The first result we establish concerns the division of a line $AB$ in a given ratio $m : n$. Suppose, as in Fig. 1.12($a$), that $\mathbf{a}$ and $\mathbf{b}$ are the position vectors of $A$ and $B$ relative to an origin $O$. We want to find the position vector of the point $C$ on $AB$ such that $AC : CB = m : n$.

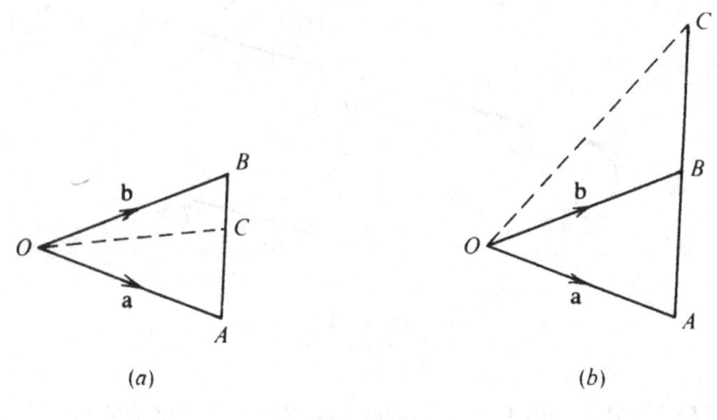


Fig. 1.12

The displacements **AC** and **AB** are therefore parallel and their lengths are in the ratio $m : m+n$. Thus

$$\mathbf{AC} = \frac{m}{m+n}\mathbf{AB}$$

$$= \frac{m}{m+n}(\mathbf{OB} - \mathbf{OA})$$

$$= \frac{m}{m+n}(\mathbf{b} - \mathbf{a}). \tag{1.24}$$

Also

$$\mathbf{OC} = \mathbf{OA} + \mathbf{AC} = \mathbf{a} + \mathbf{AC};$$

hence, using (1.24),

$$\mathbf{OC} = \mathbf{a} + \frac{m}{m+n}(\mathbf{b} - \mathbf{a})$$

$$= \frac{n\mathbf{a} + m\mathbf{b}}{m+n}. \tag{1.25}$$

Thus the position vector of $C$ is the 'weighted mean' of the position vectors of $A$ and $B$. In mechanics, $C$ is the centroid of a mass $n$ at $A$ and a mass $m$ at $B$. The coefficient of **b** in (1.25) is

$$q = \frac{m}{m+n},$$

and the coefficient of **a** is

$$\frac{n}{m+n} = 1 - \frac{m}{m+n} = 1 - q;$$

so the two ratios $m:n$ and $q:1-q$ are equal. Equation (1.25) then tells us that the point $C$ dividing $AB$ in the ratio $q:1-q$ has position vector

$$\mathbf{c}=(1-q)\mathbf{a}+q\mathbf{b}. \tag{1.26}$$

This is a useful variant of (1.25).

Equation (1.25) can be applied even if $m$ or $n$ is negative; division in the ratio $m:n$ is then called 'external division' of $AB$ in the ratio $|m|:|n|$. If $m>0$ and $0>n>-m$, for example, $B$ lies on the line between $A$ and $C$; then the condition that $AC:CB=m:n$ is better expressed *vectorially* by the equation

$$m\mathbf{CB}=n\mathbf{AC};$$

the relative positions of **A**, **B**, **C** are shown in Fig. 1.12($b$); thus (1.24), and hence (1.25) and (1.26), are still true.

We note that in deducing (1.25), we have used a notation which mixes up displacements from different origins, such as **AC** and **OC**, and position vectors such as **a** and **b**. In deducing (1.25), one *should* use abstract vectors throughout, interpreting the result in terms of displacements from $O$.

Suppose that $N$ points $A, B, \ldots, E$ have position vectors $\mathbf{a}, \mathbf{b}, \ldots, \mathbf{e}$ relative to an origin $O$. Then the **mean position** or **centroid** of $A, B, \ldots, E$ is defined as the point $G$ with position vector

$$\mathbf{OG}=\frac{1}{N}(\mathbf{a}+\mathbf{b}+\ldots+\mathbf{e}). \tag{1.27}$$

It is important to note that $G$ is a point which is fixed relative to the points $A, B, \ldots, E$. That is to say, the vectors $\mathbf{AG}, \mathbf{BG}, \ldots, \mathbf{EG}$ do not depend on the choice of the origin $O$. It is sufficient to show that **AG** is independent of the choice of origin. Now

$$\begin{aligned}\mathbf{AG}&=\mathbf{OG}-\mathbf{OA}\\&=\frac{1}{N}(\mathbf{a}+\mathbf{b}+\mathbf{c}+\ldots+\mathbf{e})-\mathbf{a}\\&=\frac{1}{N}(\mathbf{b}+\mathbf{c}+\ldots+\mathbf{e})-\frac{N-1}{N}\mathbf{a}.\end{aligned} \tag{1.28}$$

Now suppose we use a different point $P$ as origin, and that $\mathbf{OP}=\mathbf{p}$, as in Fig. 1.13. Let $\mathbf{a}', \mathbf{b}', \ldots, \mathbf{e}'$ be the position vectors of $A, B, \ldots, E$

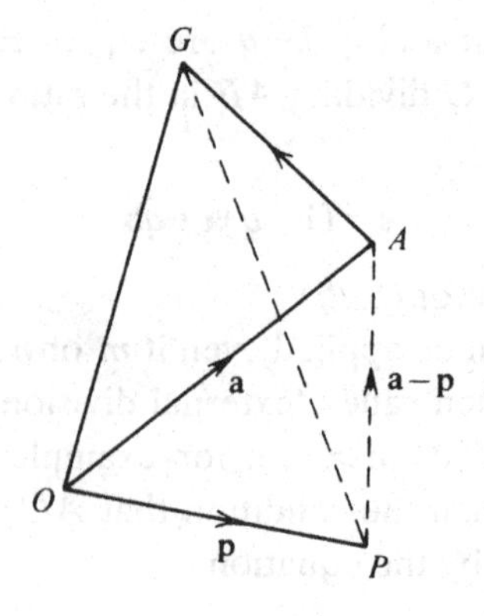


Fig. 1.13

relative to $P$; then (1.17), written in terms of position vectors, gives

$$\mathbf{a}' = \mathbf{a} - \mathbf{p},$$
$$\mathbf{b}' = \mathbf{b} - \mathbf{p},$$
$$\cdots\cdots\cdots$$
$$\mathbf{e}' = \mathbf{e} - \mathbf{p}.$$

Using $P$ as origin in the definition (1.27), we would find, instead of (1.28),

$$\mathbf{AG} = \frac{1}{N}(\mathbf{b}' + \mathbf{c}' + \ldots + \mathbf{e}') - \frac{N-1}{N}\mathbf{a}'.$$

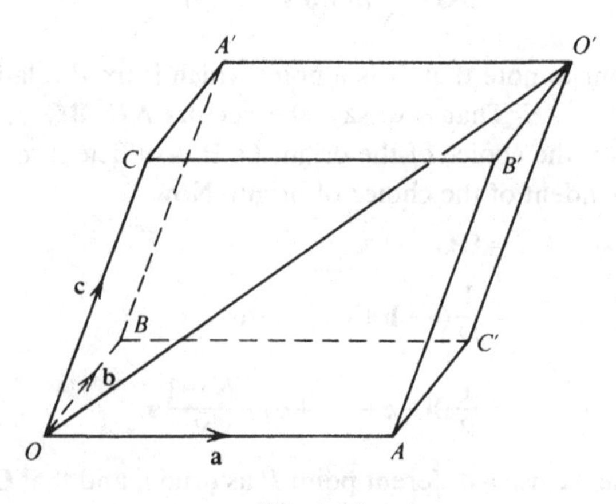


Fig. 1.14

But substituting for $\mathbf{a}', \mathbf{b}', \ldots, \mathbf{e}'$ in terms of $\mathbf{a}, \mathbf{b}, \ldots, \mathbf{e}$ and $\mathbf{p}$ gives exactly (1.28). So the same point $G$ is defined as centroid, regardless of the origin chosen.

We can define a **parallelogram** $PQRS$ by taking $P$, for example, as origin, and defining $\mathbf{PQ}=\mathbf{a}$ and $\mathbf{PS}=\mathbf{b}$ to be the displacements along two sides from $P$, as shown in Fig. 1.3; $R$ is defined by taking $\mathbf{PR}=\mathbf{a}+\mathbf{b}$. Similarly, we can define a **parallelepiped** by taking one vertex $O$ as origin and giving the position vectors **a**, **b**, **c** corresponding to displacements **OA**, **OB**, **OC** along three adjacent edges, as shown in Fig. 1.14. The remaining vertices $A'$, $B'$, $C'$ and $O'$ have position vectors $\mathbf{b}+\mathbf{c}$, $\mathbf{c}+\mathbf{a}$, $\mathbf{a}+\mathbf{b}$ and $\mathbf{a}+\mathbf{b}+\mathbf{c}$ respectively.

**Example 1.4**

Show that

(i) the medians of a triangle $ABC$ meet at the centroid $G$ of $A$, $B$ and $C$;

(ii) if $D$, $E$ and $F$ are the mid-points of $BC$, $CA$ and $AB$ respectively, then

$$AG:GD = BG:GE = CG:GF = 2:1;$$

(iii) $G$ is the centroid of the points $D$, $E$ and $F$.

Let **a**, **b**, **c** be the position vectors of $A$, $B$, $C$, relative to some origin, as in Fig. 1.15. Using (1.25) with $m=n$, the mid-point $D$ of $BC$ has position vector

$$\mathbf{d}=\tfrac{1}{2}(\mathbf{b}+\mathbf{c}).$$

Now define a point $G$ on $AD$ such that $AG:GD=2:1$. Using (1.25) with $n=1$, $m=2$, and **b** replaced by **d**, we find that $G$ has position

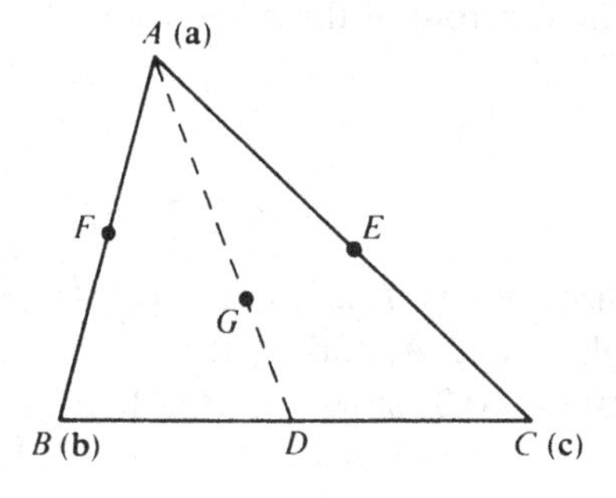


Fig. 1.15

vector

$$\tfrac{1}{3}\mathbf{a}+\tfrac{2}{3}\mathbf{d}$$

$$=\tfrac{1}{3}(\mathbf{a}+\mathbf{b}+\mathbf{c}).$$

By (1.27) with $N=3$, $G$ is the centroid of $A$, $B$, $C$. Further, symmetry between **a**, **b**, **c** shows that it is the point of trisection of the other two medians also. Thus (i) and (ii) are established.

By (1.27), the centroid of $D$, $E$, $F$ has position vector

$$\tfrac{1}{3}[\tfrac{1}{2}(\mathbf{b}+\mathbf{c})+\tfrac{1}{2}(\mathbf{c}+\mathbf{a})+\tfrac{1}{2}(\mathbf{a}+\mathbf{b})],$$

and is therefore also at $G$, proving (iii).

**Example 1.5**

Prove that the four diagonals of a parallelepiped meet and bisect each other.

Since the definition of a mid-point is independent of the origin chosen, choose the origin at $O$ as in Fig. 1.14, and define vectors **a**, **b**, **c** as shown. Then the position vectors of the pairs of points at the ends of the diagonals are

$$O, O' : \mathbf{0}, \mathbf{a}+\mathbf{b}+\mathbf{c},$$

$$A, A' : \mathbf{a}, \mathbf{b}+\mathbf{c},$$

$$B, B' : \mathbf{b}, \mathbf{c}+\mathbf{a},$$

$$C, C' : \mathbf{c}, \mathbf{a}+\mathbf{b}.$$

Using (1.25) with $m=n$ tells us that the mid-points of $OO'$, $AA'$, $BB'$ and $CC'$ all have the same position vector

$$\tfrac{1}{2}(\mathbf{a}+\mathbf{b}+\mathbf{c}).$$

So the diagonals do intersect at their mid-points. Note that the point of intersection is the centroid of the eight points $O$, $A$, $B$, $C$, $O'$, $A'$, $B'$ and $C'$.

## *Problems 1.2*

1 $O$, $A$ and $B$ are three points. $P$ and $Q$ are points lying, respectively, on the line through $O$ and $A$, and on the line through $O$ and $B$. If $\mathbf{OP}=m\mathbf{OA}$ and $\mathbf{OQ}=n\mathbf{OB}$, show that the line through $O$ and the mid-point of $AB$ divides $PQ$ in the ratio $m:n$. Draw diagrams to illustrate this result

(*a*) when $m > 1$ and $0 < n < 1$,
(*b*) when $-1 < m < 0$ and $n > 1$.

2 $L$ and $L'$ are two lines in space, and $C$, $C'$ are two fixed points on $L$, $L'$ respectively. $A$, $B$ are variable points on $L$, and $A'$, $B'$ variable points on $L'$ such that

$$\frac{AC}{CB} = \frac{A'C'}{C'B'} = \lambda,$$

where $\lambda$ is a real constant. Prove that

$$\mathbf{AA'} + \lambda \mathbf{BB'} = (1 + \lambda)\mathbf{CC'}.$$

3 Prove that the diagonals of a parallelogram bisect each other.

4 $OACB$ is a parallelogram. $D$ is a point on $OA$ such that $\mathbf{OA} = \lambda \mathbf{OD}$, and $E$ is a point on $DB$ such that $\mathbf{EB} = \lambda \mathbf{DE}$. Prove that $O$, $E$, $C$ are collinear and that $\mathbf{OC} = (\lambda + 1)\mathbf{OE}$.

5 $A$, $B$, $C$, $D$ are four points in space. $A'$ is the centroid of $B$, $C$, $D$; $B'$ is the centroid of $C$, $D$, $A$; $C'$ is the centroid of $D$, $A$, $B$, and $D'$ is the centroid of $A$, $B$, $C$. Prove that if $A \neq A'$, $B \neq B'$, $C \neq C'$ and $D \neq D'$, then $AA'$, $BB'$, $CC'$ and $DD'$ all meet at the centroid $G$ of $A$, $B$, $C$, $D$. Prove also that $G$ is the centroid of $A'$, $B'$, $C'$, $D'$.

6 $A$, $B$, $C$, $D$ are four different points in space. $P$, $Q$, $R$, $S$ are points on $AB$, $BC$, $CD$, and $DA$ respectively, dividing these lines in the same ratio $m:n$. Prove that the centroids of $A$, $B$, $C$, $D$ and of $P$, $Q$, $R$, $S$ coincide.

7 $\lambda$, $\mu$, $\nu$ are real numbers, and $\mathbf{p}$, $\mathbf{q}$, $\mathbf{r}$ are the position vectors of points $P$, $Q$, $R$ relative to a given origin. If

$$(\lambda - \mu)\mathbf{p} + (\mu - \nu)\mathbf{q} + (\nu - \lambda)\mathbf{r} = \mathbf{0},$$

and none of $\lambda$, $\mu$, $\nu$ are equal, prove that $P$, $Q$, $R$ are collinear.

8 $A$, $B$, $C$ are three non-collinear points and $\lambda$, $\mu$, $\nu$ are non-zero real numbers. $D$, $E$, $F$ are points on $BC$, $CA$, $AB$ respectively, such that

$$\frac{AF}{FB} = -\frac{\lambda}{\mu}, \qquad \frac{BD}{DC} = -\frac{\mu}{\nu}, \qquad \frac{CE}{EA} = -\frac{\nu}{\lambda},$$

with the usual sign convention. Show that $D$, $E$, $F$ are collinear. [This is Menelaus' theorem.]

# 2

# Scalar products and components

## 2.1 *Scalar products*

In the first chapter, the concept of the modulus or magnitude of a vector was used to discuss the geometric interpretation of certain axioms, but we have not yet defined it. Nor have we introduced the fundamental geometrical concept of 'angle'. We shall define both modulus and angle in terms of a more general mathematical object, the **scalar product** of two vectors, frequently called the **inner product** in mathematical works.

Before giving the formal definition of scalar product, we explain how it is related to the moduli of two vectors **a** and **b**, and the angle $\theta$ between these vectors. In Fig. 2.1 we have represented the vectors by two displacements drawn from a point, and have taken $\theta$ to be the smaller angle between the two lines, so that

$$0 \leqslant \theta \leqslant \pi. \tag{2.1}$$

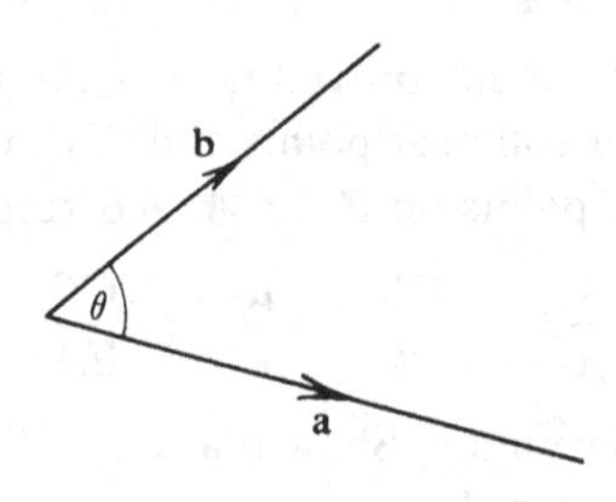


Fig. 2.1

In terms of $\theta$ and the moduli $a$ and $b$ of the vectors, the scalar product, denoted by $\mathbf{a} \cdot \mathbf{b}$ is given by

$$\mathbf{a} \cdot \mathbf{b} = ab \cos \theta. \tag{2.2}$$

When **a** and **b** denote displacements, $a$ and $b$ are the lengths of the displacements. Equation (2.2) is often regarded as the definition of $\mathbf{a} \cdot \mathbf{b}$, but this presupposes that we know what lengths and angles are – in fact, we have to assume a knowledge of Euclidean geometry. Our purpose, however, is to deduce the properties of vectors from algebraic axioms, including the axioms defining scalar products. We then *define* modulus and angle in terms of scalar products; applying these definitions to displacements in three-dimensional space, we then *deduce* Euclidean geometry.

We can, however, use (2.2) to see how 'modulus' and 'angle' are to be related to scalar products. If the vectors **a** and **b** are equal in (2.2), then in Fig. 2.1, $\theta = 0$ and $a = b$. So (2.2) becomes

$$\mathbf{a} \cdot \mathbf{a} = a^2 \cos 0 = a^2. \tag{2.3}$$

So the scalar product of a vector **a** with itself gives us the square of the modulus (or magnitude) of **a**. We always choose the modulus $a$ to be non-negative ($a \geqslant 0$); in particular, lengths of displacements are always positive or zero.

If we multiply a vector **a** by $a^{-1}$, we obtain the vector

$$\mathbf{u} = a^{-1}\mathbf{a}, \tag{2.4a}$$

called the **unit vector** corresponding to **a**. Likewise the unit vector corresponding to **b** is

$$\mathbf{u}_1 \equiv b^{-1}\mathbf{b}. \tag{2.4b}$$

The axioms that we shall write down will allow us to divide Equation (2.2) by the scalar factor $ab$ to give

$$(a^{-1}\mathbf{a}) \cdot (b^{-1}\mathbf{b}) = \cos \theta,$$

or, using (2.4),

$$\mathbf{u} \cdot \mathbf{u}_1 = \cos \theta. \tag{2.5}$$

So the scalar product of two *unit* vectors is the cosine of the angle between the vectors. We note that scalar products define *cosines* of angles, not the angles themselves; however, if we are given $\cos \theta$, with

$$|\cos \theta| \leqslant 1, \tag{2.6}$$

this determines a unique angle $\theta$ in the range (2.1). We shall see later that all cosines defined by (2.2) or (2.5) satisfy the condition (2.6), and so correspond to a definite angle $\theta$ in the range $0 \leqslant \theta \leq \pi$.

Now that we have seen how scalar products are to be related to the familiar geometrical properties of length and angle, we shall give the axioms defining them. Then we shall show that the familiar properties of length and angle do in fact follow from the axioms, when they are introduced through (2.2), or (equivalently) through (2.3) and (2.5).

The axioms defining scalar products are:

(4A) To any two vectors **a** and **b** there corresponds a unique real number, the scalar product, which is denoted by $\mathbf{a}\cdot\mathbf{b}$.

(4B) $$\mathbf{a}\cdot\mathbf{b}=\mathbf{b}\cdot\mathbf{a}, \tag{2.7}$$

(4C) $$\mathbf{a}\cdot\lambda\mathbf{b}=\lambda(\mathbf{a}\cdot\mathbf{b}),$$

(4D) $$\mathbf{a}\cdot(\mathbf{b}+\mathbf{c})=\mathbf{a}\cdot\mathbf{b}+\mathbf{a}\cdot\mathbf{c}, \tag{2.8}$$

(4E) $$\mathbf{a}\cdot\mathbf{a}>0 \text{ unless } \mathbf{a}=\mathbf{0}. \tag{2.9}$$

The last axiom (4E) tells us that, when $\mathbf{a}\neq\mathbf{0}$, we can write

$$\mathbf{a}\cdot\mathbf{a}=a^2, \tag{2.3}$$

where we can choose $a>0$; this accords with (2.2) when $\theta=0$. Then

$$a=(\mathbf{a}\cdot\mathbf{a})^{\frac{1}{2}}>0 \tag{2.10}$$

is the definition of the **modulus** of the vector **a**. It is easy to show that the modulus of the zero vector **0** is zero; we leave the proof as an example. The modulus $a$ of a vector **a** is often written as $|\mathbf{a}|$.

Axiom (4B) accords with the interpretation (2.2) of the scalar product, since

$$\mathbf{b}\cdot\mathbf{a}=ab\cos\theta$$

also; this axiom tells us that scalar multiplication of two vectors is **commutative**. Using (2.7), Axiom (4C) can be immediately extended to give

$$\mathbf{a}\cdot(\lambda\mathbf{b})=(\lambda\mathbf{a})\cdot\mathbf{b}=\lambda(\mathbf{a}\cdot\mathbf{b}); \tag{2.11}$$

so multiplication of either **a** or **b** by a real number $\lambda$ results in multiplication of their scalar product by $\lambda$. This property was used when we multiplied the scalar product $\mathbf{a}\cdot\mathbf{b}$ in (2.2) by $(ab)^{-1}$ in order to give $\mathbf{u}\cdot\mathbf{u}_1$ in (2.5); it is an **associative law**.

The unit vector **u** in the direction of **a** is defined by (2.4a); we find the modulus of **u** from (2.3), using (2.11):

$$u^2=\mathbf{u}\cdot\mathbf{u}=(a^{-1}\mathbf{a})\cdot(a^{-1}\mathbf{a})$$
$$=a^{-2}(\mathbf{a}\cdot\mathbf{a})=1.$$

So $u=1$, justifying the term ‘unit vector’.

Axiom (4D) is most easily understood geometrically if we multiply it throughout by $a^{-1}$; defining the unit vector **u** by (2.4a), (2.8) becomes

$$\mathbf{u} \cdot (\mathbf{b}+\mathbf{c}) = \mathbf{u} \cdot \mathbf{b} + \mathbf{u} \cdot \mathbf{c}. \tag{2.12}$$

Consider the term $\mathbf{u} \cdot \mathbf{b}$, for example. Equation (2.2) tells us that it is to be interpreted as

$$\mathbf{u} \cdot \mathbf{b} = b \cos \theta;$$

as in Fig. 2.2, this is the **projection** of a displacement **b** in the direction of **u**. So (2.12) can be interpreted in terms of projections, as shown in Fig. 2.3; since $\mathbf{u} \cdot \mathbf{b} = OP$, $\mathbf{u} \cdot \mathbf{c} = RS$, $\mathbf{u} \cdot (\mathbf{b}+\mathbf{c}) = OQ$, (2.12) has the geometric interpretation

$$OQ = OP + RS.$$

Axioms (4C) and (4D) express the **linearity** of the scalar product; they allow us to 'multiply out' scalar products, and to factor out scalars. Axiom (4D) is a **distributive law**.

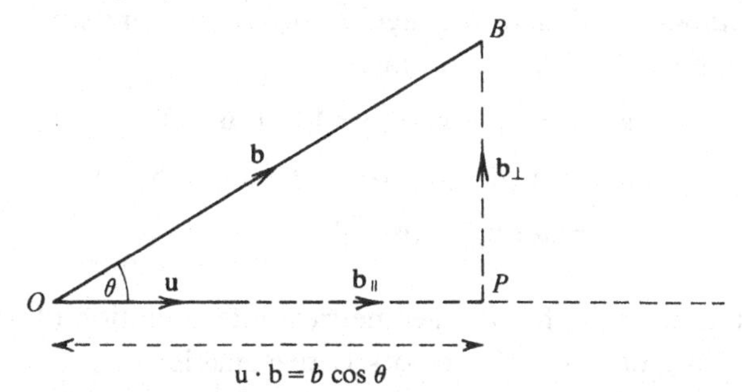


Fig. 2.2

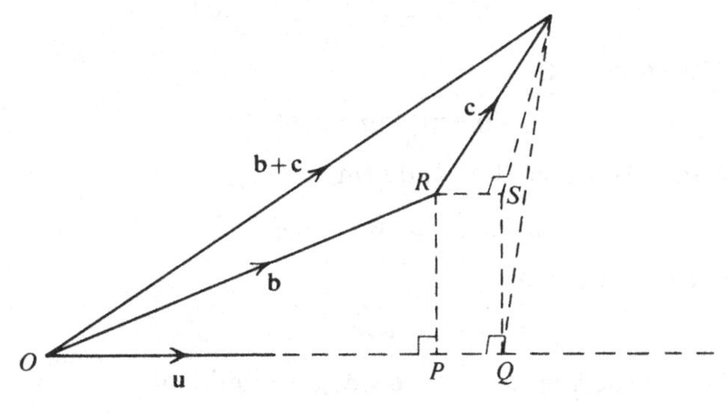


Fig. 2.3

**Example 2.1**

Using only the axioms, expand

$$(\mathbf{a}+\lambda\mathbf{b})\cdot(\mathbf{c}+\lambda\mathbf{d}).$$

The axioms used are indicated at each stage.

$$\begin{aligned}
&(\mathbf{a}+\lambda\mathbf{b})\cdot(\mathbf{c}+\lambda\mathbf{d})\\
&\quad=(\mathbf{a}+\lambda\mathbf{b})\cdot\mathbf{c}+(\mathbf{a}+\lambda\mathbf{b})\cdot(\lambda\mathbf{d}) &&\text{(4D)}\\
&\quad=\mathbf{c}\cdot(\mathbf{a}+\lambda\mathbf{b})+\lambda[\mathbf{d}\cdot(\mathbf{a}+\lambda\mathbf{b})] &&\text{(4B, 4C)}\\
&\quad=\mathbf{c}\cdot\mathbf{a}+\lambda(\mathbf{c}\cdot\mathbf{b})+\lambda(\mathbf{d}\cdot\mathbf{a})+\lambda^2(\mathbf{d}\cdot\mathbf{b}) &&\text{(4D, 4C)}\\
&\quad=\mathbf{a}\cdot\mathbf{c}+\lambda(\mathbf{b}\cdot\mathbf{c}+\mathbf{a}\cdot\mathbf{d})+\lambda^2(\mathbf{b}\cdot\mathbf{d}). &&\text{(4B)}
\end{aligned}$$

**Example 2.2**

Simplify

$$[\mathbf{a}+\mathbf{b}(1+\lambda)+\mathbf{c}]\cdot[\mathbf{a}+\mathbf{b}(1-\lambda)-\mathbf{c}].$$

Using methods familiar in the algebra of real or complex numbers, and remembering (2.3), this becomes

$$\begin{aligned}
&[(\mathbf{a}+\mathbf{b})+(\lambda\mathbf{b}+\mathbf{c})]\cdot[(\mathbf{a}+\mathbf{b})-(\lambda\mathbf{b}+\mathbf{c})]\\
&\quad=(\mathbf{a}+\mathbf{b})\cdot(\mathbf{a}+\mathbf{b})-(\lambda\mathbf{b}+\mathbf{c})\cdot(\lambda\mathbf{b}+\mathbf{c})\\
&\quad=|\mathbf{a}+\mathbf{b}|^2-|\lambda\mathbf{b}+\mathbf{c}|^2.
\end{aligned}$$

We can now show that the geometrical interpretation (2.5) of the scalar product of two unit vectors is reasonable; in particular, we shall establish (2.6). Suppose that $\mathbf{u}$ and $\mathbf{u}_1$ are any two unit vectors, so that

$$\mathbf{u}\cdot\mathbf{u}=\mathbf{u}_1\cdot\mathbf{u}_1=1. \tag{2.13}$$

Axiom (2.9) tells us that

$$(\mathbf{u}-\mathbf{u}_1)\cdot(\mathbf{u}-\mathbf{u}_1)>0$$

unless $\mathbf{u}-\mathbf{u}_1=\mathbf{0}$. Expanding and using (2.13),

$$2\mathbf{u}\cdot\mathbf{u}_1<\mathbf{u}\cdot\mathbf{u}+\mathbf{u}_1\cdot\mathbf{u}_1=2,$$

provided $\mathbf{u}\neq\mathbf{u}_1$, so that

$$\mathbf{u}\cdot\mathbf{u}_1<1.$$

Similarly, considering $(\mathbf{u}+\mathbf{u}_1)\cdot(\mathbf{u}+\mathbf{u}_1)$, we find that

$$\mathbf{u}\cdot\mathbf{u}_1>-1$$

provided $\mathbf{u} \neq \mathbf{u}_1$. Thus

$$|\mathbf{u} \cdot \mathbf{u}_1| < 1 \qquad (2.14)$$

unless $\mathbf{u} = \pm\mathbf{u}_1$. Using the word 'iff' to denote 'if and only if', we can easily see that

$$\mathbf{u} \cdot \mathbf{u}_1 = 1 \text{ iff } \mathbf{u} = \mathbf{u}_1 \qquad (2.15a)$$

and

$$\mathbf{u} \cdot \mathbf{u}_1 = -1 \text{ iff } \mathbf{u} = -\mathbf{u}_1. \qquad (2.15b)$$

Equations (2.14)–(2.15) establish the essential condition (2.6) for accepting the interpretation (2.5). They go further, however. The statement (2.15a) tells us that if $\cos\theta = 1$ in (2.5), so that the angle between $\mathbf{u}$ and $\mathbf{u}_1$ is zero, then $\mathbf{u}$ and $\mathbf{u}_1$ are the same unit vector. Likewise, (2.15b) tells us that there is only one unit vector making an angle $\pi$ with a given unit vector $\mathbf{u}$, and that this vector is $-\mathbf{u}$.

To summarise this section, we have laid down Axioms (4A)–(4E) defining the scalar product $\mathbf{a} \cdot \mathbf{b}$ of any two vectors $\mathbf{a}$ and $\mathbf{b}$; when $\mathbf{b} = \mathbf{a}$, (2.10) defines the modulus of a vector, which has been shown to have the basic properties of 'length'. The cosine of the angle $\theta$ between two vectors has been defined by (2.2) or (2.5), and has been shown to have the essential properties (2.6), and (2.15). Later in the chapter, when we have introduced Cartesian coordinates, we shall see that we have given a definition of 'cosine' which accords with its usual trigonometrical meaning.

The result (2.6) or (2.14) gives rise to an important inequality relating the moduli of vectors. If the angle between two vectors $\mathbf{a}$, $\mathbf{b}$ is $\theta$, then, using (2.2) and (2.3),

$$\begin{aligned} |\mathbf{a}+\mathbf{b}|^2 &= a^2 + b^2 + 2ab\cos\theta \\ &\leqslant a^2 + b^2 + 2ab \\ &= (a+b)^2. \end{aligned}$$

Since $a$, $b$ and $|\mathbf{a}+\mathbf{b}|$ are non-negative,

$$|\mathbf{a}+\mathbf{b}| \leqslant a + b, \qquad (2.16)$$

and by (2.15a), the equality holds only if $\theta = 0$. Equation (2.16) is known as the **triangle inequality**; its geometrical interpretation, that the sum of the lengths $a$, $b$ of two sides of a triangle is greater than the length $|\mathbf{a}+\mathbf{b}|$ of the third side, is well known.

Suppose that **a** and **b** are two non-zero vectors, so that $a \neq 0$ and $b \neq 0$. Then if their scalar product is zero,

$$\mathbf{a} \cdot \mathbf{b} = 0, \tag{2.17}$$

it follows from (2.2) that $\cos\theta = 0$. Since $0 \leqslant \theta \leqslant \pi$, $\theta = \frac{1}{2}\pi$. Two non-zero displacements, represented by **a** and **b** and satisfying (2.17), are thus **perpendicular**; non-zero *vectors* **a** and **b** satisfying (2.17) are said to be **orthogonal** or **normal**. The concept of orthogonality plays an important role, not only in geometry, but in many other applications of the algebra of linear spaces.

### *Problems 2.1*

1 Use Axiom (4C) to show that the modulus of **0** is zero.

2 Using (2.4 a) and (2.11), show that $\lambda\mathbf{a}$ and **a** have the same unit vector associated with them when $\lambda > 0$. What happens for other values of $\lambda$?

3 Assuming that $\cos\theta$ is defined for $0 \leqslant \theta \leqslant \pi$, give definitions of $\sin\theta$ and $\tan\theta$ in terms of $\cos\theta$, for the same range of angles.

4 If **a** and **b** are vectors, and $a$ and $b$ are their lengths, derive the identity

$$(\mathbf{a}+\mathbf{b}) \cdot (\mathbf{a}-\mathbf{b}) = a^2 - b^2$$

from the axioms and the definition (2.3).
If $a$ and $b$ represent displacements along adjacent sides of a parallelogram, interpret the above identity in terms of the geometry of the parallelogram.

5 Simplify the following expressions:
(i) $(\mathbf{a}+\mathbf{d}) \cdot (\mathbf{b}-\mathbf{c}) + (\mathbf{a}-\mathbf{d}) \cdot (\mathbf{b}+\mathbf{c})$,
(ii) $\mathbf{a} \cdot (\mathbf{b}-\mathbf{c}) + \mathbf{b} \cdot (\mathbf{c}-\mathbf{a}) + \mathbf{c} \cdot (\mathbf{a}-\mathbf{b})$,
(iii) $\mathbf{a} \cdot (\lambda\mathbf{b} - \mu\mathbf{c}) + \mathbf{b} \cdot (\lambda\mathbf{c} - \mu\mathbf{a}) + \mathbf{c} \cdot (\lambda\mathbf{a} - \mu\mathbf{b})$.

6 Prove the statements (2.15a) and (2.15b).

## 2.2 *Linear dependence and dimension*

Suppose that **a, b, c**, . . . , **e** are a finite number of vectors, and that they obey a relationship of the form

$$\alpha\mathbf{a} + \beta\mathbf{b} + \gamma\mathbf{c} + \ldots + \varepsilon\mathbf{e} = \mathbf{0}, \tag{2.18}$$

where the real numbers $\alpha, \beta, \gamma, \ldots, \varepsilon$ are not all zero. Then the vectors **a**, **b**, **c**, . . . , **e** are said to be **linearly dependent**. (If $\alpha=\beta=\gamma=\ldots=\varepsilon=0$, (2.18) is always trivially satisfied.) The geometric significance of linear dependence can be seen by considering two examples.

**Example 2.3**

Let **a** be any vector, and $\lambda$ any real number. Then **a** and $\mathbf{b}\equiv\lambda\mathbf{a}$ are linearly dependent, since

$$-\lambda\,\mathbf{a}+\mathbf{b}=\mathbf{0}.$$

This is a relationship of the form (2.18), and the coefficient of **b** is non-zero.

Geometrically, a displacement $\lambda\mathbf{a}$ from a point $P$ is either 'in the same line as' **a**, as in Fig. 1.11, or (if $\lambda=0$) it is the zero displacement **0**. So linear dependence of two displacements **a** and **b** from the same point $P$ means that $P$ and the two end-points, $Q$ and $R$, are **collinear**; in fact, we take this to be the *definition* of collinearity of three points. Note that **a** or **b** (or both) may be **0**. If **a** and **b** represent displacements from different points, these displacements are parallel (or possibly zero).

**Example 2.4**

Suppose that three vectors **a**, **b**, **c** are linearly dependent, with

$$\alpha\mathbf{a}+\beta\mathbf{b}+\gamma\mathbf{c}=\mathbf{0}.$$

Not all of $\alpha$, $\beta$, $\gamma$ are zero; suppose $\gamma\neq 0$, for example. Then defining finite real numbers $\lambda=-\alpha/\gamma$ and $\mu=-\beta/\gamma$,

$$\mathbf{c}=\lambda\mathbf{a}+\mu\mathbf{b}. \tag{2.19}$$

If **a** and **b** represent displacements, they are in general not parallel. Taking them to represent displacements from a single point $P$, as in Fig. 2.4, they then define a plane. If **c** is also represented as a displacement from $P$, as in the figure, then **c** 'lies in the plane' defined by **a** and **b**. We can formally define the **plane** as the set of points whose position vectors relative to a point $P$ are of the form (2.19); $\lambda$ and $\mu$ may take any real values.

If **a** and **b** in (2.19) are linearly dependent, representing displacements in the same line, they do not define a plane. Displacements **c** are then in the same direction as **a** and **b**.

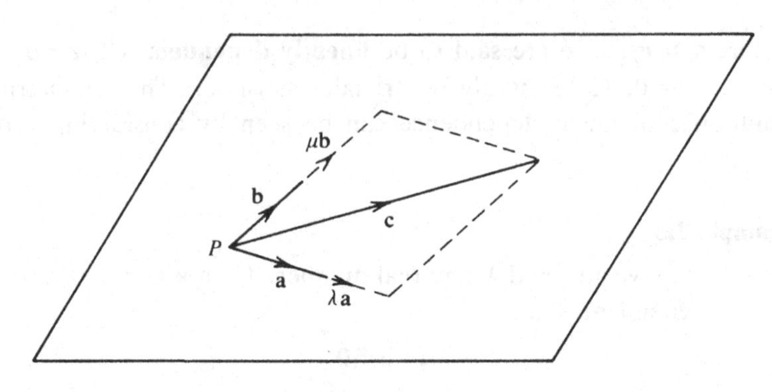


Fig. 2.4

Generally, if the set **a**, **b**, . . . , **e** are linearly dependent, then the set **a**, **b**, . . . , **e**, **f** (with *any* other vector **f** added) are also linearly dependent, for we can add the term 0**f** to the left-hand side of (2.18), obtaining a non-trivial relation of the same kind, but involving **f** as well as **a**, **b**, . . . , **e**.

A set of vectors **a**, **b**, **c**, . . . , **e** is **linearly independent** if the only solution of (2.18) is the trivial solution

$$\alpha = \beta = \gamma = \ldots = \varepsilon = 0. \tag{2.20}$$

It follows at once from Example 2.3 that two linearly independent displacements are non-zero and are *not* parallel; and from Example 2.4, that three linearly independent displacements **a**, **b**, **c** from a point do *not* lie in a plane, since no relationship of the type (2.19) exists.

We now state and prove an important theorem dealing with the 'expansion' of a vector **v**. Suppose that **a**, **b**, **c**, . . . , **e** are a linearly independent set of vectors, and that the vector **v** can be expressed as a linear combination.

$$\mathbf{v} = \alpha'\mathbf{a} + \beta'\mathbf{b} + \gamma'\mathbf{c} + \ldots + \varepsilon'\mathbf{e} \tag{2.21}$$

of this set. Then the **uniqueness of expansion theorem** states that the linear independence of **a**, **b**, . . . , **e** implies that the coefficients $\alpha', \beta', \gamma', \ldots, \varepsilon'$ in (2.21) are unique.

The proof of this theorem is simple. Suppose that **v** can also be expanded as

$$\mathbf{v} = \alpha''\mathbf{a} + \beta''\mathbf{b} + \gamma''\mathbf{c} + \ldots + \varepsilon''\mathbf{e}.$$

Subtracting this equation from (2.21) gives

$$(\alpha' - \alpha'')\mathbf{a} + (\beta' - \beta'')\mathbf{b} + \ldots + (\varepsilon' - \varepsilon'')\mathbf{e} = \mathbf{0}.$$

This is a relationship of type (2.18) with $\alpha = \alpha' - \alpha''$, $\beta = \beta' - \beta''$, and so on. The condition (2.20) of linear independence therefore gives

$$\alpha' = \alpha'', \beta' = \beta'', \ldots, \varepsilon' = \varepsilon''.$$

So the expansion (2.21) is unique.

As an example of the uniqueness of expansion theorem, consider the expansion (2.19) of a displacement **c** lying in the plane of two non-parallel displacements **a** and **b**. Since **a** and **b** correspond to linearly independent vectors, the coefficients $\lambda$ and $\mu$ in (2.19) are uniquely determined. When **a** and **b** are collinear displacements (linearly dependent), so that **b** is a multiple of **a**, then **c** is also a multiple of **a**; the coefficients $\lambda$ and $\mu$ in (2.19) are then not uniquely fixed.

In this book, we are studying vectors in three-dimensional space. So far we have not introduced the concept of the dimension of a linear vector space. The **dimension** of a linear space $\{\mathbf{a}\}$, defined by Axioms (1A)–(1D) and (2A)–(2E), is the *maximum number of linearly independent vectors* in the space. Thus a one-dimensional space consists of all vectors of the form $\lambda\mathbf{a}$, where **a** is a single non-zero vector; if these vectors $\lambda\mathbf{a}$ correspond to displacements from a point $P$, the end-points define a **line** through that point, as in Fig. 1.11.

Likewise, a plane is a two-dimensional space of displacements; if **a** and **b** are two linearly independent displacements from a given point, then all displacements **c** from the point are uniquely expressible in the form (2.19).

For brevity, we shall refer to $n$-dimensional space as '$n$-space', so that a line is a 1-space of displacements and a plane is a 2-space. In 3-space, let **a**, **b**, **c** be three linearly independent vectors. If **d** is any other vector in the 3-space, **a**, **b**, **c** and **d** must be linearly dependent; so there is a non-trivial relationship of the form

$$\alpha\mathbf{a} + \beta\mathbf{b} + \gamma\mathbf{c} + \delta\mathbf{d} = \mathbf{0}.$$

In this equation $\delta \neq 0$, since $\delta = 0$ would imply that **a**, **b** and **c** were linearly dependent. So we can divide out by $\delta$, putting $\lambda = -\alpha/\delta$, $\mu = -\beta/\delta$, $\nu = -\gamma/\delta$; this gives

$$\mathbf{d} = \lambda\mathbf{a} + \mu\mathbf{b} + \nu\mathbf{c}, \tag{2.22}$$

the **expansion** of **d** in terms of **a**, **b**, **c**. Since **a**, **b**, **c** are linearly independent, the uniqueness of expansion theorem tells us that this expansion is unique. Since any vector **d** can be expanded uniquely in the form (2.22), the vectors (**a**, **b**, **c**) are said to form a **basis** in the 3-space; they are also said to **span** the 3-space.

Quite generally, an $n$-space is (by definition) spanned by $n$ linearly independent vectors; any other vector in the $n$-space has a unique expansion in terms of them. Equations (2.19) and (2.22) are examples of these expansions, in 2-space and 3-space respectively. Some important linear spaces used in mathematics contain an arbitrarily large number of linearly independent vectors; these spaces are said to be **infinite-dimensional**. When a scalar product is defined in such a space, satisfying Axioms (4A)–(4E), the space is called a **Hilbert space**. We can if we wish look upon a finite-dimensional space, for example 3-space, as a special kind of Hilbert space.

We have now written down and discussed the full set of axioms defining a vector space of $n$ dimensions with a scalar product, and we have given further axioms defining points and displacements in geometrical or 'Euclidean' space. The algebraic axioms are known to be self-consistent, and the geometrical interpretation of the axioms provides us with an intuitive 'picture' or 'model' of the axioms. The algebraic and geometrical ideas therefore give each other mutual support: the consistency of the axiomatic algebraic scheme assures us that Euclidean geometry rests on a coherent and simple mathematical foundation, while the geometrical model in 3-space provides us with an intuitive picture of abstract vectors in terms of lengths, angles and other familiar geometrical concepts. A further advantage is that these geometrical intuitions can be used to understand the algebra of spaces of four and higher dimensions, and even of infinite-dimensional spaces. This advantage will not be apparent in this book, however, since we are restricting our study to 3-space.

## *Problems 2.2*

1 $A$, $B$, $C$ are three non-collinear points and $\lambda$, $\mu$, $\nu$ are non-zero real numbers. $D$, $E$, $F$ are points on $BC$, $CA$, $AB$ respectively, such that

$$\lambda AF = \mu FB, \qquad \mu BD = \nu DC, \qquad \nu CE = \lambda EA.$$

Show that $AD$, $BE$ and $CF$ meet at a point. Find the position vector of the point in terms of the position vectors of $A$, $B$ and $C$. (Ceva's theorem.)

2 Vectors **a** and **b** are linearly independent. A set of points $P(n_1, n_2)$ is defined to have position vectors

$$\mathbf{v}(n_1, n_2) = n_1\mathbf{a} + n_2\mathbf{b}$$

relative to an origin $O$, with the numbers $n_1$ and $n_2$ each taking all integral values $0, \pm1, \pm2, \pm3, \ldots$. Draw a diagram to indicate the positions of these points. What happens when **a** and **b** are linearly dependent?

3 A vector **c** is expressed in the form

$$\mathbf{c} = \lambda\mathbf{a} + \mu\mathbf{b},$$

where **a** and **b** are linearly independent. By taking scalar products of this equation with **a** and **b**, express $\lambda$ and $\mu$ in terms of scalar products of the three vectors. Why is linear independence necessary?

4 A vector **d** is expanded as

$$\mathbf{d} = \lambda\mathbf{a} + \mu\mathbf{b} + \nu\mathbf{c},$$

where **a**, **b**, **c** are linearly independent. Show how to find $\lambda$, $\mu$, $\nu$ in terms of scalar products of the four vectors.

5 Two unit vectors **i** and **j** satisfy $\mathbf{i} \cdot \mathbf{j} = 0$. Show that **i** and **j** are linearly independent.

If **c** is a third vector such that **i**, **j**, **c** are linearly independent, and $\mathbf{c}'$ is defined by

$$\mathbf{c}' = \mathbf{c} - (\mathbf{c} \cdot \mathbf{i})\mathbf{i} - (\mathbf{c} \cdot \mathbf{j})\mathbf{j},$$

show that
(i) $\mathbf{c}' \cdot \mathbf{i} = \mathbf{c}' \cdot \mathbf{j} = 0$,
(ii) $\mathbf{c}' \neq \mathbf{0}$.
Interpret (i) and (ii) geometrically.

## *2.3 Components of a vector*

In interpreting Equation (2.12), we introduced the projection $\mathbf{b} \cdot \mathbf{u}$ of a vector **b** in the direction of a unit vector **u**. The projection is a scalar, and is represented geometrically in Fig. 2.2. If we multiply this scalar into the unit vector **u**, we obtain the vector

$$\mathbf{b}_{\|} \equiv (\mathbf{b} \cdot \mathbf{u})\mathbf{u}; \qquad (2.23)$$

this vector is represented in Fig. 2.2 by the displacement **OP**. The vector $\mathbf{b}_{\|}$ is called the **component of b parallel to u**, explaining the notation $\mathbf{b}_{\|}$. The vector represented by the displacement **PB** in Fig.

2.2 is then

$$\mathbf{b}_\perp \equiv \mathbf{b} - \mathbf{b}_\parallel$$
$$= \mathbf{b} - (\mathbf{b} \cdot \mathbf{u})\mathbf{u}. \tag{2.24}$$

Equation (2.24) defines the **component of b orthogonal to u**. It is intuitively obvious that the displacement $\mathbf{PB} = \mathbf{b}_\perp$ in Fig. 2.2 is orthogonal to $\mathbf{u}$; it is easy to see that $\mathbf{b}_\perp$ and $\mathbf{u}$ satisfy the orthogonality condition (2.17), since from (2.24),

$$\mathbf{b}_\perp \cdot \mathbf{u} = \mathbf{b} \cdot \mathbf{u} - (\mathbf{b} \cdot \mathbf{u})(\mathbf{u} \cdot \mathbf{u}) = 0.$$

In coordinate geometry we study displacements in terms of their components along three mutually orthogonal directions, as indicated in Fig. 1.1. We are now in a position to introduce Cartesian coordinates for vectors by defining a set of three unit vectors, $(\mathbf{i}, \mathbf{j}, \mathbf{k})$, known as a **triad**, such that

$$\mathbf{i} \cdot \mathbf{j} = \mathbf{j} \cdot \mathbf{k} = \mathbf{k} \cdot \mathbf{i} = 0. \tag{2.25}$$

Geometrically, $\mathbf{i}$, $\mathbf{j}$ and $\mathbf{k}$ are mutually orthogonal and represent unit displacements along the three Cartesian axes. We must first establish that triads exist. In 3-space, we know that we can find three linearly independent (and hence non-zero) vectors $\mathbf{a}$, $\mathbf{b}$ and $\mathbf{c}$. Define the first vector of the triad as

$$\mathbf{i} = a^{-1}\mathbf{a};$$

this is the unit vector corresponding to $\mathbf{a}$, indicated in Fig. 2.5. Then, as in (2.2), define

$$\mathbf{b}_\perp = \mathbf{b} - (\mathbf{b} \cdot \mathbf{i})\mathbf{i},$$

orthogonal to $\mathbf{i}$. Clearly $\mathbf{b}_\perp \neq \mathbf{0}$; for if $\mathbf{b}_\perp = \mathbf{0}$, $\mathbf{b} = (\mathbf{b} \cdot \mathbf{i})\mathbf{i}$, meaning that $\mathbf{a}$ and $\mathbf{b}$ would not be linearly independent. Let $\mathbf{j}$ be the unit vector corresponding to $\mathbf{b}_\perp$, again indicated in Fig. 2.5. Then $\mathbf{i} \cdot \mathbf{j} = 0$, as required. To define the third vector $\mathbf{k}$ of the triad, we consider the vector (see Problems 2.2, Question 5)

$$\mathbf{c}' = \mathbf{c} - (\mathbf{c} \cdot \mathbf{i})\mathbf{i} - (\mathbf{c} \cdot \mathbf{j})\mathbf{j}. \tag{2.26}$$

Here we have subtracted from $\mathbf{c}$ its components along $\mathbf{i}$ and $\mathbf{j}$, and it is easy to check that

$$\mathbf{c}' \cdot \mathbf{i} = \mathbf{c}' \cdot \mathbf{j} = 0.$$

Further, $\mathbf{c}' \neq \mathbf{0}$, since if $\mathbf{c}' = \mathbf{0}$, the vectors $\mathbf{a}$, $\mathbf{b}$, $\mathbf{c}$ would be linearly dependent. We therefore define $\mathbf{k}$ as the unit vector corresponding to $\mathbf{c}'$; this satisfies (2.25), completing the triad $(\mathbf{i}, \mathbf{j}, \mathbf{k})$. A triad is also known as an **orthonormal basis**.

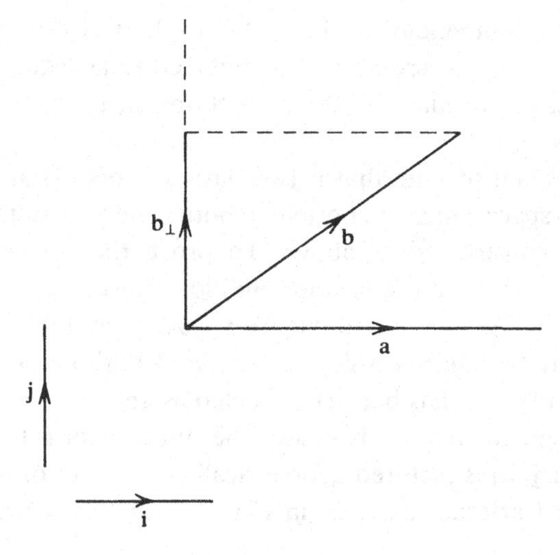


Fig. 2.5

The vectors **i**, **j**, **k** are linearly independent; for if

$$\alpha\mathbf{i}+\beta\mathbf{j}+\gamma\mathbf{k}=\mathbf{0},$$

then taking scalar products with **i**, **j** and **k** successively and using (2.25), we find

$$\alpha=\beta=\gamma=0.$$

So, using (2.22), it follows that *any* vector **v** in 3-space can be expressed in the form

$$\mathbf{v}=v_1\mathbf{i}+v_2\mathbf{j}+v_3\mathbf{k}. \tag{2.27}$$

Further, the uniqueness of expansion theorem tells us that the coefficients $v_1$, $v_2$, $v_3$ in (2.27) are unique. By forming the scalar products of (2.27) with **i**, **j** and **k** in turn, we find

$$v_1=\mathbf{v}\cdot\mathbf{i}, \qquad v_2=\mathbf{v}\cdot\mathbf{j}, \qquad v_3=\mathbf{v}\cdot\mathbf{k}. \tag{2.28}$$

Comparison with (2.17) tells us that $v_1$, $v_2$, $v_3$ are the projections of **v** along **i**, **j** and **k** respectively. The vectors in the vector sum (2.27),

$$\left.\begin{aligned} v_1\mathbf{i}&\equiv(\mathbf{v}\cdot\mathbf{i})\mathbf{i},\\ v_2\mathbf{j}&\equiv(\mathbf{v}\cdot\mathbf{j})\mathbf{j}\\ \text{and}\qquad v_3\mathbf{k}&\equiv(\mathbf{v}\cdot\mathbf{k})\mathbf{k}\end{aligned}\right\} \tag{2.29}$$

are called the **components** of the vector **v** relative to the basis **i**, **j**, **k**. The scalars $v_1$, $v_2$, $v_3$ are also often referred to as 'components'; to avoid confusion we shall call the projections $v_1$, $v_2$, $v_3$ the **Cartesian components**.

It is important to note that if two basis vectors **i**, **j** are given, *all* vectors in 3-space which are normal to both **i** and **j** are of the form $\lambda\mathbf{k}$, where **k** is constructed as above. To prove this, use the general expression (2.27) for a vector; the condition that **v** is normal to **i** and **j** gives, using (2.28), $v_1 = v_2 = 0$. Hence $\mathbf{v} = v_3\mathbf{k}$, a multiple of **k**. There are thus only two *unit* vectors orthogonal to both **i** and **j**; assuming that one of them, **k**, has been constructed as above from a particular linearly independent set **a**, **b**, **c**, the other unit vector is then $-\mathbf{k}$.

A triad (**i**, **j**, **k**) is pictured geometrically as the set of unit vectors along three Cartesian axes, as in Fig. 2.6. In this figure, we have

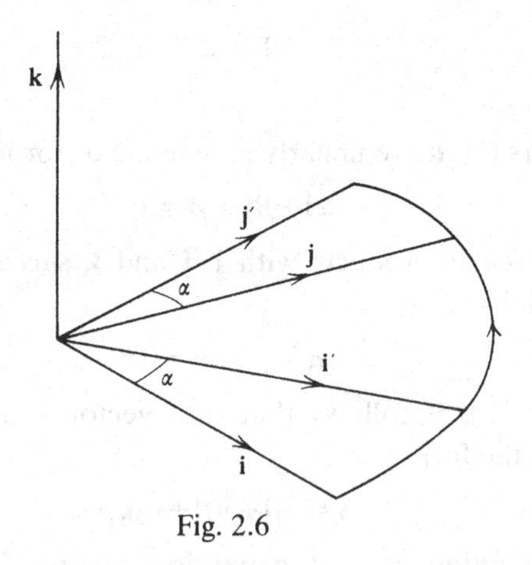


Fig. 2.6

drawn (**i**, **j**, **k**) as a **right-handed triad**, meaning that a rotation by the smaller angle from **i** to **j** (indicated by the arrow on the arc of the circle), accompanied by a translation in the direction of **k**, gives a right-hand screw motion. With this convention, we see that (**i**, **j**, $-$**k**), the only other triad containing **i** and **j**, is **left-handed**. The concept of 'right-handedness' is *purely geometrical*, and there is nothing in the *algebraic* definition of (**i**, **j**, **k**) which defines such a concept. We shall, however, follow the usual geometrical convention of associating a triad denoted by (**i**, **j**, **k**) with a right-handed set of axes, unless we specifically introduce a left-handed triad.

In the plane defined by **i** and **j**, the rotation through $\frac{1}{2}\pi$ from **i** to **j** is said to be a **rotation in the positive sense**. If **i**′ and **j**′ are two other mutually orthogonal unit vectors in the plane of **i** and **j**, as shown in Fig. 2.6, and if

$$\mathbf{i}' \cdot \mathbf{i} = \mathbf{j}' \cdot \mathbf{j}, \tag{2.30}$$

then the angles $\alpha$ marked in the figure are equal. We then say that the rotation from **i**′ to **j**′ through $\frac{1}{2}\pi$ is also in the positive sense. But if

$$\mathbf{i}' \cdot \mathbf{i} = -\mathbf{j}' \cdot \mathbf{j}, \tag{2.31}$$

then the rotation through $\frac{1}{2}\pi$ from **i**′ to **j**′ is a **rotation in the negative sense**.

More generally, suppose that **a** and **b** are any two non-zero vectors in the plane of **i** and **j**. Let **i**′ be the unit vector corresponding to **a**, and let **j**′ be the unit vector corresponding to the component $\mathbf{b}_\perp$ defined by (2.24); these replace **i** and **j** in Fig. 2.5. Then the **sense of rotation** (positive or negative) from **a** to **b**, through the smaller angle $\theta$ between them is the same as sense of rotation from **i**′ to **j**′. The sense of rotation from **a** to **b** through the reflex angle $(2\pi - \theta)$ is defined to be in the opposite sense to the rotation through the smaller angle $\theta$.

From the general expansion (2.27) of a vector in terms of its components and from (2.25), it is easy to establish certain rules of manipulation. Multiplying (2.27) by a scalar $\lambda$,

$$\lambda\mathbf{v} = (\lambda v_1)\mathbf{i} + (\lambda v_2)\mathbf{j} + (\lambda v_3)\mathbf{k}; \tag{2.32}$$

so multiplication of a vector by $\lambda$ implies that its projections or its components are each multiplied by $\lambda$. If also

$$\mathbf{w} = w_1\mathbf{i} + w_2\mathbf{j} + w_3\mathbf{k}, \tag{2.33}$$

then

$$\mathbf{v} + \mathbf{w} = (v_1 + w_1)\mathbf{i} + (v_2 + w_2)\mathbf{j} + (v_3 + w_3)\mathbf{k}, \tag{2.34}$$

so that addition of vectors implies the addition of respective components, a result closely related to Axiom (4D). In deducing (2.32) and (2.34), the Axioms (2D) and (2E) have been used.

The rules governing scalar products of basis vectors **i**, **j**, **k** are

$$\mathbf{i} \cdot \mathbf{i} = \mathbf{j} \cdot \mathbf{j} = \mathbf{k} \cdot \mathbf{k} = 1 \tag{2.35}$$

and the orthogonality conditions (2.25):

$$\mathbf{i} \cdot \mathbf{j} = \mathbf{j} \cdot \mathbf{k} = \mathbf{k} \cdot \mathbf{i} = 0. \tag{2.25}$$

If $\mathbf{v}$ and $\mathbf{w}$ are given by (2.27) and (2.33), then

$$\begin{aligned}\mathbf{v}\cdot\mathbf{w} &= (v_1\mathbf{i}+v_2\mathbf{j}+v_3\mathbf{k})\cdot(w_1\mathbf{i}+w_2\mathbf{j}+w_3\mathbf{k}) \\ &= v_1w_1(\mathbf{i}\cdot\mathbf{i})+v_2w_2(\mathbf{j}\cdot\mathbf{j})+v_3w_3(\mathbf{k}\cdot\mathbf{k}) \\ &\quad +(v_1w_2+v_2w_1)\mathbf{i}\cdot\mathbf{j}+(v_1w_3+v_3w_1)\mathbf{i}\cdot\mathbf{k} \\ &\quad +(v_2w_3+v_3w_2)\mathbf{j}\cdot\mathbf{k}.\end{aligned}$$

Using (2.35) and (2.25), this gives

$$\mathbf{v}\cdot\mathbf{w}=v_1w_1+v_2w_2+v_3w_3. \tag{2.36}$$

This is the formula for the scalar product of any two vectors $\mathbf{v}$ and $\mathbf{w}$ in terms of the components of $\mathbf{v}$ and $\mathbf{w}$. The condition that non-zero vectors $\mathbf{v}$ and $\mathbf{w}$ are orthogonal is that expression (2.36) is zero.

The modulus $v$ of $\mathbf{v}$ is given by putting $\mathbf{w}\equiv\mathbf{v}$. Then (2.36) and (2.3) give

$$v^2=\mathbf{v}\cdot\mathbf{v}=v_1^2+v_2^2+v_3^2. \tag{2.37}$$

If $\mathbf{v}$ is interpreted as a displacement from a point $P$, (2.37) is immediately identifiable as the Pythagorean relation (1.1). We have therefore established Pythagoras' theorem as a theorem in vector algebra.

**Example 2.5**

Evaluate the scalar product of the two vectors

$$\mathbf{v}=\mathbf{i}+2\mathbf{j}+3\mathbf{k},$$

$$\mathbf{w}=2\mathbf{i}-3\mathbf{j}+2\mathbf{k}.$$

Find the moduli of the two vectors.

The vectors $\mathbf{v}$ and $\mathbf{w}$ are given by (2.27) and (2.33) with

$$(v_1, v_2, v_3)=(1, 2, 3)$$

and

$$(w_1, w_2, w_3)=(2, -3, 2).$$

The scalar product, given by (2.36), is

$$\mathbf{v}\cdot\mathbf{w}=1\cdot 2+2\cdot(-3)+3\cdot 2=2.$$

The modulus of $\mathbf{v}$ is given by (2.37):

$$v^2=1^2+2^2+3^2=14.$$

Since $v>0$, $v=+\sqrt{14}$. Similarly

$$w=(2^2+3^2+2^2)^{\frac{1}{2}}=\sqrt{17}.$$

**Example 2.6**

Two vectors are given in terms of a triad (**i**, **j**, **k**) by

$$\mathbf{a} = 3\mathbf{i} - \mathbf{j} - \mathbf{k},$$

$$\mathbf{b} = 2\mathbf{i} + 2\mathbf{j} - 4\mathbf{k}.$$

Find the unit vectors corresponding to **a** and **b**, and the angle between **a** and **b**.

The moduli of **a** and **b** are given by (2.37):

$$a^2 = 3^2 + 1^2 + 1^2 = 11,$$

so that $a = \sqrt{11}$. Likewise $b = \sqrt{24} = 2\sqrt{6}$. So by (2.4), the unit vectors $\mathbf{u}$ and $\mathbf{u}_1$ corresponding to **a** and **b** are

$$\mathbf{u} = \frac{1}{\sqrt{11}}(3\mathbf{i} - \mathbf{j} - \mathbf{k}),$$

$$\mathbf{u}_1 = \frac{1}{\sqrt{6}}(\mathbf{i} + \mathbf{j} - 2\mathbf{k}).$$

The angle between **a** and **b** is given by (2.2) or, equivalently, by (2.5). We use (2.5) to give

$$\cos\theta = \frac{1}{\sqrt{66}}(3\mathbf{i} - \mathbf{j} - \mathbf{k}) \cdot (\mathbf{i} + \mathbf{j} - 2\mathbf{k}).$$

Evaluating the scalar product by using (2.36),

$$\cos\theta = \frac{1}{\sqrt{66}}(3 - 1 + 2) = \frac{4}{\sqrt{66}}.$$

Since $0 \leqslant \theta \leqslant \pi$, $\theta$ is the acute angle arccos $(4\sqrt{66})$.

## ■ *Problems 2.3*

[Throughout, (**i**, **j**, **k**) is taken to be a triad.]

1 Three vectors **a**, **b**, **c** are defined by

$$\mathbf{a} = \mathbf{i} - 2\mathbf{j} + 4\mathbf{k},$$

$$\mathbf{b} = 2\mathbf{i} + \mathbf{j},$$

$$\mathbf{c} = -\mathbf{i} + \mathbf{j} - 3\mathbf{k}.$$

Evaluate $2\mathbf{a} - \mathbf{b}$, $\mathbf{a} + \mathbf{b} + \mathbf{c}$, $\mathbf{a} + 2\mathbf{b} + 3\mathbf{c}$.

2 Vectors **a**, **b**, **c** are defined as in Question 1. Find the unit vectors corresponding to **a**, **b** and **c**, and find the angles between the three pairs of vectors.

3 $\mathbf{i}'$ and $\mathbf{j}'$ are orthogonal unit vectors. Vectors $\mathbf{a}$ and $\mathbf{b}$ are defined in terms of $\mathbf{i}'$, $\mathbf{j}'$ by

$$\mathbf{a} = 3\mathbf{i}' + 4\mathbf{j}',$$

$$\mathbf{b} = \mathbf{i}' - \mathbf{j}'.$$

Show that $\mathbf{a}$ and $\mathbf{b}$ are linearly independent. By using (2.4a) and (2.24), construct a second set of orthogonal unit vectors $\mathbf{i}$ and $\mathbf{j}$, expressed in terms of $\mathbf{i}'$ and $\mathbf{j}'$. Check that $\mathbf{i}$ and $\mathbf{j}$ are orthogonal.

4 $(\mathbf{i}', \mathbf{j}', \mathbf{k}')$ is a triad. Vectors $\mathbf{a}$ and $\mathbf{b}$ are defined by

$$\mathbf{a} = 2\mathbf{i}' + 2\mathbf{j}' + \mathbf{k}',$$

$$\mathbf{b} = 4\mathbf{i}' + \mathbf{j}' - \mathbf{k}'.$$

By using (2.4a) and (2.24), construct the unit vector $\mathbf{i}$ corresponding to $\mathbf{a}$, and a unit vector $\mathbf{j}$, normal to $\mathbf{i}$, in the plane of $\mathbf{a}$ and $\mathbf{b}$.

Find a unit vector $\mathbf{k}$ satisfying

$$\mathbf{k} \cdot \mathbf{a} = \mathbf{k} \cdot \mathbf{b} = 0.$$

Check that $(\mathbf{i}, \mathbf{j}, \mathbf{k})$ is a triad.

5 Define a rhombus in terms of position vectors. Prove that the diagonals of a rhombus are orthogonal.

6 Vectors $\mathbf{a}$, $\mathbf{b}$ and $\mathbf{c}$ satisfy the relation $(\mathbf{c} - \mathbf{a}) \cdot \mathbf{a} = (\mathbf{c} - \mathbf{b}) \cdot \mathbf{b}$. Show that $\mathbf{c} - (\mathbf{a} + \mathbf{b})$ is orthogonal to $\mathbf{a} - \mathbf{b}$. Draw a diagram to illustrate these relations when $\mathbf{a}$, $\mathbf{b}$, $\mathbf{c}$ represent displacements.

## *2.4 Geometrical applications*

Now that we have introduced scalar products and have thereby provided an axiomatic basis for Euclidean geometry, we can extend the range of geometrical theorems provable by vector methods. In §2.5 and in the next chapter we shall introduce techniques and concepts which allow a greater range of problems to be solved relatively easily. So, while in principle the whole of Euclidean geometry can now be established, we shall limit ourselves in this section to problems which can be solved easily using the methods we have established so far.

The simplest use of Cartesian components is in the geometry of a plane; for many practical purposes we can regard local travel on the earth's surface as being confined to a plane. Making this approximation, and choosing a point on the earth's surface as origin, we can take as basis two orthogonal unit displacements **i** and **j** in, say, the directions east (E) and north (N); then, as shown in Fig. 2.7, a nearby point $P$ on the earth's surface will have a position vector $\boldsymbol{\rho}$ of the form

$$\boldsymbol{\rho} = x\mathbf{i} + y\mathbf{j}. \tag{2.38}$$

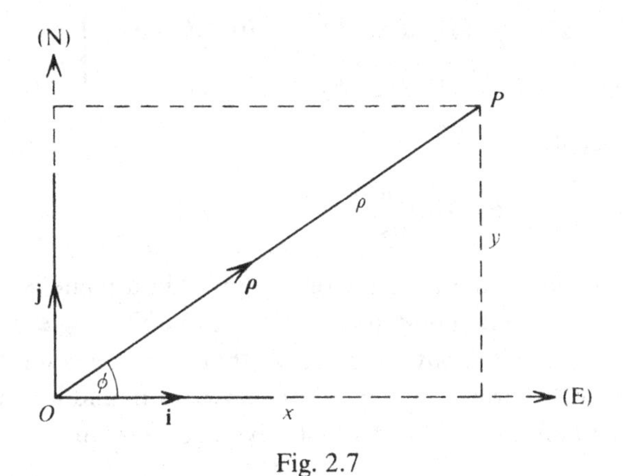


Fig. 2.7

The Cartesian components $(x, y)$ are the **rectangular coordinates** of $P$. If $P_0$ is another point with position vector

$$\boldsymbol{\rho}_0 = x_0\mathbf{i} + y_0\mathbf{j}, \tag{2.39}$$

then the displacement $\mathbf{P_0P}$ is

$$\boldsymbol{\rho} - \boldsymbol{\rho}_0 = (x - x_0)\mathbf{i} + (y - y_0)\mathbf{j}. \tag{2.40}$$

A succession of displacements $\boldsymbol{\rho}_1, \boldsymbol{\rho}_2, \ldots, \boldsymbol{\rho}_n$, where

$$\boldsymbol{\rho}_t = x_t\mathbf{i} + y_t\mathbf{j} \qquad (t = 1, 2, \ldots, n),$$

result in a total displacement, or **resultant**,

$$\begin{aligned} \boldsymbol{\rho} &\equiv \sum_{t=1}^{n} \boldsymbol{\rho}_t \\ &= \left(\sum_{t=1}^{n} x_t\right)\mathbf{i} + \left(\sum_{t=1}^{n} y_t\right)\mathbf{j}. \end{aligned} \tag{2.41}$$

This is simply an extension of the rule (2.34).

The trigonometric functions of angles are often useful for describing position vectors in a plane. The cosine of an angle, which we now denote by $\phi (0 \leq \phi \leq \pi)$, is defined by (2.5); to allow the use of the usual trigonometric formulae, we now extend the definition of $\cos \phi$

to the range $(0, 2\pi)$, and define $\sin \phi$ and $\tan \phi$ in this range. For $\pi < \phi \leqslant 2\pi$, we define

$$\cos \phi = \cos (2\pi - \phi). \tag{2.42}$$

The sine is defined by

$$\left.\begin{aligned} \sin \phi &= +(1-\cos^2 \phi)^{\frac{1}{2}} \qquad (0 \leqslant \phi \leqslant \pi), \\ \sin \phi &= -(1-\cos^2 \phi)^{\frac{1}{2}} \qquad (\pi < \phi \leqslant 2\pi), \end{aligned}\right\} \tag{2.43}$$

and the tangent by

$$\tan \phi = \frac{\sin \phi}{\cos \phi} \qquad (0 \leqslant \phi \leqslant 2\pi). \tag{2.44}$$

Suppose that the position vector of a point $P$ in a plane is given by (2.38), and that we define $\phi$ $(0 \leqslant \phi \leqslant 2\pi)$ to be the angle between $\mathbf{i}$ and $\boldsymbol{\rho}$, measured in the *positive sense*, so that $\phi = \frac{1}{2}\pi$ when $\boldsymbol{\rho} \equiv \mathbf{j}$. This is in accord with normal trigonometric convention, and it is not hard to show that definitions (2.42)–(2.44) give rise to the usual formulae

$$\cos \phi = \frac{x}{\rho}, \qquad \sin \phi = \frac{y}{\rho}, \qquad \tan \phi = \frac{y}{x}, \tag{2.45}$$

for all real values of $x$ and $y$. These relationships are shown in Fig. 2.7. We can extend the definition of trigonometric functions to all real values of $\phi$ by postulating that they are unchanged when $\phi$ is increased or decreased by $2\pi$; for example

$$\cos \phi = \cos (\phi + 2n\pi) \qquad (n + \pm 1, \pm 2, \ldots). \tag{2.46}$$

We now give some further examples of geometric applications of vectors.

**Example 2.7**

A ship makes successively the following movements:

10 kilometres north,
20 kilometres north-west,
12 kilometres 30° west of south,
6 kilometres south-east.

Find its overall change of position.

Taking **i** and **j** as unit displacements to the east and north, the four displacements are

$$\begin{aligned} &10\mathbf{j},\\ -&10\sqrt{2}\mathbf{i}+10\sqrt{2}\mathbf{j},\\ -&6\mathbf{i}-6\sqrt{3}\mathbf{j},\\ &3\sqrt{2}\mathbf{i}-3\sqrt{2}\mathbf{j}.\end{aligned}$$

The total displacement is thus

$$(-6-7\sqrt{2})\mathbf{i}+(10-6\sqrt{3}+7\sqrt{2})\mathbf{j}$$
$$\simeq -15.898\mathbf{i}+9.506\mathbf{j}.$$

This is a displacement of magnitude

$$[(15.898)^2+(9.506)^2]^{\frac{1}{2}} \simeq 18.5$$

in a direction

$$\tan^{-1}\frac{15.898}{9.506} \simeq 59°\,7'$$

west of north.

**Example 2.8**

Define an isosceles triangle vectorially. Prove that the angles at the base are equal.

Let $ABC$ be the triangle, with $AB = AC$. Choose $A$ as origin, so that it has position vector **0**. Let $B$ and $C$ have position vectors **b** and **c**, so that $\mathbf{AB} = \mathbf{b}$, $\mathbf{AC} = \mathbf{c}$ and $\mathbf{BC} = \mathbf{c}-\mathbf{b}$. Then if $b = c$, the triangle is defined to be isosceles. Let $\beta$ and $\gamma$ be the angles $ABC$ and $ACB$. Then by (2.2)

$$\mathbf{AB}\cdot\mathbf{CB} = \mathbf{b}\cdot(\mathbf{b}-\mathbf{c}) = b|\mathbf{b}-\mathbf{c}|\cos\beta,$$

so that

$$\cos\beta = \frac{b^2-\mathbf{b}\cdot\mathbf{c}}{b|\mathbf{b}-\mathbf{c}|}.$$

Likewise

$$\cos\gamma = \frac{c^2-\mathbf{b}\cdot\mathbf{c}}{c|\mathbf{b}-\mathbf{c}|}.$$

But since $b = c$, $\cos\beta = \cos\gamma$. Hence $\beta = \gamma$.

**Example 2.9**

Points $A$ and $B$ have position vectors

$$\mathbf{a} = 3\mathbf{i} - 2\mathbf{j} + \mathbf{k},$$
$$\mathbf{b} = -\mathbf{i} + 2\mathbf{j} - 7\mathbf{k}.$$

Find the position vector of the point $P$ on $AB$ such that $AP{:}PB = 3{:}1$.

The position vector of $P$ is, using (1.25),

$$\tfrac{1}{4}\mathbf{a} + \tfrac{3}{4}\mathbf{b} = \mathbf{j} - 5\mathbf{k}.$$

**Example 2.10**

$\alpha$, $\beta$, $\gamma$ are any real numbers. Show that points $P$, $Q$, $R$ with position vectors

$$\mathbf{p} = \alpha\mathbf{i} + \beta\mathbf{j} + \gamma\mathbf{k},$$
$$\mathbf{q} = \beta\mathbf{i} + \gamma\mathbf{j} + \alpha\mathbf{k},$$
$$\mathbf{r} = \gamma\mathbf{i} + \alpha\mathbf{j} + \beta\mathbf{k},$$

are the vertices of an equilateral triangle. Show that the angle $PQR$ is $\frac{1}{3}\pi$.

The displacements along the sides of the triangle are

$$\mathbf{PQ} = (\beta - \alpha)\mathbf{i} + (\gamma - \beta)\mathbf{j} + (\alpha - \gamma)\mathbf{k},$$
$$\mathbf{QR} = (\gamma - \beta)\mathbf{i} + (\alpha - \gamma)\mathbf{j} + (\beta - \alpha)\mathbf{k},$$
$$\mathbf{RP} = (\alpha - \gamma)\mathbf{i} + (\beta - \alpha)\mathbf{j} + (\gamma - \beta)\mathbf{k}.$$

So the length $l$ of each side is given by

$$l^2 = (\beta - \alpha)^2 + (\gamma - \beta)^2 + (\alpha - \gamma)^2,$$

and the triangle is equilateral. Denote the angle $QPR$ by $\theta$. Then

$$\begin{aligned} l^2 \cos\theta &= \mathbf{PQ} \cdot \mathbf{PR} \\ &= (\alpha - \beta)(\alpha - \gamma) + (\alpha - \beta)(\gamma - \beta) + (\alpha - \gamma)(\beta - \gamma) \\ &= \alpha^2 + \beta^2 + \gamma^2 - \beta\gamma - \gamma\alpha - \alpha\beta \\ &= \tfrac{1}{2}l^2. \end{aligned}$$

Hence $\cos\theta = \frac{1}{2}$ and $\theta = \frac{1}{3}\pi$.

**Example 2.11**

Prove that the perpendiculars from the vertices of a triangle onto the opposite sides are concurrent.

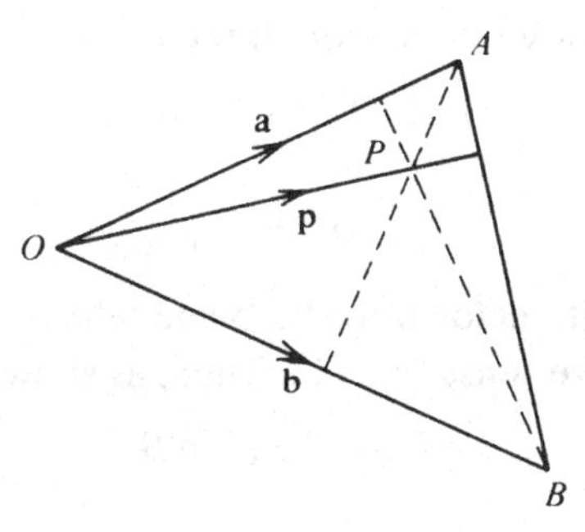


Fig. 2.8

Let $OAB$ be the triangle, as shown in Fig. 2.8. Choose $O$ as origin and let $\mathbf{OA}=\mathbf{a}$, $\mathbf{OB}=\mathbf{b}$, so that $\mathbf{AB}=\mathbf{b}-\mathbf{a}$. It is assumed that $O$, $A$, $B$ are distinct points, so that $\mathbf{a}\neq\mathbf{0}$, $\mathbf{b}\neq\mathbf{0}$, $\mathbf{a}-\mathbf{b}\neq\mathbf{0}$. If $P$ is a point such that $BP$ is perpendicular to $OA$ and $AP$ is perpendicular to $OB$, then

$$\left.\begin{aligned}(\mathbf{b}-\mathbf{p})\cdot\mathbf{a}=0,\\ (\mathbf{a}-\mathbf{p})\cdot\mathbf{b}=0.\end{aligned}\right\} \tag{2.47}$$

Subtracting these equations gives

$$\mathbf{p}\cdot(\mathbf{a}-\mathbf{b})=0,$$

showing that $\mathbf{p}$ is orthogonal to the (non-zero) vector $\mathbf{a}-\mathbf{b}$, or $OP$ is perpendicular to $AB$.

Note that (2.47) determine a unique position vector $\mathbf{p}$ in the plane. For $\mathbf{p}$ is of the form

$$\mathbf{p}=\lambda\mathbf{a}+\mu\mathbf{b},$$

so that (2.47) gives

$$\lambda a^2+\mu\mathbf{a}\cdot\mathbf{b}=\mathbf{a}\cdot\mathbf{b},$$

$$\lambda\mathbf{a}\cdot\mathbf{b}+\mu b^2=\mathbf{a}\cdot\mathbf{b},$$

which have a unique solution because $\mathbf{a}$, $\mathbf{b}$ are linearly independent. (See Problems 2.2, Question 3.)

The sine and cosine of an angle have been defined vectorially. We shall now use these definitions to establish familiar formulae for the sine and cosine of the sum of two angles. Suppose that $(\mathbf{i}, \mathbf{j}, \mathbf{k})$ is a triad, and that $\mathbf{i}'$, $\mathbf{j}'$ are two mutually orthogonal vectors in the plane defined by $\mathbf{i}$, $\mathbf{j}$, as indicated in Figs. 2.6 and 2.9, such that the sense of rotation from $\mathbf{i}'$ to $\mathbf{j}'$ is positive. Let $\alpha$ be the angle between $\mathbf{i}$ and $\mathbf{i}'$, measured in the positive sense from $\mathbf{i}$; then

$$\mathbf{i}' \cdot \mathbf{i} = \mathbf{j}' \cdot \mathbf{j} = \cos \alpha \tag{2.48a}$$

and

$$\mathbf{i}' \cdot \mathbf{j} = -\mathbf{j}' \cdot \mathbf{i} = \sin \alpha. \tag{2.48b}$$

Now consider a unit vector $\mathbf{u}$ in the plane which makes an angle $\beta$ with $\mathbf{i}'$ (in the positive sense from $\mathbf{i}'$). Then, as shown in Fig. 2.9,

$$\mathbf{u} = \mathbf{i}' \cos \beta + \mathbf{j}' \sin \beta. \tag{2.49}$$

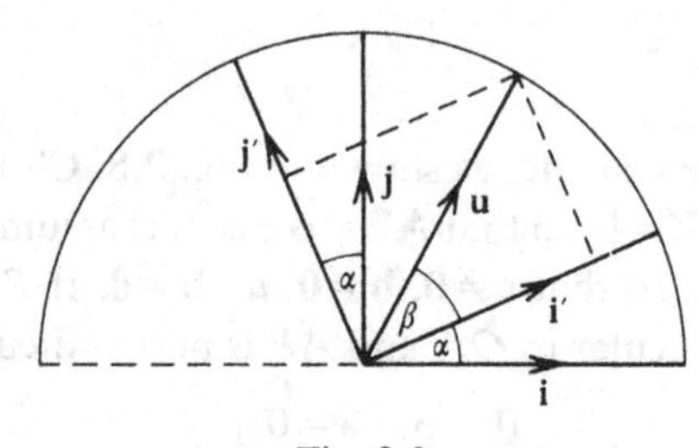


Fig. 2.9

The angle between $\mathbf{u}$ and $\mathbf{i}$ is called the **sum of the angles** $\alpha$ and $\beta$, and is denoted by $\alpha + \beta$; the cosine of this angle is, by definition,

$$\cos(\alpha + \beta) = \mathbf{u} \cdot \mathbf{i}.$$

Using (2.49) and (2.48), this becomes

$$\begin{aligned}\cos(\alpha + \beta) &= (\mathbf{i}' \cdot \mathbf{i}) \cos \beta + (\mathbf{j}' \cdot \mathbf{i}) \sin \beta \\ &= \cos \alpha \cos \beta - \sin \alpha \sin \beta. \end{aligned} \tag{2.50a}$$

Similarly it can be shown (Problems 2.4, Question 9) that

$$\sin(\alpha + \beta) = \sin \alpha \cos \beta + \cos \alpha \sin \beta. \tag{2.50b}$$

The symmetry of the formulae (2.50) between $\alpha$ and $\beta$ shows that $(\alpha + \beta)$ can be replaced by $(\beta + \alpha)$ in all trigonometric functions; this is consistent with the use of the term 'sum of angles' for the angle between $\mathbf{u}$ and $\mathbf{i}$, since we expect a 'sum' of two quantities, $\alpha$ and $\beta$, to be independent of their order. Also, putting $\beta = 0$ in (2.50) gives

$$\cos(\alpha + 0) = \cos \alpha, \qquad \sin(\alpha + 0) = \sin \alpha;$$

so we can take $\alpha + 0 = \alpha$, for all $\alpha$, ensuring that the zero angle also satisfies the usual additive property of the number 0. Addition of angles also satisfies the associative law; the use of the usual '+' sign is therefore justified.

The set of points with position vectors

$$\mathbf{r} = a \cos\phi\, \mathbf{i} + a \sin\phi\, \mathbf{j}, \qquad (2.51)$$

with $\phi$ taking values in the range $0 \leqslant \phi < 2\pi$, have Cartesian components

$$x = a\cos\phi, \qquad y = a\sin\phi,$$

which satisfy

$$x^2 + y^2 = a^2,$$

by (2.43). The points with position vectors (2.51) therefore constitute the **circle** of radius $a$, with centre at the origin. The points corresponding to $\theta = 0$ and $\theta = \pi$ have position vectors $a\mathbf{i}$ and $-a\mathbf{i}$ respectively; these are the end-points of a diameter of the circle.

**Example 2.12**

If $A$, $B$ are the end-points of the diameter of a circle, and $R$ is any other point on the circle, show that **AR** and **BR** are orthogonal.

Let the circle have radius $a$. Choose the centre of the circle as origin, and the basis vector $\mathbf{i}$ so that $A$, $B$ have position vectors $a\mathbf{i}$ and $-a\mathbf{i}$. Let $R$ have position vector (2.51), corresponding to an angle $\phi$. Then

$$\mathbf{AR} = a(\cos\phi - 1)\mathbf{i} + a\sin\phi\, \mathbf{j},$$
$$\mathbf{BR} = a(\cos\phi + 1)\mathbf{i} + a\sin\phi\, \mathbf{j}.$$

Then

$$\mathbf{AR}\cdot\mathbf{BR} = a^2[(\cos\phi - 1)(\cos\phi + 1) + \sin^2\phi]$$
$$= a^2[\cos^2\phi - 1 + \sin^2\phi] = 0.$$

But since $\phi \neq 0$ and $\phi \neq \pi$, the moduli of **AR** and **BR** are not zero. Therefore **AR** and **BR** are orthogonal.

## *Problems 2.4*

1 The position vectors of points $A$, $B$ relative to an origin $O$ are

$$\mathbf{a} = 3\mathbf{i} - \mathbf{j} - \mathbf{k},$$
$$\mathbf{b} = -2\mathbf{i} - 5\mathbf{j} + 3\mathbf{k}.$$

$P$ and $Q$ are the points of trisection of $AB$, with $AP{:}PB = QB{:}AQ = 1{:}2$. Find the position vectors of $P$ and $Q$, the lengths of

$OA$, $OP$ and $AP$, and the angles $OAB$ and $OQB$. Check that

$$OA < OP + PA.$$

2 Unit vectors **i**, **j** are parallel to adjacent edges of a large square table; the directions of **i** and **j** are referred to as 'east' and 'north'. An ant, walking on the table, makes the following movements:

4 centimetres 30° east of south,

12 centimetres south-west,

6 centimetres east,

9 centimetres 60° west of north.

Find the magnitude and direction of the ant's resultant displacement.

3 The position vectors of non-collinear points $A$, $B$, $C$ are **a**, **b**, **c**. Show that the position vector **r** of any point $R$ in the plane of $A$, $B$ and $C$ can be expressed in the form

$$\mathbf{r} = \frac{\alpha\mathbf{a} + \beta\mathbf{b} + \gamma\mathbf{c}}{\alpha + \beta + \gamma},$$

with $\alpha$, $\beta$, $\gamma$ suitably chosen. Interpret the ratios $\alpha:\beta:\gamma$ in terms of the ratios in which the lines $AR$, $BR$ and $CR$ cut the sides $BC$, $CA$ and $AB$ respectively. Prove that there is a point $P$ (the orthocentre) such that $AP$, $BP$, $CP$ intersect $BC$, $CA$ and $AB$ at right-angles, and that its position vector is

$$\cot A \cot B \cot C \{\mathbf{a} \tan A + \mathbf{b} \tan B + \mathbf{c} \tan C\}.$$

4 If $A$, $B$, $C$, $D$ are any four points in space, show that

$$\mathbf{DA} \cdot \mathbf{BC} + \mathbf{DB} \cdot \mathbf{CA} + \mathbf{DC} \cdot \mathbf{AB} = 0.$$

5 If $G$ is the centroid of $n$ points $A_1, A_2, A_3, \ldots, A_n$, and $P$ is any other point, show that

$$A_1P^2 + A_2P^2 + \ldots + A_nP^2$$
$$= A_1G^2 + A_2G^2 + \ldots + A_nG^2 + nGP^2.$$

6 Points $A$, $B$ have position vectors **a**, **b**. Points $C$ and $D$ divide $AB$ internally and externally in the ratio $p:q$. Show that

$$\mathbf{OC} \cdot \mathbf{OD} = \frac{q^2a^2 - p^2b^2}{q^2 - p^2}.$$

Deduce that **OC** and **OD** are perpendicular if $pb = qa$; interpret this result in terms of the geometry of the circle with $CD$ as diameter.

7 One pair of opposite edges of a tetrahedron are equal in length and both of these edges are perpendicular to the line joining their mid-points. Show that these properties are then true for the other pairs of opposite edges.

8 A line passes through the centre of a cube, making angles $\alpha$, $\beta$, $\gamma$, $\delta$ with the four diagonals of the cube. Show that

$$\cos^2\alpha+\cos^2\beta+\cos^2\gamma+\cos^2\delta=\tfrac{4}{3}.$$

9 Use the definition (2.43) to establish (2.48b). Use vectorial methods to establish (2.50b). Find a formula for $\tan(\alpha+\beta)$ in terms of $\tan\alpha$ and $\tan\beta$.

## *2.5 Coordinate systems*

In the previous section, we introduced rectangular coordinates $(x, y)$ of a point in a plane. There are a number of useful methods of denoting points in a plane and in 3-space by a set of numbers, referred to generally as **coordinate systems**; we shall introduce some of the most familiar coordinate systems in this section.

Position vectors in a plane relative to an origin in the plane, are all of the form

$$\boldsymbol{\rho}=x\mathbf{i}+y\mathbf{j}; \tag{2.38}$$

the axioms of addition and scalar multiplication imply that we must allow $x$ and $y$ each to have any real value, usually expressed as

$$-\infty<x<\infty, \qquad -\infty<y<\infty. \tag{2.52}$$

**Polar coordinates** $(\rho, \phi)$ in a plane are defined by the relations

$$\left.\begin{aligned} x&=\rho\cos\phi,\\ y&=\rho\sin\phi,\end{aligned}\right\} \tag{2.53}$$

or by the inverse relations

$$\left.\begin{aligned} \rho&=(x^2+y^2)^{\frac{1}{2}},\\ \phi&=\tan^{-1}(y/x).\end{aligned}\right\} \tag{2.54}$$

The coordinate pairs $(x, y)$ and $(\rho, \phi)$ are shown in Fig. 2.7; comparing (2.35) and (2.54), we see that $\rho=|\boldsymbol{\rho}|$, so that the notation $\rho$ for the polar coordinate accords with the usual convention for the modulus of the vector $\boldsymbol{\rho}$.

It is not difficult to show (see Problems 2.5, Question 1) that to any set of Cartesian coordinates $(x, y)$, satisfying (2.52), there is normally a unique set of polar coordinates $(\rho, \phi)$, provided that the allowed ranges of $\rho$ and $\phi$ are

$$0 \leqslant \rho < \infty, \qquad 0 \leqslant \phi < 2\pi. \tag{2.55}$$

The only exception to this rule is the origin, given by $\rho = 0$ for any value of $\phi$. We may, if we wish, choose $\phi$ to vary over any range of length $2\pi$; another common choice of ranges is

$$0 \leqslant \rho < \infty, \qquad -\pi < \phi \leqslant \pi. \tag{2.56}$$

We now discuss several coordinate systems that are useful in 3-space. The vectors we study, as in 2-space, represent displacements. It is important to note, though, that other types of vector quantity, such as velocity and force, can also be described by using these coordinate systems.

Corresponding to the representation (2.38) of a displacement $\boldsymbol{\rho}$ in a plane, we frequently express a displacement $\mathbf{r}$ from an origin $O$ in 3-space in the form

$$\mathbf{r} = x\mathbf{i} + y\mathbf{j} + z\mathbf{k}, \tag{2.57}$$

where $(\mathbf{i}, \mathbf{j}, \mathbf{k})$ is a triad. Then, as in Fig. 2.10, the Cartesian components $(x, y, z)$ of $\mathbf{r}$ are the **rectangular coordinates** of the point $P$ with position vector $\mathbf{r}$. Equations (2.32) and (2.34) are then the familiar laws of scalar multiplication and of addition of components. In particular, (2.34) tells us that the displacement to $P$ from a point $P_1$ with position vector

$$\mathbf{r}_1 = x_1\mathbf{i} + y_1\mathbf{j} + z_1\mathbf{k} \tag{2.58}$$

is

$$\mathbf{r} - \mathbf{r}_1 = (x - x_1)\mathbf{i} + (y - y_1)\mathbf{j} + (z - z_1)\mathbf{k}. \tag{2.59}$$

The unit vector corresponding to $\mathbf{r} = \mathbf{OP}$ is, using (2.4),

$$\mathbf{u} \equiv r^{-1}\mathbf{r} = \left(\frac{x}{r}\right)\mathbf{i} + \left(\frac{y}{r}\right)\mathbf{j} + \left(\frac{z}{r}\right)\mathbf{k}. \tag{2.60}$$

The coordinates of this unit vector,

$$l_1 = \frac{x}{r}, \qquad l_2 = \frac{y}{r}, \qquad l_3 = \frac{z}{r}, \tag{2.61}$$

are called the **direction cosines** of $\mathbf{OP}$, since they determine the direction of $\mathbf{OP}$ in space, but not its length. From (2.5), we see (for example) that $l_1$ is the cosine of the angle between $\mathbf{OP}$ and $\mathbf{i}$. The three direction cosines (2.61) are not independent, since the

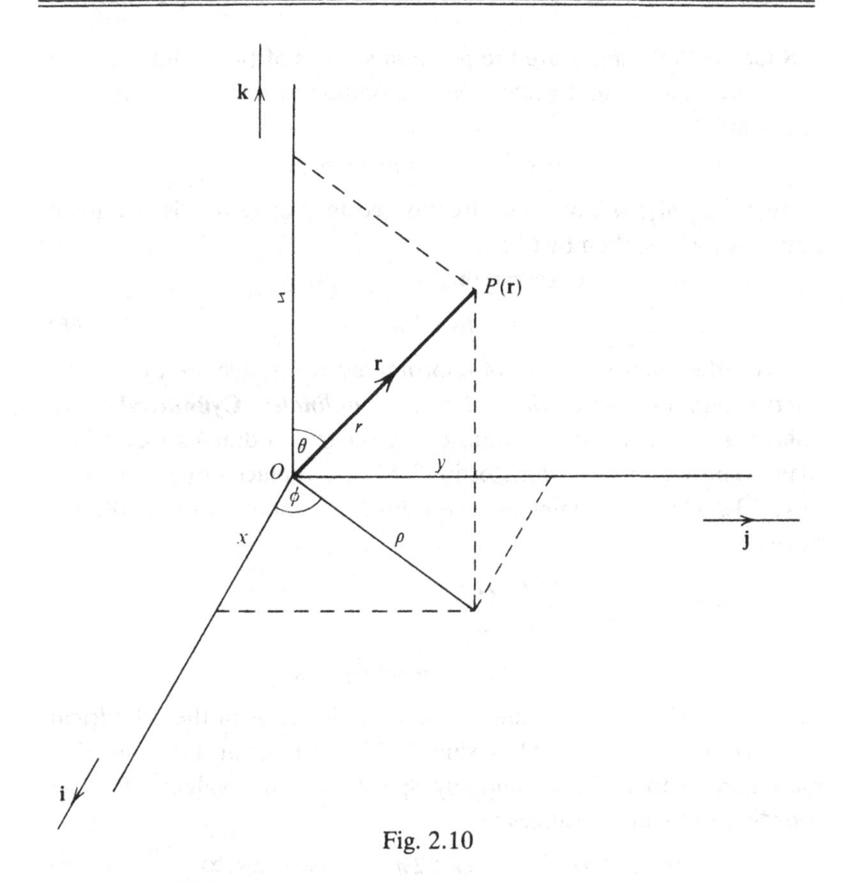


Fig. 2.10

Pythagorean relation (2.37) tells us that

$$l_1^2 + l_2^2 + l_3^2 = 1. \tag{2.62}$$

The unit vector in the direction of **r** is

$$\mathbf{u} = l_1\mathbf{i} + l_2\mathbf{j} + l_3\mathbf{k}. \tag{2.63}$$

The displacement **r** is therefore specified by giving its length $r$ and its direction cosines.

For a general displacement (2.59), with length $|\mathbf{r}-\mathbf{r}_1|$ given by

$$|\mathbf{r}-\mathbf{r}_1|^2 = (x-x_1)^2 + (y-y_1)^2 + (z-z_1)^2, \tag{2.64}$$

the direction cosines are defined as

$$l_1 = \frac{x-x_1}{|\mathbf{r}-\mathbf{r}_1|}, \qquad l_2 = \frac{y-y_1}{|\mathbf{r}-\mathbf{r}_1|}, \qquad l_3 = \frac{z-z_1}{|\mathbf{r}-\mathbf{r}_1|}. \tag{2.65}$$

They again satisfy the relation (2.62).

Suppose that $\mathbf{r}$ and $\mathbf{s}$ are the position vectors of two points relative to $O$, and that the unit vectors corresponding to $\mathbf{r}$ and $\mathbf{s}$ are given by (2.59) and by

$$\mathbf{u}_1 \equiv s^{-1}\mathbf{s} = m_1\mathbf{i} + m_2\mathbf{j} + m_3\mathbf{k},$$

so that $(m_1, m_2, m_3)$ are the direction cosines of $\mathbf{s}$. If $\theta$ is the angle between $\mathbf{r}$ and $\mathbf{s}$, then by (2.5),

$$\begin{aligned} \cos\theta &= \mathbf{u} \cdot \mathbf{u}_1 \\ &= l_1m_1 + l_2m_2 + l_3m_3. \end{aligned} \tag{2.66}$$

Two other well-used sets of coordinates in 3-space are *cylindrical polar coordinates* and *spherical polar coordinates*. **Cylindrical polar coordinates** are a simple extension of polar coordinates $(\rho, \phi)$ in a plane. The $x$ and $y$ coordinates in (2.57) are replaced by $\rho$ and $\phi$, as in (2.53), while $z$ remains as a coordinate. Defining $\boldsymbol{\rho}$ by (2.38), we have

$$\begin{aligned} \mathbf{r} &= x\mathbf{i} + y\mathbf{j} + z\mathbf{k} \\ &= \boldsymbol{\rho} + z\mathbf{k} \\ &= \rho\cos\phi\mathbf{i} + \rho\sin\phi\mathbf{j} + z\mathbf{k}. \end{aligned} \tag{2.67}$$

This expresses the rectangular coordinates in terms of the cylindrical polar coordinates $(\rho, \phi, z)$. Using (2.55), it is clear that any displacement with $\rho \neq 0$ is uniquely specified if the values of these coordinates lie in the ranges

$$0 \leqslant \rho < \infty, \qquad 0 \leqslant \phi < 2\pi, \qquad -\infty < z < \infty. \tag{2.68}$$

Two of the three **spherical polar coordinates** are $r$, the length of $\mathbf{r}$, and $\phi$, the *cylindrical* polar angle. The third coordinate is the angle $\theta$ between the vectors $\mathbf{r}$ and $\mathbf{k}$ (the '$z$-axis'). The spherical polar coordinates $(r, \theta, \phi)$, and their geometrical relationships with $x$, $y$, $z$ and $\rho$ are shown in Fig. 2.10; the position vector $\mathbf{r}$ of the point $P$ is shown as a thickened line, and is of length $r$. From the diagram it is clear that

$$\left.\begin{aligned} \rho &= r\sin\theta, \\ z &= r\cos\theta. \end{aligned}\right\} \tag{2.69}$$

Using (2.53), we find the relationship between $(x, y, z)$ and $(r, \theta, \phi)$ to be

$$\left.\begin{aligned} x &= r\sin\theta\cos\phi, \\ y &= r\sin\theta\sin\phi, \\ z &= r\cos\theta. \end{aligned}\right\} \tag{2.70}$$

We can find $r$ and trigonometrical functions of $\theta$ and $\phi$ in terms of $(x, y, z)$:

$$\left.\begin{aligned} r &= (x^2+y^2+z^2)^{\frac{1}{2}}, & \cos\theta &= z/(x^2+y^2+z^2)^{\frac{1}{2}}, \\ \cos\phi &= x/(x^2+y^2)^{\frac{1}{2}}, & \sin\phi &= y/(x^2+y^2)^{\frac{1}{2}}. \end{aligned}\right\} \tag{2.71}$$

Almost all displacements can be uniquely described if $(r, \theta, \phi)$ vary in the ranges

$$0 \leqslant r < \infty, \qquad 0 \leqslant \theta \leqslant \pi, \qquad 0 \leqslant \phi < 2\pi. \tag{2.72}$$

We can allow $\phi$ to vary over *any* range of length $2\pi$, for example

$$-\pi < \phi \leqslant \pi,$$

if we wish. The proof of (2.71) and (2.72) is left to the reader (Problems 2.5, Question 3).

## ■ *Problems 2.5*

1 Show that each point in a plane, except the origin of coordinates, corresponds to a unique set of polar coordinates $(\rho, \phi)$ lying in the ranges

$$0 < \rho < \infty, \qquad 0 \leqslant \phi < 2\pi.$$

2 Show that the distance $d$ between two points in a plane with polar coordinates $(\rho_1, \phi_1)$ and $(\rho_2, \phi_2)$ is given by

$$d^2 = \rho_1^2 + \rho_2^2 - 2\rho_1\rho_2 \cos(\phi_1 - \phi_2).$$

3 Show that the relations (2.71) between spherical polar and rectangular coordinates follow from the relations (2.70), and that each set of rectangular coordinates $(x, y, z)$ normally corresponds to exactly one set of spherical polar coordinates $(r, \theta, \phi)$ lying in the ranges

$$0 \leqslant r < \infty, \qquad 0 \leqslant \theta \leqslant \pi, \qquad 0 \leqslant \phi < 2\pi.$$

Discuss the relationship between the two sets of coordinates when $r = 0$ and when $\theta = 0$ or $\theta = \pi$.

4 If $(\rho, \phi, z)$ are cylindrical polar coordinates, write down in terms of rectangular coordinates the equations of surfaces
(i) $\rho = \rho_0$,
(ii) $\phi = \phi_0$,
where $\rho_0$ and $\phi_0$ are constants. Draw a diagram showing surfaces of types (i) and (ii).

5 If $(r, \theta, \phi)$ are spherical polar coordinates, write down in terms of rectangular coordinates the equations of surfaces

(i) $r = r_0$,

(ii) $\theta = \theta_0$,

(iii) $\phi = \phi_0$,

where $r_0$, $\theta_0$ and $\phi_0$ are constants. Draw a diagram showing surfaces of types (i), (ii) and (iii).

6 Write down the rectangular coordinates of the two unit vectors with spherical polar angles $(\theta_1, \phi_1)$ and $(\theta_2, \phi_2)$. If $\alpha$ is the angle between these two unit vectors, show that

$$\cos \alpha = \cos \theta_1 \cos \theta_2 + \sin \theta_1 \sin \theta_2 \cos (\phi_1 - \phi_2).$$

7 Two points have spherical polar coordinates $(r_1, \theta_1, \phi_1)$ and $(r_2, \theta_2, \phi_2)$; show that the distance $d$ between the points is given by

$$d^2 = r_1^2 + r_2^2 - 2r_1r_2[\cos \theta_1 \cos \theta_2 + \sin \theta_1 \sin \theta_2 \cos (\phi_1 - \phi_2)].$$

Simplify this formula by using the result of Question 6.

# 3
# Other products of vectors

## 3.1 *The vector product of two vectors*

If we are given two linearly independent vectors **a** and **b**, they define a 2-space of vectors through (2.19). We proved in §2.3 that there were only two unit vectors in 3-space orthogonal to both **a** and **b**; if we denote one of these by **n**, the other is −**n**. If $\mathbf{b}_\perp$ is the component of **b** orthogonal to **a**, the vectors **a**, $\mathbf{b}_\perp$, **n** are mutually orthogonal. We shall see in Chapter 4 that, if any given triad (**i**, **j**, **k**) is taken to be right-handed, the 'handedness' of any three mutually orthogonal vectors is determined. We can choose the labelling of the two vectors (**n**, −**n**) to ensure either that ($\mathbf{a}, \mathbf{b}_\perp, \mathbf{n}$) is right-handed *or* that ($\mathbf{a}, \mathbf{b}_\perp, \mathbf{n}$) is left-handed. We choose to define **n** so that ($\mathbf{a}, \mathbf{b}_\perp, \mathbf{n}$) is right-handed, as indicated in Fig. 3.1; then, of course, ($\mathbf{a}, \mathbf{b}_\perp, -\mathbf{n}$) is left-

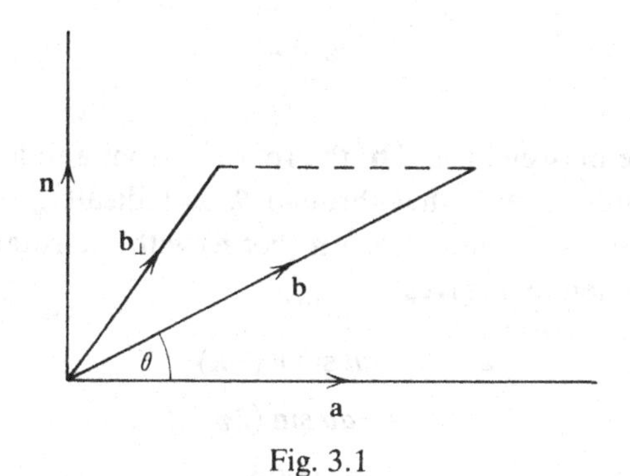

Fig. 3.1

handed. So a rotation from **a** to **b** through the smaller angle $\theta$, combined with motion in the direction **n**, produces a right-handed screw motion. The **vector product** of the two vectors **a** and **b** is defined in terms of the moduli $a$ and $b$, the angle $\theta$ $(0<\theta<\pi)$

between the vectors, and the vector **n** specified above. The vector product is denoted by $\mathbf{a} \wedge \mathbf{b}$, and has definition

$$\mathbf{a} \wedge \mathbf{b} = ab \sin \theta \mathbf{n}. \tag{3.1}$$

Since we have assumed that $0<\theta<\pi$, (3.1) defines $\mathbf{a} \wedge \mathbf{b}$ for this range of $\theta$ only. It is easy, however, to extend the definition to all real values of $\theta$. First, if $\theta=0$ or $\theta=\pi$, $\sin\theta=0$, and we can take (3.1) to mean $\mathbf{a} \wedge \mathbf{b}=\mathbf{0}$; **n** is undefined for these values of $\theta$, since **a** and **b** are then linearly dependent; the result **0** is obtained if we take the limit of (3.1) as $\theta \to 0$ or $\theta \to \pi$.

Next, if $\psi=2\pi-\theta$ is the reflex angle between **a** and **b**, shown in Fig. 3.2, then $\pi<\psi<2\pi$. Let us apply the definition (3.1), regarding

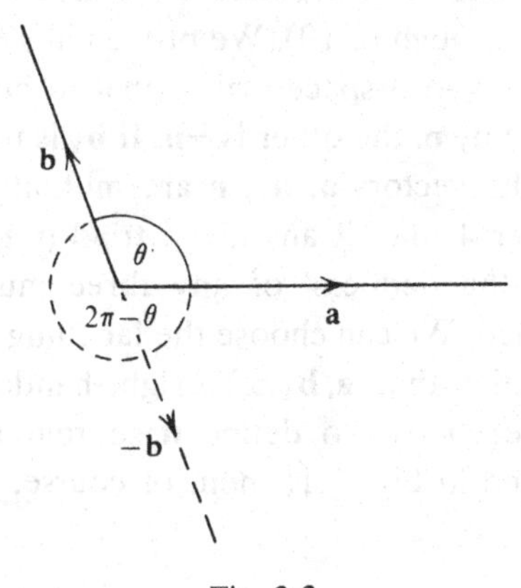


Fig. 3.2

$\psi$ as the angle between **a** and **b**; the rotation from **a** to **b** through $\psi$ is in the opposite sense to that through $\theta$, as indicated, so the 'right-hand rule' associates the vector $-\mathbf{n}$ (*not* **n**) with the rotation through the angle $\psi$. Then (3.1) gives

$$\begin{aligned}\mathbf{a} \wedge \mathbf{b} &= ab \sin \psi(-\mathbf{n}) \\ &= -ab \sin(2\pi-\theta)\mathbf{n} \\ &= ab \sin \theta \mathbf{n},\end{aligned}$$

consistent with the definition (3.1) for $0<\theta<\pi$. So (3.1) is valid for all angles in the range $0\leqslant\theta\leqslant 2\pi$. Since any angle is of the form

$$\theta \pm 2\pi n \qquad (n=0, 1, 2, \ldots; 0\leqslant\theta<2\pi),$$

and $\sin(\theta \pm 2\pi n) = \sin\theta$, the definition (3.1) can be used for any angle.

In (3.1), $\mathbf{n}$ is a unit vector; so the magnitude of $\mathbf{a} \wedge \mathbf{b}$ is

$$|ab \sin\theta|.$$

If $\mathbf{a}$ and $\mathbf{b}$ correspond to displacements from a point $P$, as in Fig. 1.3, this is the **area of the parallelogram** ***PQRS*** defined by $\mathbf{a}$ and $\mathbf{b}$.

Certain simple properties follow from the definition (3.1). First note that

$$\mathbf{a} \wedge (-\mathbf{b}) = -\mathbf{a} \wedge \mathbf{b}; \tag{3.2}$$

for, if $\theta$ is the angle between $\mathbf{a}$ and $\mathbf{b}$, as in Fig. 3.2, then $(\theta + \pi)$ is the angle between $\mathbf{a}$ and $-\mathbf{b}$, in the same sense as $\theta$; since (3.1) is valid for all angles,

$$\begin{aligned}\mathbf{a} \wedge (-\mathbf{b}) &= ab \sin(\theta + \pi)\mathbf{n} \\ &= -ab \sin\theta\mathbf{n},\end{aligned}$$

establishing (3.2). Next, if $\lambda > 0$,

$$\mathbf{a} \wedge (\lambda\mathbf{b}) = (\lambda\mathbf{a}) \wedge \mathbf{b} = \lambda(\mathbf{a} \wedge \mathbf{b}); \tag{3.3}$$

this follows at once from the definition (3.1), remembering that the lengths of $\lambda\mathbf{a}$ and $\lambda\mathbf{b}$ are $\lambda a$ and $\lambda b$ respectively. Combining (3.2) and (3.3) we see that (3.3) is valid for *any* scalar $\lambda$, positive or negative.

The effect of interchanging the order of vectors in a vector product is to change its sign; thus

$$\mathbf{b} \wedge \mathbf{a} = -\mathbf{a} \wedge \mathbf{b}. \tag{3.4}$$

This result follows by noting that, in Fig. 3.2, a rotation from $\mathbf{b}$ to $\mathbf{a}$ in the *positive* sense is through the angle $\psi = 2\pi - \theta$, so that

$$\begin{aligned}\mathbf{b} \wedge \mathbf{a} &= ba \sin(2\pi - \theta)\mathbf{n} \\ &= -ab \sin\theta\mathbf{n}.\end{aligned}$$

Putting $\mathbf{b} = \mathbf{a}$ in (3.4), we find

$$\mathbf{a} \wedge \mathbf{a} = \mathbf{0}; \tag{3.5}$$

this equation is valid, since we have already taken $\mathbf{a} \wedge \mathbf{b} = \mathbf{0}$ when $\mathbf{b}$ is parallel to $\mathbf{a}$. Its deduction from (3.4) is not valid, however; it is correctly deduced by taking the limit $\mathbf{b} \to \mathbf{a}$ with $\theta \to 0$ in (3.1). For similar reasons, we define $\mathbf{a} \wedge \mathbf{0} = \mathbf{0}$, for all vectors $\mathbf{a}$, although a unique vector $\mathbf{n}$ in (3.1) is not determined when $\mathbf{b} = \mathbf{0}$.

A very common notation for the vector product is $\mathbf{a} \times \mathbf{b}$, and the vector product is frequently called the **cross product**. The use of the ordinary multiplication sign can be confusing, since by (3.4) the vector product does not obey the commutative law of multiplication. We have therefore chosen to use the other standard notation $\mathbf{a} \wedge \mathbf{b}$ for the vector product.

## 3.2 *The distributive law for vector products; components*

Suppose that $\mathbf{b}_\perp$ is the component of $\mathbf{b}$ orthogonal to $\mathbf{a}$, as represented in Fig. 3.1. Then we shall show that

$$\mathbf{a} \wedge \mathbf{b} = \mathbf{a} \wedge \mathbf{b}_\perp. \tag{3.6}$$

The product $\mathbf{a} \wedge \mathbf{b}$ is defined by (3.1), with $0 < \theta < \pi$ and with $\mathbf{n}$ and the rotation $\mathbf{a} \to \mathbf{b}$ associated by the right-hand rule. The angle between $\mathbf{a}$ and $\mathbf{b}_\perp$ is $\frac{1}{2}\pi$; the vector $\mathbf{n}$ is orthogonal to both $\mathbf{a}$ and $\mathbf{b}_\perp$, and is associated with the rotation $\mathbf{a} \to \mathbf{b}_\perp$ by the right-hand rule. So the vector product $\mathbf{a} \wedge \mathbf{b}_\perp$ is a vector in the direction of $\mathbf{n}$. Thus the definition (3.1), with $\mathbf{b}$ replaced by $\mathbf{b}_\perp$, gives

$$\begin{aligned} \mathbf{a} \wedge \mathbf{b}_\perp &= a|\mathbf{b}_\perp| \sin \tfrac{1}{2}\pi \, \mathbf{n} \\ &= a|\mathbf{b}_\perp|\mathbf{n}. \end{aligned} \tag{3.7}$$

However, the modulus of $\mathbf{b}_\perp$ is

$$b_\perp \equiv |\mathbf{b}_\perp| = b \sin \theta,$$

so that (3.7) becomes

$$\mathbf{a} \wedge \mathbf{b}_\perp = a(b \sin \theta)\mathbf{n}.$$

Comparing with (3.1), we see that (3.6) is established for $0 < \theta < \pi$. When $\theta = 0$ or $\pi$, both sides of (3.6) are zero, so that the equation holds for all vectors $\mathbf{a}$ and $\mathbf{b}$.

If we resolve $\mathbf{b}$ into components parallel and orthogonal to $\mathbf{a}$, by taking $\mathbf{u} = a^{-1}\mathbf{a}$ in (2.23) and (2.24), then

$$\mathbf{b} = \mathbf{b}_\parallel + \mathbf{b}_\perp. \tag{3.8}$$

Equation (3.6) tells us that

$$\mathbf{a} \wedge (\mathbf{b}_\parallel + \mathbf{b}_\perp) = \mathbf{a} \wedge \mathbf{b}_\perp. \tag{3.9}$$

Since $\mathbf{a} \wedge \mathbf{b}_\parallel = \mathbf{0}$, we might regard this equation as obvious. It is not obvious, however, because we do not know that we can 'multiply out' the left-hand member of (3.9) to give

$$(\mathbf{a} \wedge \mathbf{b}_\parallel) + (\mathbf{a} \wedge \mathbf{b}_\perp).$$

Put more generally, we have not yet established the **distributive law of multiplication**

$$\mathbf{a} \wedge (\mathbf{b}+\mathbf{c}) = \mathbf{a} \wedge \mathbf{b} + \mathbf{a} \wedge \mathbf{c} \tag{3.10}$$

for vector products. This law is not an axiom, like the distributive law (2.8) for scalar products; the vector product has been defined by (3.1), and all of its properties, including (3.10), must follow from this definition.

To establish (3.10), we shall express each of the vector products in terms of a right-handed triad (**i**, **j**, **k**), chosen so that

$$\mathbf{a} = a\mathbf{k}. \tag{3.11}$$

Consider, for example, the term $\mathbf{a} \wedge \mathbf{b}$ in (3.10), which equals $\mathbf{a} \wedge \mathbf{b}_\perp$, by (3.6). The vector $\mathbf{b}_\perp$ is orthogonal to **k**, and so is of the form

$$\mathbf{b}_\perp = b_1\mathbf{i} + b_2\mathbf{j}, \tag{3.12}$$

as indicated in Fig. 3.3. We shall now show that the vector product is given by

$$\mathbf{a} \wedge \mathbf{b} = \mathbf{a} \wedge \mathbf{b}_\perp = -ab_2\mathbf{i} + ab_1\mathbf{j}. \tag{3.13}$$

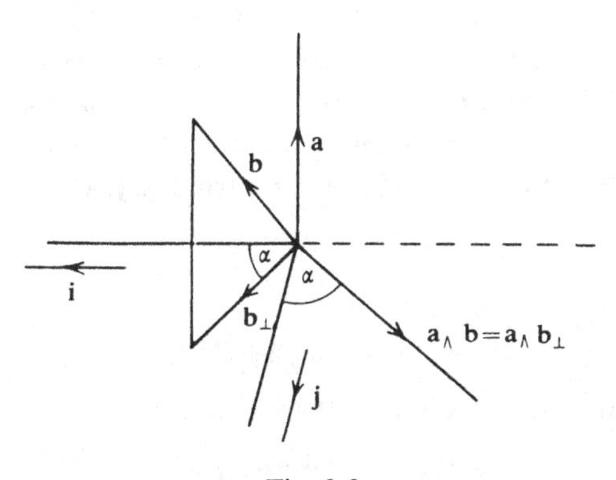


Fig. 3.3

The vector $\mathbf{a} \wedge \mathbf{b}_\perp$, given by (3.7), is generally determined by the properties:

(i) Its modulus is

$$ab_\perp \equiv a(b_1^2 + b_2^2)^{\frac{1}{2}}, \tag{3.14}$$

using (3.1) with $\theta = \frac{1}{2}\pi$, and (3.12).

(ii) $\mathbf{a} \wedge \mathbf{b}_\perp$ is orthogonal to both **a** and $\mathbf{b}_\perp$.

(iii) $(\mathbf{a}, \mathbf{b}_\perp, \mathbf{a} \wedge \mathbf{b}_\perp)$ form a right-handed set.

If it happens that $b_1 = b_2 = 0$, so that $\mathbf{b}_\perp = \mathbf{0}$, the modulus (3.14) is also zero, and (3.13) gives $\mathbf{a} \wedge \mathbf{b}_\perp = \mathbf{0}$, which is correct. We therefore check that the vector on the right of (3.13) satisfies the properties (i), (ii) and (iii) when $\mathbf{b}_\perp \neq \mathbf{0}$. Its modulus is $[(ab_2)^2 + (ab_1)^2]^{\frac{1}{2}}$, agreeing with (3.14). It is clearly orthogonal to $\mathbf{a} = a\mathbf{k}$, and its scalar product with $\mathbf{b}_\perp$ is, using (3.12) and (2.36),

$$b_1(-ab_2) + b_2(ab_1) = 0.$$

It is thus also orthogonal to $\mathbf{b}_\perp$. So we have only to establish property (iii); if $\mathbf{i}'$ and $\mathbf{j}'$ are the unit vector corresponding to the vectors (3.12) and (3.13), we need to establish Equation (2.30),

$$\mathbf{i}' \cdot \mathbf{i} = \mathbf{j}' \cdot \mathbf{j}. \tag{3.15}$$

The unit vectors corresponding to (3.14) and (3.15) are, using (2.4) and (3.14),

$$\mathbf{i}' = b_\perp^{-1}(b_1\mathbf{i} + b_2\mathbf{j})$$

and

$$\mathbf{j}' = b_\perp^{-1}(-b_2\mathbf{i} + b_1\mathbf{j}).$$

It is clear that each side of (3.15) is equal to $b_\perp^{-1}b_1$, establishing property (iii). We have therefore proved that $\mathbf{a} \wedge \mathbf{b}$ is given by (3.13). The relationships between $\mathbf{ab}$, $\mathbf{b}_\perp$, $\mathbf{i}$, $\mathbf{j}$ and $\mathbf{a} \wedge \mathbf{b}$ are represented geometrically in Fig. 3.3; the angles marked '$\alpha$' are equal, by (3.15). In a similar way, if in terms of the same triad $(\mathbf{i}, \mathbf{j}, \mathbf{k})$,

$$\mathbf{c}_\perp = c_1\mathbf{i} + c_2\mathbf{j},$$

so that

$$(\mathbf{b} + \mathbf{c})_\perp = (b_1 + c_1)\mathbf{i} + (b_2 + c_2)\mathbf{j},$$

the other vector products in (3.10) are given by

$$\mathbf{a} \wedge \mathbf{c} = -ac_2\mathbf{i} + ac_1\mathbf{j}$$

and

$$\mathbf{a} \wedge (\mathbf{b} + \mathbf{c}) = -a(b_2 + c_2)\mathbf{i} + a(b_1 + c_1)\mathbf{j}.$$

Taken together with (3.13), these equations show that the distributive law (3.10) is satisfied. The essential point of the argument is that $\mathbf{a} \wedge \mathbf{b}$, given by (3.13), is linear in the components $b_1$, $b_2$ of $\mathbf{b}$.

The distributive law (3.10) is algebraically very simple, but its geometrical significance is quite complex. The vectors $\mathbf{b}$, $\mathbf{c}$ and $\mathbf{b} + \mathbf{c}$ can be represented by displacements which form a closed triangle in space. The three vector products in (3.10) represent displacements perpendicular to **b**, **c** and **b**+**c** respectively, all lying in a plane perpendicular to **a**. The result (3.10) tells us that this second set of displacements also form a closed triangle.

The distributive law is extremely useful in practice, since it allows 'multiplying out' and 'factorisation' of vector products. Using (3.4), (3.10) becomes

$$(\mathbf{b}+\mathbf{c}) \wedge \mathbf{a}=\mathbf{b} \wedge \mathbf{a}+\mathbf{c} \wedge \mathbf{a}, \tag{3.16}$$

so that these properties also apply to the first vector in a vector product. We must always remember, though, to preserve the order of vectors in vector products, since a change of order results in a change of sign.

**Example 3.1**

Simplify

$$(\mathbf{a}+\lambda\mathbf{b}) \wedge (\mathbf{a}+\mu\mathbf{b}).$$

Using (3.3), (3.10) and (3.16), we multiply out to obtain

$$\mathbf{a} \wedge \mathbf{a}+\lambda\mathbf{b} \wedge \mathbf{a}+\mu\mathbf{a} \wedge \mathbf{b}+\lambda\mu\mathbf{b} \wedge \mathbf{b}.$$

Using (3.4) and (3.5), this becomes

$$\lambda\mathbf{b} \wedge \mathbf{a}-\mu\mathbf{b} \wedge \mathbf{a}=(\lambda-\mu)\mathbf{b} \wedge \mathbf{a}.$$

From the definition (3.1), and from (3.4) and (3.5), it follows that a right-handed triad (**i**, **j**, **k**) satisfies the following relations:

$$\left.\begin{aligned} \mathbf{j} \wedge \mathbf{k}&=-\mathbf{k} \wedge \mathbf{j}=\mathbf{i},\\ \mathbf{k} \wedge \mathbf{i}&=-\mathbf{i} \wedge \mathbf{k}=\mathbf{j},\\ \mathbf{i} \wedge \mathbf{j}&=-\mathbf{j} \wedge \mathbf{i}=\mathbf{k},\end{aligned}\right\} \tag{3.17}$$

$$\mathbf{i} \wedge \mathbf{i}=\mathbf{j} \wedge \mathbf{j}=\mathbf{k} \wedge \mathbf{k}=\mathbf{0}. \tag{3.18}$$

If vectors **v** and **w** are expressed in component form

$$\mathbf{v}=v_1\mathbf{i}+v_2\mathbf{j}+v_3\mathbf{k},$$

$$\mathbf{w}=w_1\mathbf{i}+w_2\mathbf{j}+w_3\mathbf{k},$$

then, using the fact that we can multiply out vector products,

$$\begin{aligned}\mathbf{v} \wedge \mathbf{w}=\;&v_1w_1\mathbf{i} \wedge \mathbf{i}+v_1w_2\mathbf{i} \wedge \mathbf{j}+v_1w_3\mathbf{i} \wedge \mathbf{k}\\ &+v_2w_1\mathbf{j} \wedge \mathbf{i}+v_2w_2\mathbf{j} \wedge \mathbf{j}+v_2w_3\mathbf{j} \wedge \mathbf{k}\\ &+v_3w_1\mathbf{k} \wedge \mathbf{i}+v_3w_2\mathbf{k} \wedge \mathbf{j}+v_3w_3\mathbf{k} \wedge \mathbf{k};\end{aligned}$$

using now the relations (3.17) and (3.18),

$$\mathbf{v} \wedge \mathbf{w} = (v_2 w_3 - v_3 w_2)\mathbf{i} + (v_3 w_1 - v_1 w_3)\mathbf{j} + (v_1 w_2 - v_2 w_1)\mathbf{k}. \quad (3.19)$$

**Example 3.2**

Find a unit vector orthogonal to $\mathbf{v} = \mathbf{i} + 2\mathbf{j} + 3\mathbf{k}$ and to $\mathbf{w} = \mathbf{i} - \mathbf{j} + \mathbf{k}$.
By definition, the vector product $\mathbf{v} \wedge \mathbf{w}$ is orthogonal to both $\mathbf{v}$ and $\mathbf{w}$. The Cartesian components of $\mathbf{v}$ and $\mathbf{w}$ are $(1, 2, 3)$ and $(1, -1, 1)$. Using (3.19),

$$\begin{aligned} \mathbf{v} \wedge \mathbf{w} &= [2 \cdot 1 - 3(-1)]\mathbf{i} + [3 \cdot 1 - 1 \cdot 1]\mathbf{j} + [1 \cdot (-1) - 2 \cdot 1]\mathbf{k} \\ &= 5\mathbf{i} + 2\mathbf{j} - 3\mathbf{k}. \end{aligned}$$

We can check that this vector *is* orthogonal to $\mathbf{v}$ and $\mathbf{w}$ by forming scalar products. For example, the scalar product with $\mathbf{v}$ is

$$5 \cdot 1 + 2 \cdot 2 + (-3) \cdot 3 = 0.$$

The modulus of $\mathbf{v} \wedge \mathbf{w}$ is given by

$$|\mathbf{v} \wedge \mathbf{w}|^2 = 5^2 + 2^2 + 3^2 = 38.$$

So the unit vector corresponding to $\mathbf{v} \wedge \mathbf{w}$ is

$$(5\mathbf{i} + 2\mathbf{j} - 3\mathbf{k})/\sqrt{38}.$$

We have already noted that, when $\mathbf{a}$ and $\mathbf{b}$ are displacements, the modulus $|\mathbf{a} \wedge \mathbf{b}|$ of the vector product equals the area of the parallelogram with displacements $\mathbf{a}$ and $\mathbf{b}$ along adjacent sides. Geometrically, vector products are often useful for describing areas.

**Example 3.3**

The vertices of a triangle have position vectors $\mathbf{a}$, $\mathbf{b}$, $\mathbf{c}$. Show that the area of the triangle is

$$\tfrac{1}{2}|\mathbf{b} \wedge \mathbf{c} + \mathbf{c} \wedge \mathbf{a} + \mathbf{a} \wedge \mathbf{b}|.$$

The area of the triangle is half the area of the parallelogram with displacements $\mathbf{b} - \mathbf{a}$ and $\mathbf{c} - \mathbf{a}$ along adjacent sides. The area of the parallelogram is

$$\begin{aligned} &|(\mathbf{b} - \mathbf{a}) \wedge (\mathbf{c} - \mathbf{a})| \\ &\quad = |\mathbf{b} \wedge \mathbf{c} - \mathbf{a} \wedge \mathbf{c} - \mathbf{b} \wedge \mathbf{a} + \mathbf{a} \wedge \mathbf{a}| \\ &\quad = |\mathbf{b} \wedge \mathbf{c} + \mathbf{c} \wedge \mathbf{a} + \mathbf{a} \wedge \mathbf{b}|, \end{aligned}$$

using (3.4) and (3.5). The result follows.

### Problems 3.1

1 Vectors $\mathbf{a}$, $\mathbf{b}$, $\mathbf{c}$ are defined by

$$\mathbf{a}=\mathbf{i}+4\mathbf{j}+2\mathbf{k},$$
$$\mathbf{b}=\mathbf{i}+2\mathbf{j}+\mathbf{k},$$
$$\mathbf{c}=\mathbf{i}-\mathbf{j}+\mathbf{k}.$$

Calculate $\mathbf{a} \wedge \mathbf{b}$ and $\mathbf{a} \wedge \mathbf{c}$. Show that $\mathbf{b}$ and $\mathbf{c}$ are orthogonal, but that $\mathbf{a} \wedge \mathbf{b}$ and $\mathbf{a} \wedge \mathbf{c}$ are not orthogonal. Find a vector $\mathbf{d}$ such that $\mathbf{d} \wedge \mathbf{b}$ and $\mathbf{d} \wedge \mathbf{c}$ are orthogonal.

2 If vectors $\mathbf{a}$, $\mathbf{b}$, $\mathbf{c}$ satisfy $\mathbf{a}+\mathbf{b}+\mathbf{c}=\mathbf{0}$, prove that $\mathbf{a} \wedge \mathbf{b}=\mathbf{b} \wedge \mathbf{c}=\mathbf{c} \wedge \mathbf{a}$. Interpret this result geometrically.

3 Vectors $\mathbf{a}$, $\mathbf{b}$, $\mathbf{c}$ are defined by

$$\mathbf{a}=2\mathbf{i}-\mathbf{j}+\mathbf{k},$$
$$\mathbf{b}=3\mathbf{i}+2\mathbf{j}-\mathbf{k},$$
$$\mathbf{c}=\mathbf{i}+2\mathbf{j}+3\mathbf{k}.$$

Evaluate $\mathbf{a} \wedge \mathbf{b}$ and $\mathbf{b} \wedge \mathbf{c}$, and hence evaluate the vectors $(\mathbf{a} \wedge \mathbf{b}) \wedge \mathbf{c}$ and $\mathbf{a} \wedge (\mathbf{b} \wedge \mathbf{c})$.

## *3.3 Products of more than two vectors*

The **scalar triple product** of vectors $\mathbf{a}$, $\mathbf{b}$ and $\mathbf{c}$ is defined to be

$$\mathbf{a} \cdot (\mathbf{b} \wedge \mathbf{c}). \tag{3.20}$$

We shall now establish some basic properties of this product:

(i) $\mathbf{a}$, $\mathbf{b}$, $\mathbf{c}$ are linearly dependent if and only if

$$\mathbf{a} \cdot (\mathbf{b} \wedge \mathbf{c})=0. \tag{3.21}$$

(ii) $\mathbf{a} \cdot (\mathbf{b} \wedge \mathbf{c})=\mathbf{b} \cdot (\mathbf{c} \wedge \mathbf{a})=\mathbf{c} \cdot (\mathbf{a} \wedge \mathbf{b}).$ (3.22)

First, consider property (i). If $\mathbf{a}$, $\mathbf{b}$, $\mathbf{c}$ are linearly dependent, then a relation

$$\alpha\mathbf{a}+\beta\mathbf{b}+\gamma\mathbf{c}=\mathbf{0}$$

holds, with at least one of $\alpha$, $\beta$, $\gamma$ non-zero. If $\alpha=0$, $\mathbf{b}$ and $\mathbf{c}$ are then linearly dependent (or parallel), and $\mathbf{b} \wedge \mathbf{c}=\mathbf{0}$, establishing (3.29). If $\alpha \neq 0$, then $\mathbf{a} \cdot \mathbf{b} \wedge \mathbf{c}$ can be expressed as a linear combination of $\mathbf{b} \cdot (\mathbf{b} \wedge \mathbf{c})$ and $\mathbf{c} \cdot (\mathbf{b} \wedge \mathbf{c})$; but

$$\mathbf{b} \cdot (\mathbf{b} \wedge \mathbf{c})=\mathbf{c} \cdot (\mathbf{b} \wedge \mathbf{c})=0, \tag{3.23}$$

since $\mathbf{b} \wedge \mathbf{c}$ is orthogonal to both $\mathbf{b}$ and $\mathbf{c}$. So linear dependence always implies (3.21). Conversely, if (3.21) holds, either $\mathbf{a}=\mathbf{0}$, or $\mathbf{a}$ is orthogonal to $\mathbf{b} \wedge \mathbf{c}$, or $\mathbf{b} \wedge \mathbf{c}=0$; that is, either $\mathbf{a}$ is linearly dependent on $\mathbf{b}$ and $\mathbf{c}$, or $\mathbf{b}$ and $\mathbf{c}$ are linearly dependent. In either case, $(\mathbf{a}, \mathbf{b}, \mathbf{c})$ are linearly dependent. So (i) is established.

To prove (3.22), replace $\mathbf{b}$ by $\mathbf{a}+\mathbf{b}$ in (3.23). Then

$$(\mathbf{a}+\mathbf{b}) \cdot [(\mathbf{a}+\mathbf{b}) \wedge \mathbf{c}]=0.$$

Expanding,

$$\mathbf{a} \cdot (\mathbf{a} \wedge \mathbf{c})+\mathbf{b} \cdot (\mathbf{a} \wedge \mathbf{c})+\mathbf{a} \cdot (\mathbf{b} \wedge \mathbf{c})+\mathbf{b} \cdot (\mathbf{b} \wedge \mathbf{c})=0.$$

Using (3.23) again, and writing $\mathbf{a} \wedge \mathbf{c}=-\mathbf{c} \wedge \mathbf{a}$, we obtain the result

$$\mathbf{a} \cdot (\mathbf{b} \wedge \mathbf{c})=\mathbf{b} \cdot (\mathbf{c} \wedge \mathbf{a}).$$

Since

$$\mathbf{a} \cdot (\mathbf{b} \wedge \mathbf{c})=(\mathbf{b} \wedge \mathbf{c}) \cdot \mathbf{a}, \tag{3.24}$$

(3.22) implies that *all scalar triple products with* $\mathbf{a}$, $\mathbf{b}$, $\mathbf{c}$ *in the same cyclic order are equal.* For this reason we introduce the notation

$$[\mathbf{a}, \mathbf{b}, \mathbf{c}] \equiv \mathbf{a} \cdot (\mathbf{b} \wedge \mathbf{c}) \tag{3.25}$$

for the scalar triple product.

When $\mathbf{a}$, $\mathbf{b}$, $\mathbf{c}$ denote displacements, the scalar triple product has an important geometrical significance. Let $\mathbf{a}$, $\mathbf{b}$, $\mathbf{c}$ be displacements along adjacent edges of a parallelepiped, as in Fig. 3.4. Then

$$\mathbf{d}=\mathbf{b} \wedge \mathbf{c}$$

is perpendicular to the base of the parallelepiped, and

$$d=|\mathbf{b} \wedge \mathbf{c}| \tag{3.26}$$

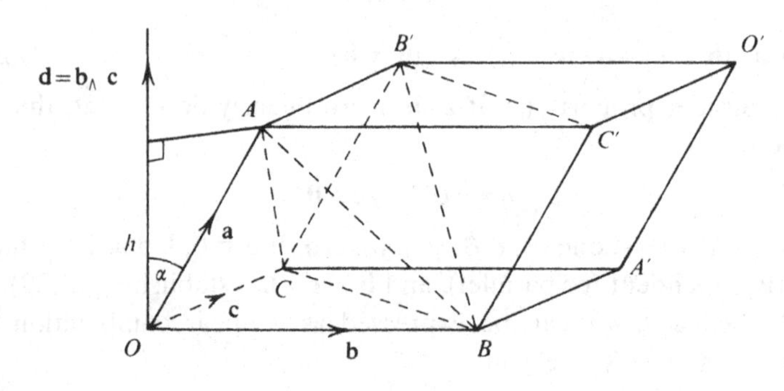


Fig. 3.4

is equal to the area of the base. If $\alpha$ is the angle between $\mathbf{a}$ and $\mathbf{d}$,

$$\begin{aligned}\mathbf{a}\cdot(\mathbf{b}\wedge\mathbf{c})&=\mathbf{a}\cdot\mathbf{d}\\&=ad\cos\alpha.\end{aligned}$$

But, as shown, $|a\cos\alpha|=h$, the height of the parallelepiped, so that

$$\begin{aligned}|\mathbf{a}\cdot(\mathbf{b}\wedge\mathbf{c})|&=dh\\&=\text{area of base}\times\text{height},\end{aligned}$$

or

$$\text{volume of parallelepiped}=|[\mathbf{a},\mathbf{b},\mathbf{c}]|. \tag{3.27}$$

The cyclic symmetry of $[\mathbf{a},\mathbf{b},\mathbf{c}]$ means that the volume is symmetric between the vectors $\mathbf{a}$, $\mathbf{b}$, $\mathbf{c}$.

The geometric figure with vertices at $O$, $A$, $B$, $C$ in Fig. 3.4 is a **tetrahedron**, and its volume is

$$\tfrac{1}{6}|[\mathbf{a},\mathbf{b},\mathbf{c}]|,$$

or one-sixth of the volume (3.27) of the parallelepiped. We justify this statement by noting that the parallelepiped can be exactly divided up into six non-overlapping tetrahedra; three of these, with vertices $(O, A, B, C)$, $(A, C, B, B')$ and $(A, B, C', B')$ are shown in Fig. 3.4. If we consider the tetrahedron $(A, C, B, B')$, for example, the displacements along three edges are

$$\begin{aligned}\mathbf{AB}&=\mathbf{b}-\mathbf{a},\\\mathbf{AC}&=\mathbf{c}-\mathbf{a},\\\mathbf{AB}'&=\mathbf{c}.\end{aligned}$$

Our assumption then gives the volume of this tetrahedron to be

$$\begin{aligned}&\tfrac{1}{6}|[\mathbf{b}-\mathbf{a},\mathbf{c}-\mathbf{a},\mathbf{c}]|\\&=\tfrac{1}{6}|[\mathbf{a},\mathbf{b},\mathbf{c}]|.\end{aligned}$$

In a similar way, the assumed formula gives the same result for the volumes of all six tetrahedra. Since the tetrahedra are non-overlapping, the volume of the parallelepiped should be the sum of the volumes of the tetrahedra; the factor $\frac{1}{6}$ ensures that this is so.

If $\mathbf{a}$, $\mathbf{b}$, $\mathbf{c}$ are given in component form, with

$$\begin{aligned}\mathbf{a}&=a_1\mathbf{i}+a_2\mathbf{j}+a_3\mathbf{k},\\\mathbf{b}&=b_1\mathbf{i}+b_2\mathbf{j}+b_3\mathbf{k},\\\mathbf{c}&=c_1\mathbf{i}+c_2\mathbf{j}+c_3\mathbf{k},\end{aligned}$$

then (3.19) and (2.36) give

$$[\mathbf{a}, \mathbf{b}, \mathbf{c}] = a_1(b_2c_3 - b_3c_2) + a_2(b_3c_1 - b_1c_3) + a_3(b_1c_2 - b_2c_1). \quad (3.28)$$

Once again, the cyclic symmetry between **a**, **b** and **c** is evident.

**Example 3.4**

If $\mathbf{a} = 3\mathbf{i} - 2\mathbf{j}$, $\mathbf{b} = 2\mathbf{i} + \mathbf{j} + 2\mathbf{k}$, $\mathbf{c} = -\mathbf{i} - 3\mathbf{j} + \mathbf{k}$, evaluate $[\mathbf{a}, \mathbf{b}, \mathbf{c}]$.

Using (3.27) with $(a_1, a_2, a_3) = (3, -2, 0)$, $(b_1, b_2, b_3) = (2, 1, 2)$ and $(c_1, c_2, c_3) = (-1, -3, 1)$,

$$\begin{aligned}[\mathbf{a}, \mathbf{b}, \mathbf{c}] &= 3(1 \cdot 1 + 2 \cdot 3) - 2(-2 \cdot 1 - 2 \cdot 1) \\ &= 3 \cdot 7 + 2 \cdot 4 = 29.\end{aligned}$$

The **vector triple product** of three vectors **a**, **b** and **c** is

$$\mathbf{a} \wedge (\mathbf{b} \wedge \mathbf{c}). \quad (3.29)$$

This product does not possess symmetry between the three vectors **a**, **b** and **c**: it is orthogonal to **a**; but since it is orthogonal to $\mathbf{b} \wedge \mathbf{c}$, it must be linearly dependent on **b** and **c** (geometrically, (3.29) lies in the plane of **b** and **c**); thus

$$\mathbf{a} \wedge (\mathbf{b} \wedge \mathbf{c}) = \lambda\mathbf{b} + \mu\mathbf{c},$$

for some values of the scalars $\lambda$ and $\mu$. We shall now find the values of the scalars $\lambda$ and $\mu$. For any particular vectors **b** and **c**, we define a triad (**i**, **j**, **k**) such that $\mathbf{i} = b^{-1}\mathbf{b}$, and **j** lies in the plane of **b** and **c**. Then **a**, **b**, **c** are of the form

$$\left.\begin{aligned}\mathbf{a} &= a_1\mathbf{i} + a_2\mathbf{j} + a_3\mathbf{k}, \\ \mathbf{b} &= b\mathbf{i}, \\ \mathbf{c} &= c_1\mathbf{i} + c_2\mathbf{j}.\end{aligned}\right\} \quad (3.30)$$

Using (3.17) and (3.18)

$$\mathbf{b} \wedge \mathbf{c} = b\mathbf{i} \wedge c_2\mathbf{j} = bc_2\mathbf{k},$$

and hence

$$\begin{aligned}\mathbf{a} \wedge (\mathbf{b} \wedge \mathbf{c}) &= (a_1\mathbf{i} + a_2\mathbf{j}) \wedge bc_2\mathbf{k} \\ &= -a_1bc_2\mathbf{j} + a_2bc_2\mathbf{i}.\end{aligned}$$

Adding and subtracting $a_1bc_1\mathbf{i}$, we find

$$\begin{aligned}\mathbf{a} \wedge (\mathbf{b} \wedge \mathbf{c}) &= (a_1c_1 + a_2c_2)b\mathbf{i} - a_1b(c_1\mathbf{i} + c_2\mathbf{j}) \\ &= (a_1c_1 + a_2c_2)\mathbf{b} - a_1b\mathbf{c}.\end{aligned}$$

But from (3.30),

$$\mathbf{a}\cdot\mathbf{b}=a_1 b,$$

$$\mathbf{a}\cdot\mathbf{c}=a_1 c_1+a_2 c_2.$$

So we have finally

$$\mathbf{a}\wedge(\mathbf{b}\wedge\mathbf{c})=(\mathbf{a}\cdot\mathbf{c})\mathbf{b}-(\mathbf{a}\cdot\mathbf{b})\mathbf{c}. \tag{3.31}$$

Although we have introduced a particular triad in establishing (3.31), the result does not depend on this triad, but only on the original vectors **a**, **b** and **c**. The formula (3.31) is extremely important in practice; with the other formulae already established, any complicated product of vectors can be reduced to a simpler form. We note that an interchange of vectors in (3.31) leads to the formula

$$(\mathbf{a}\wedge\mathbf{b})\wedge\mathbf{c}=(\mathbf{a}\cdot\mathbf{c})\mathbf{b}-(\mathbf{b}\cdot\mathbf{c})\mathbf{a}. \tag{3.32}$$

**Example 3.5**

If $\mathbf{a}=2\mathbf{i}-\mathbf{j}-2\mathbf{k}$, $\mathbf{b}=3\mathbf{i}+2\mathbf{k}$ and $\mathbf{c}=-3\mathbf{i}+\mathbf{j}+\mathbf{k}$, find $\mathbf{a}\wedge(\mathbf{b}\wedge\mathbf{c})$ and $(\mathbf{a}\wedge\mathbf{b})\wedge\mathbf{c}$.

First evaluate the scalar products occurring in (3.31) and (3.32):

$$\mathbf{a}\cdot\mathbf{b}=2\cdot 3-2\cdot 2=2,$$

$$\mathbf{a}\cdot\mathbf{c}=-2\cdot 3-1\cdot 1-2\cdot 1=-9,$$

$$\mathbf{b}\cdot\mathbf{c}=-3\cdot 3+2\cdot 1=-7.$$

Then (3.31) gives

$$\mathbf{a}\wedge(\mathbf{b}\wedge\mathbf{c})=-9(3\mathbf{i}+2\mathbf{k})-2(-3\mathbf{i}+\mathbf{j}+\mathbf{k})$$
$$=-21\mathbf{i}-2\mathbf{j}-20\mathbf{k},$$

while (3.32) gives

$$(\mathbf{a}\wedge\mathbf{b})\wedge\mathbf{c}=-9(3\mathbf{i}+2\mathbf{k})+7(2\mathbf{i}-\mathbf{j}-2\mathbf{k})$$
$$=-13\mathbf{i}-7\mathbf{j}-32\mathbf{k}.$$

Note that these vector triple products are quite different vectors.

**Example 3.6**

Show that

$$(\mathbf{a}\wedge\mathbf{b})\cdot(\mathbf{c}\wedge\mathbf{d})=(\mathbf{a}\cdot\mathbf{c})(\mathbf{b}\cdot\mathbf{d})-(\mathbf{a}\cdot\mathbf{d})(\mathbf{b}\cdot\mathbf{c}), \tag{3.33}$$

for any four vectors **a**, **b**, **c**, **d**.

Putting $\mathbf{e}=\mathbf{a}\wedge\mathbf{b}$ in the equality

$$\mathbf{e}\cdot(\mathbf{c}\wedge\mathbf{d})=(\mathbf{e}\wedge\mathbf{c})\cdot\mathbf{d}$$

gives

$$(\mathbf{a} \wedge \mathbf{b}) \cdot (\mathbf{c} \wedge \mathbf{d}) = [(\mathbf{a} \wedge \mathbf{b}) \wedge \mathbf{c}] \cdot \mathbf{d};$$

using (3.32), this becomes

$$\begin{aligned}[(\mathbf{a} \cdot \mathbf{c})\mathbf{b} - (\mathbf{b} \cdot \mathbf{c})\mathbf{a}] \cdot \mathbf{d} \\ = (\mathbf{a} \cdot \mathbf{c})(\mathbf{b} \cdot \mathbf{d}) - (\mathbf{b} \cdot \mathbf{c})(\mathbf{a} \cdot \mathbf{d}).\end{aligned}$$

An interesting identity between any four vectors **a**, **b**, **c**, **d** follows by considering the product

$$(\mathbf{a} \wedge \mathbf{b}) \wedge (\mathbf{c} \wedge \mathbf{d}).$$

Treating $\mathbf{a} \wedge \mathbf{b}$ as a single vector, and using (3.31), we find

$$\begin{aligned}(\mathbf{a} \wedge \mathbf{b}) \wedge (\mathbf{c} \wedge \mathbf{d}) &= \{(\mathbf{a} \wedge \mathbf{b}) \cdot \mathbf{d}\}\mathbf{c} - \{(\mathbf{a} \wedge \mathbf{b}) \cdot \mathbf{c}\}\mathbf{d} \\ &= [\mathbf{a}, \mathbf{b}, \mathbf{d}]\mathbf{c} - [\mathbf{a}, \mathbf{b}, \mathbf{c}]\mathbf{d}. \qquad (3.34)\end{aligned}$$

Similarly, the product may be expanded as a linear combination of **a** and **b**:

$$(\mathbf{a} \wedge \mathbf{b}) \wedge (\mathbf{c} \wedge \mathbf{d}) = [\mathbf{a}, \mathbf{c}, \mathbf{d}]\mathbf{b} - [\mathbf{b}, \mathbf{c}, \mathbf{d}]\mathbf{a}. \qquad (3.35)$$

From (3.34) and (3.35) follows the identity

$$[\mathbf{b}, \mathbf{c}, \mathbf{d}]\mathbf{a} - [\mathbf{c}, \mathbf{d}, \mathbf{a}]\mathbf{b} + [\mathbf{d}, \mathbf{a}, \mathbf{b}]\mathbf{c} - [\mathbf{a}, \mathbf{b}, \mathbf{c}]\mathbf{d} = \mathbf{0}. \qquad (3.36a)$$

We know that any four vectors **a**, **b**, **c** and **d** in 3-space are linearly dependent. Equation (3.36a) is a linear relationship that they satisfy; it can also be written

$$[\mathbf{a}, \mathbf{b}, \mathbf{c}]\mathbf{d} = [\mathbf{d}, \mathbf{b}, \mathbf{c}]\mathbf{a} + [\mathbf{d}, \mathbf{c}, \mathbf{a}]\mathbf{b} + [\mathbf{d}, \mathbf{a}, \mathbf{b}]\mathbf{c}. \qquad (3.36b)$$

If $[\mathbf{a}, \mathbf{b}, \mathbf{c}] \neq 0$, so that **a**, **b** and **c** are linearly independent, we can divide through by $[\mathbf{a}, \mathbf{b}, \mathbf{c}]$ in (3.36b). We then have an expression for any vector **d** in terms of **a**, **b**, **c**.

**Example 3.7**

Simplify

$$\{(\mathbf{a} \wedge \mathbf{b}) \wedge (\mathbf{c} \wedge \mathbf{a})\} \wedge (\mathbf{b} \wedge \mathbf{c}).$$

By (3.34), the expression in curly brackets is equal to

$$[\mathbf{a}, \mathbf{b}, \mathbf{a}]\mathbf{c} - [\mathbf{a}, \mathbf{b}, \mathbf{c}]\mathbf{a} = -[\mathbf{a}, \mathbf{b}, \mathbf{c}]\mathbf{a}.$$

Thus the whole expression equals

$$\begin{aligned}-[\mathbf{a}, \mathbf{b}, \mathbf{c}]\{\mathbf{a} \wedge (\mathbf{b} \wedge \mathbf{c})\} \\ = [\mathbf{a}, \mathbf{b}, \mathbf{c}]\{(\mathbf{a} \cdot \mathbf{b})\mathbf{c} - (\mathbf{a} \cdot \mathbf{c})\mathbf{b}\},\end{aligned}$$

using (3.31).

## Problems 3.2

1 Vectors **a**, **b**, **c** are defined by

$$\mathbf{a}=4\mathbf{i}-2\mathbf{j}+\mathbf{k},$$
$$\mathbf{b}=\mathbf{i}-\mathbf{j}-3\mathbf{k},$$
$$\mathbf{c}=2\mathbf{i}+\mathbf{j}+2\mathbf{k}.$$

Evaluate $\mathbf{a}\cdot(\mathbf{b}\wedge\mathbf{c})$, $(\mathbf{a}\wedge\mathbf{c})\cdot\mathbf{b}$, $(\mathbf{a}\wedge\mathbf{c})\wedge\mathbf{b}$, $\mathbf{a}\wedge(\mathbf{c}\wedge\mathbf{b})$ and $\{(\mathbf{a}\wedge\mathbf{b})\wedge\mathbf{c}\}\wedge\mathbf{c}$.

2 If **a**, **b**, **c**, **d**, **e**, **f** are any vectors, show that
(i) $\mathbf{a}\wedge\{\mathbf{b}\wedge(\mathbf{c}\wedge\mathbf{d})\}=(\mathbf{b}\cdot\mathbf{d})(\mathbf{a}\wedge\mathbf{c})-(\mathbf{b}\cdot\mathbf{c})(\mathbf{a}\wedge\mathbf{d})$,
(ii) $(\mathbf{a}\wedge\mathbf{b})\cdot\{(\mathbf{b}\wedge\mathbf{c})\wedge(\mathbf{c}\wedge\mathbf{a})\}=[\mathbf{a},\mathbf{b},\mathbf{c}]^2$,
(iii) $|\mathbf{a}\wedge\mathbf{b}|^2=a^2b^2-(\mathbf{a}\cdot\mathbf{b})^2$,
(iv) $(\mathbf{a}\wedge\mathbf{b})\cdot\{(\mathbf{c}\wedge\mathbf{d})\wedge(\mathbf{e}\wedge\mathbf{f})\}$
$=[\mathbf{a},\mathbf{b},\mathbf{d}][\mathbf{c},\mathbf{e},\mathbf{f}]-[\mathbf{a},\mathbf{b},\mathbf{c}][\mathbf{d},\mathbf{e},\mathbf{f}]$.

3 The corners of a tetrahedron have position vectors **a**, **b**, **c** and **d**. Show that the volume of the tetrahedron is

$$\tfrac{1}{6}|[\mathbf{b},\mathbf{c},\mathbf{d}]-[\mathbf{c},\mathbf{d},\mathbf{a}]+[\mathbf{d},\mathbf{a},\mathbf{b}]-[\mathbf{a},\mathbf{b},\mathbf{c}]|.$$

4 The lengths of two opposite edges of a tetrahedron are $a$ and $b$; the acute angle between these edges is $\theta$, and the perpendicular distance between them is $h$. Show that the volume of the tetrahedron is $\frac{1}{6}abh\sin\theta$.

5 $A$, $B$, $C$, $D$ are four points in a plane; lines through the four points, perpendicular to a plane $\pi$, meet $\pi$ in points $P$, $Q$, $R$, $S$. Show that the tetrahedra $AQRS$ and $PBCD$ have equal volumes.

6 $ABCD$ is a tetrahedron with vertices at non-coplanar points $A$, $B$, $C$, $D$, and $O$ is another point. The lines $AO$, $BO$, $CO$, $DO$ through the four vertices and $O$ meet the faces $BCD$, $CDA$, $DAB$, and $ABC$, respectively, at points $P$, $Q$, $R$ and $S$. Prove that

$$\frac{AO}{AP}+\frac{BO}{BQ}+\frac{CO}{CR}+\frac{DO}{DS}=3.$$

7 If **a**, **b**, **c** and **p** are four vectors, show that

$$[\mathbf{a},\mathbf{b},\mathbf{c}]\mathbf{p}=(\mathbf{a}\cdot\mathbf{p})(\mathbf{b}\wedge\mathbf{c})+(\mathbf{b}\cdot\mathbf{p})(\mathbf{c}\wedge\mathbf{a})+(\mathbf{c}\cdot\mathbf{p})(\mathbf{a}\wedge\mathbf{b}).$$

Explain why **p** can be expanded in terms of $\mathbf{b}\wedge\mathbf{c}$, $\mathbf{c}\wedge\mathbf{a}$ and $\mathbf{a}\wedge\mathbf{b}$ only if $[\mathbf{a},\mathbf{b},\mathbf{c}]\neq 0$.

## 3.4 Further geometry of planes and lines

Suppose that **a** and **b** are linearly independent vectors corresponding to displacements **QA** and **QB** from a point $Q$; then (by definition) the displacement from $Q$ to a point $R$ in the plane containing **QA** and **QB** is of the form

$$\mathbf{r}_Q \equiv \mathbf{QR} = \lambda\mathbf{a} + \mu\mathbf{b}. \tag{3.37}$$

If an arbitrary point $O$ in space is chosen as origin, with

$$\mathbf{OQ} = \mathbf{q}, \tag{3.38}$$

as in Fig. 3.5, then the position vector of $R$ relative to $O$ is, using (1.22),

$$\begin{aligned}\mathbf{r} &= \mathbf{q} + \mathbf{r}_Q \\ &= \mathbf{q} + \lambda\mathbf{a} + \mu\mathbf{b}.\end{aligned} \tag{3.39}$$

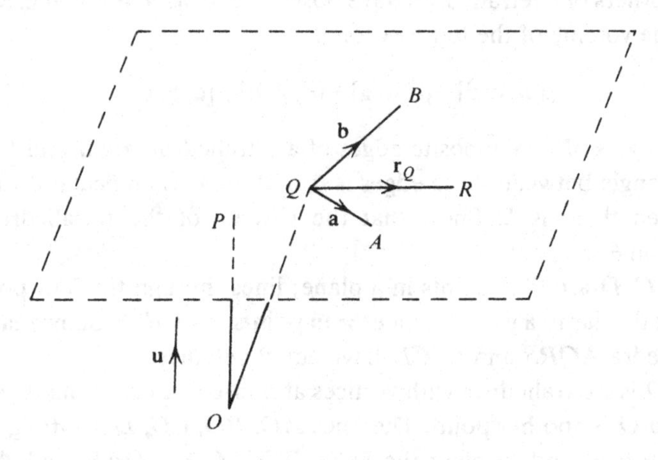


Fig. 3.5

This is the **parametric equation of a plane**, relative to origin $O$; all points in the plane are described by allowing $\lambda$ and $\mu$ to take every value in the range $(-\infty, \infty)$. The variable vector **r** is called the **running vector** in the plane.

Now suppose that **u** is one of the two unit vectors normal to the plane defined by **a** and **b** (the other such vector is $-\mathbf{u}$). Then

$$\mathbf{u} \cdot \mathbf{a} = \mathbf{u} \cdot \mathbf{b} = 0. \tag{3.40}$$

The projection of the vector $\mathbf{q}$ in direction $\mathbf{u}$ is

$$p = \mathbf{q} \cdot \mathbf{u}; \tag{3.41}$$

equation (2.2) tells us that this is the length of $OP$, the perpendicular from $O$ onto the plane, shown in Fig. 3.5. Also, by (2.23), the displacement $\mathbf{OP}$ corresponds to the vector

$$\mathbf{p} = (\mathbf{q} \cdot \mathbf{u})\mathbf{u} = p\mathbf{u}. \tag{3.42}$$

Taking the scalar product of Equation (3.39) with $\mathbf{u}$ and using (3.41), we find

$$\mathbf{r} \cdot \mathbf{u} = p. \tag{3.43}$$

This is another form of the equation of a plane; it tells us that the projection perpendicular to the plane, of the position vector of any point $R$ in the plane, is equal to $p$.

We note that any equation of the form

$$\mathbf{r} \cdot \mathbf{v} = m \qquad (\mathbf{v} \neq \mathbf{0}) \tag{3.44}$$

can be written as

$$\mathbf{r} \cdot \mathbf{u} = v^{-1}m,$$

where $\mathbf{u} = v^{-1}\mathbf{v}$ is a unit vector; this is of the form (3.43), so that (3.44) also defines a plane.

Also, using (3.42), the equation of the plane can be written

$$(\mathbf{r} - \mathbf{p}) \cdot \mathbf{u} = 0. \tag{3.45}$$

**Example 3.8**

Find the equation of the plane through the point with position vector $2\mathbf{i} - \mathbf{j} - 2\mathbf{k}$ which is orthogonal to the vector $\mathbf{i} + \mathbf{j} + 3\mathbf{k}$.

If $\mathbf{r}$ is the position vector of any point in the plane, then $\mathbf{r} - (2\mathbf{i} - \mathbf{j} - 2\mathbf{k})$ is perpendicular to $\mathbf{i} + \mathbf{j} + 3\mathbf{k}$. So

$$\mathbf{r} \cdot (\mathbf{i} + \mathbf{j} + 3\mathbf{k}) = (2\mathbf{i} - \mathbf{j} - 2\mathbf{k}) \cdot (\mathbf{i} + \mathbf{j} + 3\mathbf{k}) = -5.$$

If $\mathbf{r} = x\mathbf{i} + y\mathbf{j} + z\mathbf{k}$, the equation can be written

$$x + y + 3z = -5.$$

If three non-collinear points $R_1$, $R_2$, $R_3$ with position vectors $\mathbf{r}_1$, $\mathbf{r}_2$ and $\mathbf{r}_3$ lie in plane, the plane is determined. We can identify $R_1$, say, with $Q$ in Fig. 3.5, so that

$$\mathbf{q} = \mathbf{r}_1;$$

also, we can take the linearly independent displacements in the plane to be

$$\mathbf{a}=\mathbf{r}_2-\mathbf{r}_1$$

and

$$\mathbf{b}=\mathbf{r}_3-\mathbf{r}_1,$$

so that $R_2$ and $R_3$ coincide with $A$ and $B$ respectively. Substituting for $\mathbf{q}$, $\mathbf{a}$ and $\mathbf{b}$ in (3.39), and writing $\lambda=\lambda_2$, $\mu=\lambda_3$, the parametric equation of the plane becomes

$$\mathbf{r}=\mathbf{r}_1+\lambda_2(\mathbf{r}_2-\mathbf{r}_1)+\lambda_3(\mathbf{r}_3-\mathbf{r}_1)$$

or

$$\mathbf{r}=(1-\lambda_2-\lambda_3)\mathbf{r}_1+\lambda_2\mathbf{r}_2+\lambda_3\mathbf{r}_3. \tag{3.46}$$

Defining $\lambda_1$ by

$$\lambda_1+\lambda_2+\lambda_3=1, \tag{3.47}$$

Equation (3.46) can be written in the symmetrical form

$$\mathbf{r}=\lambda_1\mathbf{r}_1+\lambda_2\mathbf{r}_2+\lambda_3\mathbf{r}_3. \tag{3.48}$$

We note that this result gives a partial answer to Question 3 of Problems 2.4.

**Example 3.9**

Find the equation of the plane through the points with position vectors

$$\mathbf{r}_1=2\mathbf{i}-\mathbf{k},$$
$$\mathbf{r}_2=3\mathbf{i}+2\mathbf{j}+\mathbf{k},$$
$$\mathbf{r}_3=-\mathbf{i}+4\mathbf{j}+2\mathbf{k}.$$

The equation can be written in the parametric form (3.47) and (3.48) directly. Another form of the equation can be given: if $\mathbf{r}$ is a point in the plane, then $\mathbf{r}-\mathbf{r}_1$, $\mathbf{r}_2-\mathbf{r}_1$ and $\mathbf{r}_3-\mathbf{r}_1$ all lie in the plane. The condition for these vectors to be coplanar is, by (3.21),

$$[\mathbf{r}-\mathbf{r}_1, \mathbf{r}_2-\mathbf{r}_1, \mathbf{r}_3-\mathbf{r}_1]=0.$$

Since

$$(\mathbf{r}_2-\mathbf{r}_1)\wedge(\mathbf{r}_3-\mathbf{r}_1)=-2\mathbf{i}-9\mathbf{j}+10\mathbf{k},$$

the equation is

$$\mathbf{r}\cdot(-2\mathbf{i}-9\mathbf{j}+10\mathbf{k})=(2\mathbf{i}-\mathbf{k})\cdot(-2\mathbf{i}-9\mathbf{j}+10\mathbf{k})$$

or

$$2x+9y-10z=14.$$

If $A$ is a given point, any point $R$ on a line through $A$ has position vector relative to $A$ of the form

$$\mathbf{r}_A = s\mathbf{u}, \tag{3.49}$$

where $\mathbf{u}$ is a fixed unit vector and $s$ is a variable parameter. Changing origin to a point $O$, with

$$\mathbf{OA} = \mathbf{a}, \tag{3.50}$$

the position vector of $R$ relative to $O$ is

$$\mathbf{r} = \mathbf{a} + \mathbf{r}_A = \mathbf{a} + s\mathbf{u}. \tag{3.51}$$

This is the **parametric equation of a line** through $A$ in direction $\mathbf{u}$, and $\mathbf{r}$ is the running vector. If the rectangular components of $\mathbf{r}$, $\mathbf{a}$ and $\mathbf{u}$ are $(x, y, z)$, $(a_1, a_2, a_3)$ and $(l_1, l_2, l_3)$ respectively, (3.51) can be written

$$x = a_1 + sl_1,$$
$$y = a_2 + sl_2,$$
$$z = a_3 + sl_3.$$

We can eliminate $s$ from these equations to give

$$\frac{x-a_1}{l_1} = \frac{y-a_2}{l_2} = \frac{z-a_3}{l_3}, \tag{3.52}$$

an equation satisfied by the Cartesian coordinates $(x, y, z)$ of every point on the line. This is the usual **coordinate equation of a line** through a point $(a_1, a_2, a_3)$, with direction cosines $(l_1, l_2, l_3)$.

If the point $B$ with coordinates $(b_1, b_2, b_3)$ also lies on the line, then

$$\frac{b_1-a_1}{l_1} = \frac{b_2-a_2}{l_2} = \frac{b_3-a_3}{l_3}.$$

Eliminating $(l_1, l_2, l_3)$ between this equation and (3.52) gives

$$\frac{x-a_1}{a_1-b_1} = \frac{y-a_2}{a_2-b_2} = \frac{z-a_3}{a_3-b_3}. \tag{3.53}$$

This is the **equation of the line through two given points** $A$ and $B$ with position vectors $\mathbf{a}$ and $\mathbf{b}$.

Two non-parallel planes intersect in a line; we now show how to determine their line of intersection. Suppose that the equations of the planes are given in the form (3.44) as

$$\left.\begin{aligned}\mathbf{r}\cdot\mathbf{v}&=m,\\ \mathbf{r}\cdot\mathbf{w}&=n.\end{aligned}\right\}\tag{3.54}$$

Since $\mathbf{v}$ and $\mathbf{w}$ respectively are perpendicular to the two planes, the condition that the planes are not parallel is equivalent to assuming that $\mathbf{v}$ and $\mathbf{w}$ are linearly independent. Then $\mathbf{v}$, $\mathbf{w}$ and $\mathbf{v}\wedge\mathbf{w}$ form a linearly independent set of vectors, and we can therefore express the position vector $\mathbf{r}$ of any point in 3-space in the form

$$\mathbf{r}=\alpha\mathbf{v}+\beta\mathbf{w}+t\mathbf{v}\wedge\mathbf{w}.\tag{3.55}$$

For points common to the two planes, $\mathbf{r}$ satisfies (3.54). Substituting (3.55) in (3.54), and remembering that $[\mathbf{v},\mathbf{v},\mathbf{w}]=[\mathbf{w},\mathbf{v},\mathbf{w}]=0$, we find

$$\begin{aligned}\alpha v^2+\beta\mathbf{v}\cdot\mathbf{w}&=m,\\ \alpha\mathbf{v}\cdot\mathbf{w}+\beta w^2&=n.\end{aligned}\tag{3.56}$$

Since $\mathbf{v}\cdot\mathbf{w}\neq vw$ for linearly independent vectors $\mathbf{v}$ and $\mathbf{w}$, (3.56) can be solved to determine $\alpha$ and $\beta$ uniquely. So, in (3.55), only $t$ is undetermined; the equation is therefore of the form (3.51), with $\mathbf{u}$ the unit vector corresponding to $\mathbf{v}\wedge\mathbf{w}$. The parameters $s$ and $t$ in (3.51) and (3.55) are related by

$$s=t|\mathbf{v}\wedge\mathbf{w}|.$$

The direction $\mathbf{u}$ of the line is thus orthogonal to both $\mathbf{v}$ and $\mathbf{w}$, in accord with geometrical intuition.

**Example 3.10**

Find the equation of the line common to the two planes

$$\begin{aligned}\mathbf{r}\cdot(\mathbf{i}+\mathbf{j}+\mathbf{k})&=4,\\ \mathbf{r}\cdot(2\mathbf{i}-\mathbf{j}+\mathbf{k})&=-2.\end{aligned}$$

The direction of the common line is parallel to $\mathbf{v}\wedge\mathbf{w}$, using the notation of (3.54). This vector is

$$(\mathbf{i}+\mathbf{j}+\mathbf{k})\wedge(2\mathbf{i}-\mathbf{j}+\mathbf{k})=2\mathbf{i}+\mathbf{j}-3\mathbf{k}.$$

In (3.55), $\alpha$ and $\beta$ are determined by (3.56):

$$\left.\begin{aligned}3\alpha+2\beta&=4,\\ 2\alpha+6\beta&=-2,\end{aligned}\right\}$$

so that $\alpha=2$, $\beta=-1$. So the parametric equation of the line of intersection is

$$\begin{aligned}\mathbf{r}&=2(\mathbf{i}+\mathbf{j}+\mathbf{k})-(2\mathbf{i}-\mathbf{j}+\mathbf{k})+t(2\mathbf{i}+\mathbf{j}-3\mathbf{k})\\&=3\mathbf{j}+\mathbf{k}+t(2\mathbf{i}+\mathbf{j}-3\mathbf{k}).\end{aligned}$$

The Equations (3.54) for two planes may be written

$$\mathbf{r}\cdot\mathbf{v}-m=0,$$
$$\mathbf{r}\cdot\mathbf{w}-n=0.$$

Consider an equation of the form

$$(\mathbf{r}\cdot\mathbf{v}-m)+\alpha(\mathbf{r}\cdot\mathbf{w}-n)=0, \tag{3.57}$$

where $\alpha$ is some real number. We assume that the planes (3.54) are not parallel, so that the terms $\mathbf{r}\cdot(\mathbf{v}+\alpha\mathbf{w})$ in (3.57) cannot be identically zero. Then (3.57) is the equation of a plane of the form (3.44); further, if a vector $\mathbf{r}$ satisfies both of the Equations (3.54), it satisfies (3.57). So the plane (3.57) contains the line of intersection of the planes (3.54); if we allow $\alpha$ to vary, (3.57) represents the **family of planes**, all of which contain this line of intersection. One of the original planes corresponds to $\alpha=0$; dividing (3.57) by $\alpha$ and then putting $\alpha^{-1}=0$ gives the second plane.

**Example 3.11**

Find the condition for the lines

$$\left.\begin{aligned}&\mathbf{r}=\mathbf{a}_1+s_1\mathbf{u}_1,\\ \text{and}\qquad&\\&\mathbf{r}=\mathbf{a}_2+s_2\mathbf{u}_2\end{aligned}\right\} \tag{3.58}$$

to intersect.

If there are values of $s_1$ and $s_2$ corresponding to the same position vector $\mathbf{r}$, then

$$\mathbf{a}_1-\mathbf{a}_2=s_1\mathbf{u}_1-s_2\mathbf{u}_2.$$

Forming the scalar product with $\mathbf{u}_1\wedge\mathbf{u}_2$ gives

$$[\mathbf{a}_1-\mathbf{a}_2,\mathbf{u}_1,\mathbf{u}_2]=0, \tag{3.59}$$

a condition on $\mathbf{a}_1$, $\mathbf{a}_2$, $\mathbf{u}_1$ and $\mathbf{u}_2$ only. This is the condition for the two lines (3.58) to intersect.

**Example 3.12**
Find the perpendicular distance from the point $\mathbf{p}=2\mathbf{i}+\mathbf{j}+3\mathbf{k}$ to the line

$$\mathbf{r}=3\mathbf{i}+2\mathbf{j}+12\mathbf{k}+t(3\mathbf{i}-\mathbf{j}+\mathbf{k}),$$

and find the position vector $\mathbf{r}_1$ of the foot of the perpendicular.

Let $\mathbf{r}=\mathbf{r}_1$ correspond to $t=t_1$. Then

$$(\mathbf{r}_1-\mathbf{p})\cdot(3\mathbf{i}-\mathbf{j}+\mathbf{k})=0$$

or

$$[\mathbf{i}+\mathbf{j}+9\mathbf{k}+t_1(3\mathbf{i}-\mathbf{j}+\mathbf{k})]\cdot(3\mathbf{i}-\mathbf{j}+\mathbf{k})=0,$$

giving $t_1=-1$. So the foot of the perpendicular has position vector

$$\mathbf{r}_1=3\mathbf{j}+11\mathbf{k}.$$

The length of the perpendicular is

$$|\mathbf{r}_1-\mathbf{p}|=|-2\mathbf{i}+2\mathbf{j}+8\mathbf{k}|=6\sqrt{2}.$$

It is often convenient to define a line in terms of its **vector moment** $\mathbf{m}$ about the origin $O$. This is defined by taking the vector product of the running vector (3.51) with $\mathbf{u}$; since $\mathbf{u}\wedge\mathbf{u}=\mathbf{0}$, this gives

$$\mathbf{r}\wedge\mathbf{u}=\mathbf{a}\wedge\mathbf{u}\equiv\mathbf{m}. \tag{3.60}$$

Since $s$ has been eliminated, $\mathbf{m}$ is the same vector for all points $\mathbf{r}$ on the line. We shall now show that the line is determined if $\mathbf{u}$ and $\mathbf{m}$ are given; they must of course satisfy the condition

$$\mathbf{u}\cdot\mathbf{m}=0. \tag{3.61}$$

The position vectors $\mathbf{r}$ are constructed to satisfy (3.60), with given $\mathbf{u}$ and $\mathbf{m}$; this is a **vector equation** for $\mathbf{r}$. We solve the equation by taking the vector product of (3.60) with $\mathbf{u}$, giving

$$\mathbf{u}\wedge(\mathbf{r}\wedge\mathbf{u})=\mathbf{u}\wedge\mathbf{m}$$

or, using (3.32),

$$\mathbf{r}-(\mathbf{r}\cdot\mathbf{u})\mathbf{u}=\mathbf{u}\wedge\mathbf{m}. \tag{3.62}$$

The left-hand member of this equation is, by (2.24), just the component $\mathbf{r}_\perp$ or $\mathbf{r}$ orthogonal to $\mathbf{u}$, and $(\mathbf{r}\cdot\mathbf{u})$ is the projection of $\mathbf{r}$ in direction $\mathbf{u}$, as indicated in Fig. 3.6. Thus $\mathbf{u}$, $\mathbf{m}$, $\mathbf{r}_\perp$ are a mutually

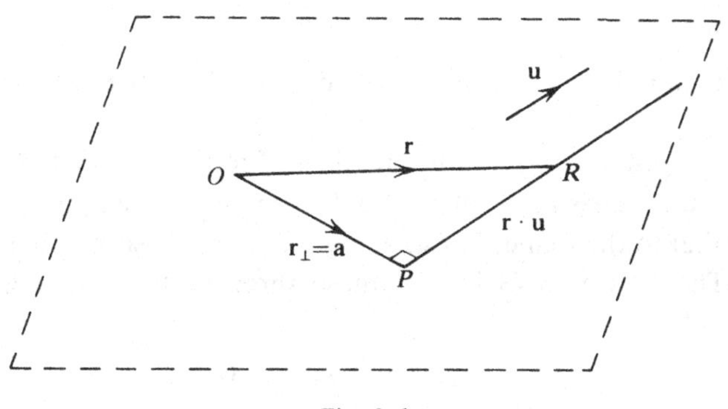


Fig. 3.6

orthogonal right-handed set of vectors. In (3.62), $(\mathbf{r}\cdot\mathbf{u})$ is just a scalar parameter defining the distance $PR$, and can thus be identified with $s$ in (3.51). Defining the vector $\mathbf{a}$ by

$$\mathbf{a}=\mathbf{r}_{\perp}=\mathbf{u}\wedge\mathbf{m}, \qquad (3.63)$$

we see that (3.62) is exactly of the form (3.51); the displacement $\mathbf{OP}=\mathbf{a}$ is given in terms of $(\mathbf{u},\mathbf{m})$ by (3.63). Thus (3.60) is another form of the equation of a line. The pair of vectors $(\mathbf{u},\mathbf{m})$ are called the **fundamental vectors** of the line.

If a line is defined as the intersection of the planes (3.54), we have shown that the unit vector $\mathbf{u}$ along the line is given by

$$\mathbf{u}=\frac{\mathbf{v}\wedge\mathbf{w}}{|\mathbf{v}\wedge\mathbf{w}|}. \qquad (3.64)$$

The vector moment is thus

$$\mathbf{m}=\frac{\mathbf{r}\wedge(\mathbf{v}\wedge\mathbf{w})}{|\mathbf{v}\wedge\mathbf{w}|};$$

using (3.32) and the Equations (3.54) of the planes, this becomes

$$\mathbf{m}=\frac{(\mathbf{r}\cdot\mathbf{w})\mathbf{v}-(\mathbf{r}\cdot\mathbf{v})\mathbf{w}}{|\mathbf{v}\wedge\mathbf{w}|}$$

$$=\frac{n\mathbf{v}-m\mathbf{w}}{|\mathbf{v}\wedge\mathbf{w}|}. \qquad (3.65)$$

Equations (3.64) and (3.65) define the fundamental vectors $(\mathbf{u},\mathbf{m})$ for the line of intersection of the two planes (3.54).

**Example 3.13**

Find the equation of the plane containing the point $\mathbf{r}_1$ and the line $(\mathbf{u}, \mathbf{m})$.

Let $\mathbf{r}$ represent any point in the plane. From (3.62) or (3.63), the point $\mathbf{a} = \mathbf{u} \wedge \mathbf{m}$ lies in the plane. So the vectors $\mathbf{r} - \mathbf{r}_1$ and $(\mathbf{r}_1 - \mathbf{u} \wedge \mathbf{m})$ are parallel to the plane; but $\mathbf{u}$, along the line, is also parallel to the plane. The condition (3.21) for these three vectors to be linearly dependent is

$$[\mathbf{r} - \mathbf{r}_1, \mathbf{r}_1 - \mathbf{u} \wedge \mathbf{m}, \mathbf{u}] = 0.$$

Since $(\mathbf{u} \wedge \mathbf{m}) \wedge \mathbf{u} = \mathbf{m}$, this reduces to

$$\mathbf{r} \cdot (\mathbf{r}_1 \wedge \mathbf{u} + \mathbf{m}) = \mathbf{r}_1 \cdot \mathbf{m}.$$

**Example 3.14**

Find the equation of the plane through the line $(\mathbf{u}_1, \mathbf{m}_1)$ which is parallel to the line $(\mathbf{u}_2, \mathbf{m}_2)$.

The point $\mathbf{a} = \mathbf{u}_1 \wedge \mathbf{m}_1$ lies in the plane, so $\mathbf{r} - \mathbf{a}$ is parallel to the plane for all points $\mathbf{r}$ in the plane. But $\mathbf{u}_1$ and $\mathbf{u}_2$ are also parallel to the plane. Hence

$$[\mathbf{r} - \mathbf{a}, \mathbf{u}_1, \mathbf{u}_2] = 0,$$

which reduces to

$$\mathbf{r} \cdot (\mathbf{u}_1 \wedge \mathbf{u}_2) = \mathbf{u}_2 \cdot \mathbf{m}_1.$$

## *3.5 Vector equations*

In §3.4 we encountered the vector equation (3.60) for the vector $\mathbf{r}$, subject to the condition (3.61). We showed that its solution was of the form (3.51), with $\mathbf{a}$ given by (3.63). This is one example of a vector equation; more generally, **vector equations** are mathematical statements about an unknown vector $\mathbf{x}$ which determine, or partly determine, $\mathbf{x}$. Equation (3.60) does not determine $\mathbf{r}$ exactly, since the scalar $s = (\mathbf{r} \cdot \mathbf{u})$ in (3.51) is undetermined; thus the component of $\mathbf{r}$ in the direction of $\mathbf{u}$ is arbitrary.

We now consider some other types of vector equation. First, suppose that $\mathbf{x}$ satisfies

$$\lambda \mathbf{x} + \mathbf{x} \wedge \mathbf{a} = \mathbf{b}, \tag{3.66}$$

where $\lambda$, $\mathbf{a}$ and $\mathbf{b}$ are given, with $\lambda \neq 0$. If $\lambda = 0$, Equation (3.66) reduces to the form (3.60), which we have already solved. Taking the vector product of (3.66) with **a**, and using (3.32), we obtain

$$\lambda(\mathbf{x} \wedge \mathbf{a}) + (\mathbf{x} \cdot \mathbf{a})\mathbf{a} - a^2\mathbf{x} = \mathbf{b} \wedge \mathbf{a}. \tag{3.67}$$

Also, taking the scalar product of (3.66) with **a** gives

$$\lambda(\mathbf{x} \cdot \mathbf{a}) = \mathbf{b} \cdot \mathbf{a}$$

or, since $\lambda \neq 0$,

$$\mathbf{x} \cdot \mathbf{a} = \lambda^{-1}(\mathbf{b} \cdot \mathbf{a}). \tag{3.68}$$

Now, from (3.66) itself,

$$\lambda(\mathbf{x} \wedge \mathbf{a}) = \lambda(\mathbf{b} - \lambda\mathbf{x}). \tag{3.69}$$

Substituting (3.68) and (3.69) into (3.67), we find

$$\lambda(\mathbf{b} - \lambda\mathbf{x}) + \lambda^{-1}(\mathbf{b} \cdot \mathbf{a})\mathbf{a} - a^2\mathbf{x} = \mathbf{b} \wedge \mathbf{a},$$

or

$$\mathbf{x} = [\lambda\mathbf{b} + \lambda^{-1}(\mathbf{b} \cdot \mathbf{a})\mathbf{a} - \mathbf{b} \wedge \mathbf{a}]/(\lambda^2 + a^2). \tag{3.70}$$

So when $\lambda \neq 0$, (3.66) has the unique solution (3.70); but when $\lambda = 0$, we know that provided $\mathbf{a} \cdot \mathbf{b} = 0$, there is a solution **x** which is not uniquely determined. These facts can also be understood as properties of linear equations, which we discuss in Chapter 4: if we express **x**, **a** and **b** in terms of components, then the three components of the vector equation (3.66) become three linear equations for $x_1$, $x_2$, $x_3$. If the determinant of coefficients in this equation is non-zero, **x** is uniquely determined; it is not hard to show that this condition is just $\lambda \neq 0$ (see Problems 3.3, Question 15). When $\lambda = 0$, the determinant of coefficients is zero; then there is either no solution for **x** or an infinite number of solutions.

Another type of vector equation is

$$\lambda\mathbf{x} + (\mathbf{x} \cdot \mathbf{b})\mathbf{a} = \mathbf{c}, \tag{3.71}$$

in which we assume that $\lambda \neq 0$. If we form the scalar product of each side with **b**, we find

$$(\mathbf{x} \cdot \mathbf{b})[\lambda + \mathbf{a} \cdot \mathbf{b}] = \mathbf{b} \cdot \mathbf{c}. \tag{3.72}$$

So if

$$\lambda + \mathbf{a} \cdot \mathbf{b} \neq 0, \tag{3.73}$$

$$\mathbf{x} \cdot \mathbf{b} = \frac{\mathbf{b} \cdot \mathbf{c}}{\lambda + \mathbf{a} \cdot \mathbf{b}}.$$

Substituting into (3.71), we obtain the solution

$$\mathbf{x}=\lambda^{-1}\left[\mathbf{c}-\mathbf{a}\left(\frac{\mathbf{b}\cdot\mathbf{c}}{\lambda+\mathbf{a}\cdot\mathbf{b}}\right)\right], \tag{3.74}$$

since $\lambda \neq 0$. There is therefore a unique solution (3.74) provided (3.73) holds; again, this condition ensures that, in component form, the three linear equations (3.71) have a non-zero determinant of coefficients, and so have a unique solution.

If, however,

$$\lambda+\mathbf{a}\cdot\mathbf{b}=0,$$

then (3.72) implies that $\mathbf{b}\cdot\mathbf{c}=0$; if $\mathbf{b}$ and $\mathbf{c}$ are not orthogonal, (3.71) cannot then be satisfied. Provided $\mathbf{b}\cdot\mathbf{c}=0$, (3.72) is then trivially satisfied, and $\mathbf{x}\cdot\mathbf{b}$ is not determined. Writing $t=\mathbf{x}\cdot\mathbf{b}$, (3.71) then becomes

$$\mathbf{x}=\lambda^{-1}[\mathbf{c}-t\mathbf{a}], \tag{3.75}$$

with $t$ arbitrary. So there is an infinity of solutions $\mathbf{x}$; in geometric terms, $\mathbf{x}$ lies on a line through $\lambda^{-1}\mathbf{c}$ and parallel to the vector $\mathbf{a}$.

## *Problems 3.3*

1 A plane $\pi$ is normal to the vector $2\mathbf{i}-\mathbf{j}-2\mathbf{k}$ and passes through the point with position vector $3\mathbf{i}+2\mathbf{j}-\mathbf{k}$. Find the perpendicular distance from the origin to the plane $\pi$; find also the position vector of the foot of the perpendicular.

2 Show that the points with position vectors $\mathbf{i}-\mathbf{j}+\mathbf{k}$, $2\mathbf{i}+3\mathbf{k}$, $-\mathbf{i}+2\mathbf{j}+4\mathbf{k}$ and $6\mathbf{i}-\mathbf{j}+4\mathbf{k}$ lie in a plane.

3 Find the equation of the plane which contains the line

$$\mathbf{r}=\mathbf{i}-\mathbf{j}+2\mathbf{k}+t(4\mathbf{i}-\mathbf{j}-\mathbf{k}),$$

and to which the vector $2\mathbf{i}+\mathbf{j}-3\mathbf{k}$ is parallel.

4 Find the plane $\pi$ containing the line of intersection of the planes $\mathbf{r}\cdot(2\mathbf{i}+3\mathbf{j}-\mathbf{k})=4$ and $\mathbf{r}\cdot(\mathbf{i}-\mathbf{j}+2\mathbf{k})=3$, and passing through the point with position vector $-\mathbf{i}+2\mathbf{j}+4\mathbf{k}$.

5 Show that the equation of the line passing through the point with position vector $\mathbf{r}_2$, and meeting the line $\mathbf{r}=\mathbf{r}_1+\lambda\mathbf{u}$ at right-angles, is

$$\mathbf{r}=\mathbf{r}_2+\mu[\mathbf{r}_2-\mathbf{r}_1-\{(\mathbf{r}_2-\mathbf{r}_1)\cdot\mathbf{u}\}\mathbf{u}].$$

6 Show that the lines

$$2(x-2)=-(3y+1)=z-2$$

and

$$\tfrac{1}{2}(x+3)=y+4=\tfrac{1}{3}(2z+1)$$

intersect, and find the point of intersection.

7 Find the equation of the line containing the point $\mathbf{r}_2$, which is parallel to the plane $\mathbf{r}\cdot\mathbf{v}=m$ and perpendicular to the line $\mathbf{r}=\mathbf{r}_1+s\mathbf{u}$; the vectors $\mathbf{u}$ and $\mathbf{v}$ are not parallel.

8 If $\mathbf{v}_1$ and $\mathbf{v}_2$ are not parallel, show that there is a unique line (the common perpendicular) which intersects and is perpendicular to the two lines

$$\mathbf{r}=\mathbf{r}_1+t\mathbf{v}_1,$$

$$\mathbf{r}=\mathbf{r}_2+t\mathbf{v}_2.$$

Show that the perpendicular distance between the lines is

$$|[\mathbf{r}_2-\mathbf{r}_1,\mathbf{v}_1,\mathbf{v}_2]|/|\mathbf{v}_1\wedge\mathbf{v}_2|.$$

9 Find the length and the equation of the common perpendicular of the two lines

$$x-3=y-4=-z-1$$

and

$$2(x+6)=y+5=-4(z+1).$$

[See Question 8 above.]

10 Find the equation of the line of intersection of the planes

$$\mathbf{r}\cdot(\mathbf{i}+2\mathbf{j}-\mathbf{k})=9,$$

$$\mathbf{r}\cdot(3\mathbf{i}-4\mathbf{j}-\mathbf{k})=-2,$$

and find the fundamental vectors of the line.

11 Find the angle between the line $\mathbf{r}=\mathbf{a}+t\mathbf{v}$ and the plane $\mathbf{r}\cdot\mathbf{w}=n$.

12 Find the equation of the plane containing the line $(\mathbf{u},\mathbf{m})$ and perpendicular to the plane $\mathbf{r}\cdot\mathbf{w}=n$.

13 The line $(\mathbf{u}_1,\mathbf{m})$ is projected orthogonally onto the plane $\mathbf{r}\cdot\mathbf{u}_2=p$, where $\mathbf{u}_1$ and $\mathbf{u}_2$ are non-parallel unit vectors. Show that the fundamental vectors of the resulting line are $k\mathbf{u}_2\wedge(\mathbf{u}_1\wedge\mathbf{u}_2)$, $k[p(\mathbf{u}_2\wedge\mathbf{u}_1)+(\mathbf{u}_2\cdot\mathbf{m})\mathbf{u}_2]$, where $k|\mathbf{u}_1\wedge\mathbf{u}_2|=1$.

14 The plane $\pi$ passes through the point with position vector $\mathbf{a}$, and is perpendicular to the vector $\mathbf{a}$; $\pi$ meets the coordinate axes at the points $A$, $B$, $C$. Show that the area of the triangle $ABC$ is

$$\tfrac{1}{2}a^5/(\mathbf{a}\cdot\mathbf{i})(\mathbf{a}\cdot\mathbf{j})(\mathbf{a}\cdot\mathbf{k}).$$

15 Write Equation (3.66) in terms of Cartesian coordinates of the vectors $\mathbf{x}$, $\mathbf{a}$, $\mathbf{b}$. If $\Delta$ is the determinant of coefficients of these equations for the components of $\mathbf{x}$, show that $\Delta = 0$ if and only if $\lambda = 0$.

## 3.6 *Spherical trigonometry*

The formulae of spherical trigonometry are basic to the study of astronomy and astro-navigation. We shall establish the two fundamental formulae, the *sine formula* and the *cosine formula.* These formulae relate certain angles defined in terms of points on the unit sphere whose centre is at the origin 0; the position vectors of all points on this sphere are then unit vectors. The cosine formula is the result to be established in Problems 2.5, Question 6; it is expressed in terms of spherical polar coordinates $(r, \theta, \phi)$ relative to a given frame of reference. Since $r = 1$ for all points on the unit sphere, the spherical polar coordinates of two points $P_1$, $P_2$ on the sphere are of the form $(1, \theta_1, \phi_1)$ and $(1, \theta_2, \phi_2)$. So the unit position vectors of $P_1$ and $P_2$ are, from (2.70),

$$\mathbf{u}_1 = \sin\theta_1 \cos\phi_1\, \mathbf{i} + \sin\theta_1 \sin\phi_1\, \mathbf{j} + \cos\theta_1\, \mathbf{k},$$

$$\mathbf{u}_2 = \sin\theta_2 \cos\phi_2\, \mathbf{i} + \sin\theta_2 \sin\phi_2\, \mathbf{j} + \cos\theta_2\, \mathbf{k}.$$

Using (2.66) and (2.50a), the angle $\alpha$ between the vectors $\mathbf{u}_1$ and $\mathbf{u}_2$ is given by

$$\cos\alpha = \sin\theta_1 \cos\phi_1 \sin\theta_2 \cos\phi_2 + \sin\theta_1 \sin\phi_1 \sin\theta_2 \sin\phi_2 + \cos\theta_1 \cos\theta_2$$

or

$$\cos\alpha = \cos\theta_1 \cos\theta_2 + \sin\theta_1 \sin\theta_2 \cos(\phi_1 - \phi_2). \tag{3.76}$$

This is the **cosine formula**.

The **sine formula** concerns the angles of a **spherical triangle**, which we now define. As in Fig. 3.7, let $A$, $B$, $C$ be three points on the unit sphere, centre $O$, and let $\mathbf{a}$, $\mathbf{b}$, $\mathbf{c}$ be their unit position vectors. The planes $OAB$, $OBC$ and $OCA$ intersect the sphere in the circular arcs $AB$, $BC$, $CA$, as shown; these three circular arcs form a spherical triangle. The angles between the pairs of planes are denoted by $A$, $B$, $C$, as shown. The angles subtended at $O$ by the arcs $BC$, $CA$, $AB$ are $\alpha$, $\beta$, $\gamma$ respectively, and these angles are assumed to be different from 0 and $\pi$; two of these angles are shown. The sine formula states

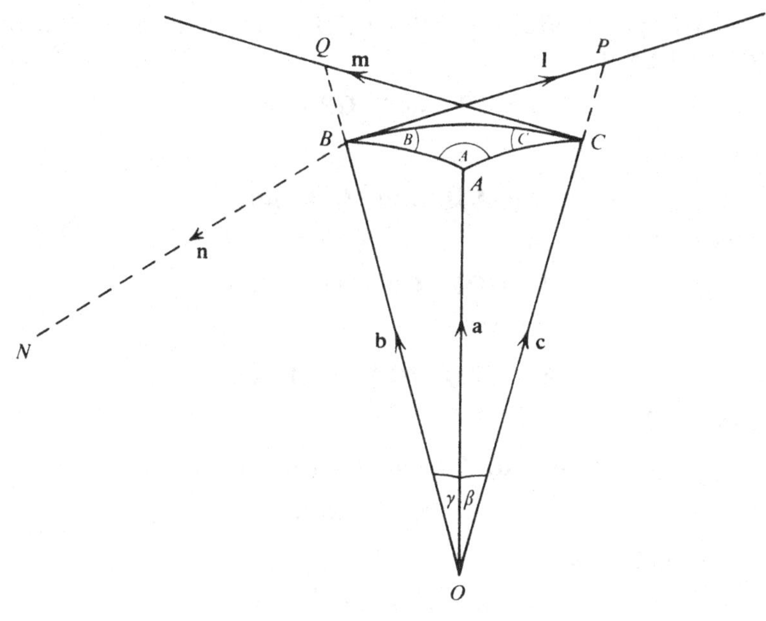


Fig. 3.7

that

$$\frac{\sin A}{\sin \alpha}=\frac{\sin B}{\sin \beta}=\frac{\sin C}{\sin \gamma}. \tag{3.77}$$

We now establish this formula.

Let **n** be the unit normal to the plane $OAB$, represented by the displacement **BN**; then

$$\mathbf{a} \wedge \mathbf{b}=\mathbf{n} \sin \gamma, \tag{3.78}$$

since **a** and **b** are unit vectors. If the perpendiculars to $OB$, $OC$ at $B$, $C$ respectively, in the plane $OBC$, meet the lines $OC$, $OB$ at $P$, $Q$ respectively, then the angle $NBP$ equals $B+\frac{1}{2}\pi$. So if **l** is the unit vector parallel to **BP**,

$$\mathbf{n} \cdot \mathbf{l}=\cos(B+\tfrac{1}{2}\pi)=-\sin B.$$

Using (3.78), this gives

$$(\mathbf{a} \wedge \mathbf{b}) \cdot \mathbf{l}=-\sin \gamma \sin B. \tag{3.79}$$

Likewise, if **m** is the unit vector parallel to **CQ**,

$$(\mathbf{c} \wedge \mathbf{a}) \cdot \mathbf{m}=-\sin \beta \sin C. \tag{3.80}$$

Now, in the plane $OBC$, since $OB = OC$, it follows that $BP = CQ$ and $OP = OQ$. So, if

$$\mathbf{l} = \lambda \mathbf{BP} \quad \text{and} \quad \mathbf{OP} = \mu \mathbf{c},$$

then

$$\mathbf{m} = \lambda \mathbf{CQ} \quad \text{and} \quad \mathbf{OQ} = \mu \mathbf{b}.$$

It follows that

$$\mathbf{l} = \lambda(\mathbf{OP} - \mathbf{OB}) = \lambda(\mu \mathbf{c} - \mathbf{b}),$$

and

$$\mathbf{m} = \lambda(\mathbf{OQ} - \mathbf{OC}) = \lambda(\mu \mathbf{b} - \mathbf{c}).$$

Hence, in (3.79),

$$\begin{aligned}(\mathbf{a} \wedge \mathbf{b}) \cdot \mathbf{l} &= \lambda(\mathbf{a} \wedge \mathbf{b}) \cdot (\mu \mathbf{c} - \mathbf{b}),\\ &= \lambda\mu(\mathbf{a} \wedge \mathbf{b}) \cdot \mathbf{c};\end{aligned}$$

similarly, in (3.80),

$$(\mathbf{c} \wedge \mathbf{a}) \cdot \mathbf{m} = \lambda\mu(\mathbf{c} \wedge \mathbf{a}) \cdot \mathbf{b}.$$

But

$$(\mathbf{a} \wedge \mathbf{b}) \cdot \mathbf{c} = (\mathbf{c} \wedge \mathbf{a}) \cdot \mathbf{b},$$

so that (3.79) and (3.80) give

$$\sin \gamma \sin B = \sin \beta \sin C.$$

Since we are assuming that $\sin \beta$ and $\sin \gamma$ are non-zero,

$$\frac{\sin B}{\sin \beta} = \frac{\sin C}{\sin \gamma}.$$

The proof can clearly be extended to establish in full the sine formula (3.77).

# 4

# Transformations of vectors

## 4.1 Vectors and matrices

To study rotations in 3-space, we need to use matrix algebra. The fundamentals of matrix algebra are laid down in a variety of textbooks [Reference 4.1]. We are only interested in matrices of dimensions 1, 2 and 3, arising in the study of the geometry of the plane and of 3-space. Many of the basic properties of this class of matrices are closely connected with properties of vectors that we have already established. So, in the first two sections of this chapter, we shall establish several basic properties of this class of matrices and their determinants by using vector methods and notation; this approach to matrix algebra can be generalised to $(n \times n)$ matrices through the algebra of **exterior forms** [Reference 4.2]. Our approach complements the normal approach to matrix algebra by demonstrating the close linkage of the theory of $(3 \times 3)$ matrices and determinants with the vector algebra developed in the first three chapters, and thus with familiar concepts of the geometry of 3-space.

We shall assume that the reader has a familiarity with the basic matrix operations of addition, subtraction and multiplication; nevertheless we state these properties, for completeness. Matrices are denoted by capital letters $A, B, C, \ldots$, and are rectangular arrays of numbers; for example, the matrix

$$A = \begin{pmatrix} a_{11} & a_{12} & \cdots & a_{1n} \\ a_{21} & g_{22} & \cdots & a_{2n} \\ \cdots & \cdots & \cdots & \cdots \\ a_{m1} & a_{m2} & \cdots & a_{mn} \end{pmatrix} \tag{4.1}$$

is an $(m \times n)$ matrix whose general element is $a_{pq}(p = 1, 2, \ldots, m;$ $q = 1, 2, \ldots, n)$; (4.1) can also be written

$$A = (a_{pq}). \tag{4.2}$$

The matrix $A$ has $m$ **rows** and $n$ **columns**. For example, the second row of $A$ is

$$a_{21} \quad a_{22} \ldots a_{2n},$$

every element having its first suffix equal to 2. If $B$ is another $(m \times n)$ matrix, then the matrix sum $A+B$ exists, and is the $(m \times n)$ matrix formed by adding corresponding elements of $A$ and $B$; thus

$$A+B=(a_{pq}+b_{pq}). \tag{4.3}$$

Likewise the matrix difference $A-B$ is the $(m \times n)$ matrix

$$A-B=(a_{pq}-b_{pq}). \tag{4.4}$$

The matrix product $AB$ of two matrices $A$ and $B$ exists only if the number of rows of $B$ is equal to the number of columns of $A$; if $A=(a_{pq})$ is an $(m \times n)$ matrix and $B=(b_{pq})$ is an $(n \times k)$ matrix, then the $(p, r)$ element of $AB$ is defined to be

$$\sum_{q=1}^{n} a_{pq}b_{qr}. \tag{4.5}$$

It is important to note the relationship between this expression and (2.36) for the scalar product of two vectors. Let us set out the matrix product in full, displaying rows of $A$ and columns of $B$:

$$\begin{pmatrix} a_{11} & a_{12} & \ldots & a_{1n} \\ \ldots & \ldots & \ldots & \ldots \\ a_{p1} & a_{p2} & \ldots & a_{pn} \\ \ldots & \ldots & \ldots & \ldots \\ a_{m1} & a_{m2} & \ldots & a_{mn} \end{pmatrix} \begin{pmatrix} b_{11} & \ldots & b_{1r} & \ldots & b_{1k} \\ b_{21} & \ldots & b_{2r} & \ldots & b_{2k} \\ \vdots & & \vdots & & \vdots \\ b_{n1} & \ldots & b_{nr} & \ldots & b_{nk} \end{pmatrix}. \tag{4.6}$$

Equation (2.36) can be written

$$\mathbf{v} \cdot \mathbf{w} = \sum_{q=1}^{3} v_q w_q.$$

If we looked upon the row $a_{p1}\ a_{p2}\ \ldots\ a_{pn}$ as the $n$ components of a 'vector', and likewise the column $b_{1r}\ b_{2r}\ \ldots\ b_{nr}$ as the $n$ components of another 'vector', then the sum (4.5) has exactly the form of a 'scalar product' between these two vectors, the sum over the suffix $q$ being from 1 to $n$, rather than from 1 to 3. (We shall in fact be dealing only with matrices for which $n=3$ or $n=2$.) So if we regard the rows of $A$ as $m$ 'vectors' in component form, and likewise the columns of $B$ as $k$ 'vectors', then the elements (4.5) of the $(m \times k)$ matrix $AB$ can be regarded as the 'scalar products' of the $m$ rows of $A$ with the $k$ columns of $B$.

We have used the phrases 'vectors' and 'scalar products' in inverted commas because the rows and columns of matrices are not in fact vectors, but simply have the same form as the set of components of a vector in $n$-space. Later in this chapter, we shall define certain matrices by identifying their rows or columns with certain vectors in component form; the 'scalar products' we have just discussed will then be true scalar products of vectors.

An $(n \times 1)$ matrix, consisting of a single column, is called a **column matrix**. It is customary to omit the second suffix from the elements of a column matrix, since it takes only one value. If the elements of a column matrix are $v_p$ $(p = 1, 2, \ldots, m)$, the matrix is written

$$V = \begin{pmatrix} v_1 \\ v_2 \\ \cdot \\ \cdot \\ \cdot \\ v_m \end{pmatrix}. \tag{4.7a}$$

It is often convenient to write the three components of a vector $\mathbf{v}$ in this form, with $m = 3$:

$$V = \begin{pmatrix} v_1 \\ v_2 \\ v_3 \end{pmatrix}; \tag{4.7b}$$

we then call $V$ a **component matrix**. A clear distinction should be made between the vector

$$\mathbf{v} = v_1\mathbf{i} + v_2\mathbf{j} + v_3\mathbf{k}$$

itself and the component matrix (4.7b), since the components of $\mathbf{v}$ depend upon the choice of the basis $(\mathbf{i}, \mathbf{j}, \mathbf{k})$. If a different triad $(\mathbf{i}', \mathbf{j}', \mathbf{k}')$ is chosen as basis, the *same* vector $\mathbf{v}$ is expressible as

$$\mathbf{v} = v_1'\mathbf{i}' + v_2'\mathbf{j}' + v_3'\mathbf{k}',$$

with component matrix

$$V' = \begin{pmatrix} v_1' \\ v_2' \\ v_3' \end{pmatrix},$$

which is not in general equal to $V$.

A triad $(\mathbf{i}, \mathbf{j}, \mathbf{k})$ is frequently called a **reference frame**, and is denoted by $F$. One of our main tasks in this chapter is to investigate the relationship between the component matrices $V$ and $V'$ of a vector $\mathbf{v}$ in different reference frames $F$ and $F'$.

The **transposed matrix**, or **transpose**, of a given matrix $A$ is formed by interchanging the rows and columns of $A$. So if $A=(a_{pq})$ is an $(m \times n)$ matrix, the transpose $A^{\mathrm{T}}$ is an $(n \times m)$ matrix $(a^{\mathrm{T}}_{pq})$; its $(p, q)$ element is given by

$$a^{\mathrm{T}}_{pq}=a_{qp}. \tag{4.8}$$

It is easy to show (Problems 4.1, Question 2) that $(AB)^{\mathrm{T}}=B^{\mathrm{T}}A^{\mathrm{T}}$ whenever $AB$ exists with the row and column suffixes interchanged. The transpose of the column matrix (4.7b) is the **row matrix**

$$V^{\mathrm{T}}=(v_1 \quad v_2 \quad v_3). \tag{4.9}$$

If $w_p$ $(p=1,2,3)$ are the components of a vector $\mathbf{w}$ relative to the triad $(\mathbf{i}, \mathbf{j}, \mathbf{k})$, then the product of the row matrix (4.9) into the column matrix

$$W=\begin{pmatrix} w_1 \\ w_2 \\ w_3 \end{pmatrix}$$

is the $(1 \times 1)$ matrix

$$V^{\mathrm{T}}W=(v_1w_1+v_2w_2+v_3w_3). \tag{4.10}$$

The single element of this matrix is equal to the scalar product $\mathbf{v}\cdot\mathbf{w}$ of the vectors $\mathbf{v}$ and $\mathbf{w}$ given by (2.36). In particular, the square of the modulus of $\mathbf{v}$, given by (2.37), is equal to the single element of the matrix product $V^{\mathrm{T}}V$. Since we are interested in the relationship of component matrices $V$ and $V^{\mathrm{T}}$ in different reference frames, it is important to note that the definition (in Chapter 2) of a scalar product $\mathbf{v}\cdot\mathbf{w}$ did not depend upon any reference frame; it follows that its value in terms of the components of $\mathbf{v}$ and $\mathbf{w}$ is independent of the reference frame used. Using components in two frames $F$ and $F'$,

$$\begin{aligned}\mathbf{v}\cdot\mathbf{w}&=v_1w_1+v_2w_2+v_3w_3\\&=v_1'w_1'+v_2'w_2'+v_3'w_3',\end{aligned} \tag{4.11}$$

or, expressed in terms of the matrix form (4.10),

$$V^{\mathrm{T}}W=V'^{\mathrm{T}}W'. \tag{4.12a}$$

The scalar product is also the single element of

$$W^{\mathrm{T}}V=W'^{\mathrm{T}}V' \tag{4.12b}$$

Equations (4.11) and (4.12) express the **invariance of scalar products** under a change of reference frame; this invariance is in fact the meaning of the word 'scalar'. Since the modulus of a vector and the angle between two vectors have been defined by (2.10) and (2.2) in terms of scalar products, it follows that all moduli and angles are independent of the choice of reference frame.

A matrix with the same number of rows and columns ($m=n$) is called a **square matrix**. We shall deal only with $(2\times 2)$ and $(3\times 3)$ square matrices. If a $(3\times 3)$ square matrix $A$ is multiplied into the column matrix (4.7b), we obtain the product

$$AV=\begin{pmatrix} \sum_{s=1}^{3} a_{1s}v_s \\ \sum_{s=1}^{3} a_{2s}v_s \\ \sum_{s=1}^{3} a_{3s}v_s \end{pmatrix}; \tag{4.13}$$

like $V$, this is a column matrix, with elements

$$\sum_{s=1}^{3} a_{rs}v_s \qquad (r=1,2,3). \tag{4.14}$$

We say that $V$ 'is transformed into' $AV$ by multiplication by $A$; we also say that $A$ 'operates on' $V$ to give $AV$. Since the components (4.14) of $AV$ are linear combinations of $v_1$, $v_2$, $v_3$, we refer to $A$ as a **linear operator**; multiplication of $V$ by $A$ is called a **linear transformation** of $V$. We shall be especially interested in linear transformations of component matrices of a vector $\mathbf{v}$; for the present, we simply note that these transformations are of two distinct types:

($a$) transformations due to the change of the vector $\mathbf{v}$, relative to a *fixed reference frame*; these are **active transformations**;

($b$) transformations of the components of a *given* vector $\mathbf{v}$, due to a change of the reference frame; these are **passive transformations**.

The matrix equivalent of the number 1 is a **unit matrix**. This is defined in terms of a two-suffix quantity $\delta_{pq}$, known as the **Kronecker delta**; the values of $\delta_{pq}$ for $p, q=1,2,3,\ldots$ are

$$\left.\begin{aligned} \delta_{pq}&=1 \qquad (p=q), \\ \delta_{pq}&=0 \qquad (p\neq q). \end{aligned}\right\} \tag{4.15}$$

The $(m\times m)$ unit matrix $I$ is the matrix whose elements are $\delta_{pq}$

$(p, q = 1, 2, \ldots, m)$. If $A$ is an $(m \times n)$ matrix, and $I$ is $(m \times m)$, then using (4.5), we find

$$IA = A; \tag{4.16a}$$

If $I$ is $(n \times n)$, then

$$AI = A. \tag{4.16b}$$

So when multiplication of a matrix $A$ by a unit matrix is possible, it leaves $A$ unchanged.

A **diagonal matrix** is a square matrix $(a_{pq})$ whose 'non-diagonal' elements (those with $p \neq q$) are all zero. The $(m \times m)$ diagonal matrix with elements $d_1, d_2, \ldots, d_m$ along the leading diagonal is then

$$D = (d_p \delta_{pq}). \tag{4.17}$$

The unit $(m \times m)$ matrix is therefore the diagonal matrix whose diagonal elements $d_p$ $(p = 1, \ldots, m)$ are all unity.

If $E = (e_q \delta_{qr})$ is a second $(m \times m)$ diagonal matrix, then the $(p, r)$ element of the product $DE$ is

$$\textstyle\sum_q d_p \delta_{pq} e_q \delta_{qr} = d_p e_p \delta_{pr},$$

using (4.15). So $DE$ is the diagonal matrix whose $p$th diagonal element is $d_p e_p$, the product of the corresponding elements of $D$ and $E$; since we are simply multiplying corresponding diagonal elements, the product $ED$ will give the same result. In general, two square matrices $D$, $E$ for which $DE = ED$ are **commuting matrices**. So *all pairs of diagonal matrices commute.* In order to commute, a pair of square matrices do not have to be diagonal; for example, $I$ and $A$ commute, where $A$ is *any* square matrix.

## 4.2 *Determinants; inverse of a square matrix*

The **determinant** of a square matrix $A$, denoted by $\Delta(A)$, is a very important single number defined from the elements $a_{pq}$ of $A$. We shall only be concerned with the determinants of $(2 \times 2)$ and $(3 \times 3)$ matrices. If

$$A = \begin{pmatrix} a_{11} & a_{12} \\ a_{21} & a_{22} \end{pmatrix},$$

then we define

$$\Delta(A) = a_{11}a_{22} - a_{12}a_{21}. \tag{4.18}$$

Let us now treat the columns

$$A_1 = \begin{pmatrix} a_{11} \\ a_{21} \end{pmatrix}, \qquad A_2 = \begin{pmatrix} a_{12} \\ a_{22} \end{pmatrix} \tag{4.19a}$$

of $A$ as though they were component matrices of two vectors, $\mathbf{a}_1$ and $\mathbf{a}_2$, lying in the $(\mathbf{i}, \mathbf{j})$ plane. If we introduce the third vector $\mathbf{k}$ of a right-handed triad, $\mathbf{a}_1$ and $\mathbf{a}_2$ will have zero components along $\mathbf{k}$; so, using (3.19), $\mathbf{a}_1 \wedge \mathbf{a}_2$ has a single component, in the $\mathbf{k}$-direction, equal to $\Delta(A)$ in (4.18); that is,

$$\mathbf{a}_1 \wedge \mathbf{a}_2 = \Delta(A)\mathbf{k}. \tag{4.20a}$$

Therefore $\Delta(A)$ is in magnitude equal to the area of the parallelogram with adjacent sides corresponding to vectors $\mathbf{a}_1$ and $\mathbf{a}_2$.

We have chosen to regard the matrix

$$A = (A_1 \quad A_2) \tag{4.21a}$$

as composed of the two column matrices (4.19a). We could equally write

$$A = \begin{pmatrix} \tilde{A}_1 \\ \tilde{A}_2 \end{pmatrix}, \tag{4.21b}$$

where $\tilde{A}_1$ and $\tilde{A}_2$ are the row matrices

$$\tilde{A}_1 = (a_{11} \quad a_{12}), \qquad \tilde{A}_2 = (a_{21} \quad a_{22}); \tag{4.19b}$$

$\tilde{A}_1$ and $\tilde{A}_2$ could again be regarded as the component matrices of vectors $\tilde{\mathbf{a}}_1$, $\tilde{\mathbf{a}}_2$ lying in the $(\mathbf{i}, \mathbf{j})$ plane. Exactly as before,

$$\tilde{\mathbf{a}}_1 \wedge \tilde{\mathbf{a}}_2 = \Delta(A)\mathbf{k}, \tag{4.20b}$$

so that $\Delta(A)$ is in magnitude also equal to the area of the parallelogram with sides corresponding to $\tilde{\mathbf{a}}_1$ and $\tilde{\mathbf{a}}_2$.

The determinant $\Delta(A)$ of the $(3 \times 3)$ matrix

$$A = \begin{pmatrix} a_{11} & a_{12} & a_{13} \\ a_{21} & a_{22} & a_{23} \\ a_{31} & a_{32} & a_{33} \end{pmatrix} \tag{4.22}$$

can also be defined in terms of column matrices and their associated vectors. We write

$$A_1 = \begin{pmatrix} a_{11} \\ a_{21} \\ a_{31} \end{pmatrix}, \qquad A_2 = \begin{pmatrix} a_{12} \\ a_{22} \\ a_{32} \end{pmatrix}, \qquad A_3 = \begin{pmatrix} a_{13} \\ a_{23} \\ a_{33} \end{pmatrix}; \tag{4.23a}$$

in terms of these three columns,

$$A = (A_1 \quad A_2 \quad A_3). \tag{4.24a}$$

Let us now regard $A_1$, $A_2$, $A_3$ as the component matrices of three vectors $\mathbf{a}_1$, $\mathbf{a}_2$, $\mathbf{a}_3$, relative to a reference frame $(\mathbf{i}, \mathbf{j}, \mathbf{k})$. Then we define the **determinant** of $A$ to be the scalar triple product

$$\Delta(A) = [\mathbf{a}_1, \mathbf{a}_2, \mathbf{a}_3]; \tag{4.25a}$$

substituting the components (4.23a) into (3.28),

$$\Delta(A) = a_{11}(a_{22}a_{33} - a_{23}a_{32}) + a_{21}(a_{32}a_{13} - a_{12}a_{33}) + a_{31}(a_{12}a_{23} - a_{22}a_{13}). \tag{4.26}$$

We can also choose to write $A$ in terms of its rows

$$\left.\begin{aligned} \tilde{A}_1 &= (a_{11} \quad a_{12} \quad a_{13}), \\ \tilde{A}_2 &= (a_{21} \quad a_{22} \quad a_{23}), \\ \tilde{A}_3 &= (a_{31} \quad a_{32} \quad a_{33}), \end{aligned}\right\} \tag{4.23b}$$

giving

$$A = \begin{pmatrix} \tilde{A}_1 \\ \tilde{A}_2 \\ \tilde{A}_3 \end{pmatrix}. \tag{4.24b}$$

Just as before, we can regard the row matrices (4.23b) as the component matrices of three vectors $\tilde{\mathbf{a}}_1$, $\tilde{\mathbf{a}}_2$, $\tilde{\mathbf{a}}_3$; then

$$\Delta(A) = [\tilde{\mathbf{a}}_1, \tilde{\mathbf{a}}_2, \tilde{\mathbf{a}}_3], \tag{4.25b}$$

since this scalar triple product, given by (4.23b) and (3.28), equals (4.26).

Since $\Delta(A)$ has the same definition in terms of the rows and columns of $A$, it follows that $\Delta(A)$ is unchanged if the rows and columns of $A$ are interchanged. Thus we have established the first of a number of properties of determinants.

(i) $$\Delta(A) = \Delta(A^{\mathrm{T}}), \tag{4.27}$$

where $A^{\mathrm{T}}$ is the transpose of $A$.

From the definition (4.18) it is clear that (4.27) is also true for $(2 \times 2)$ matrices. We shall only establish this and other properties of determinants for $(3 \times 3)$ matrices, since the proofs for $(2 \times 2)$ matrices are very simple. All of these properties have generalisations which are true for $(n \times n)$ matrices; general proofs can be found in standard texts on linear algebra [Reference 4.1]. The proofs which we give have generalisations in the algebra of exterior forms [Reference 4.2].

The next three properties of $\Delta(A)$ have already been established in §3.3 as properties of scalar triple products, and follow through (4.25a) and (4.25b).

(ii) Property (i) of scalar triple products (p. 65) tells us that the following three statements are equivalent to each other:

($a$) $\Delta(A)=0$,

($b$) the columns (4.23a) of $A$ are linearly dependent,

($c$) the rows (4.23b) of $A$ are linearly dependent.

Statement ($c$), for example, means that there exist constants $\alpha_1$, $\alpha_2$, $\alpha_3$ (not all zero) such that $\alpha_1\tilde{\mathbf{a}}_1+\alpha_2\tilde{\mathbf{a}}_2+\alpha_3\tilde{\mathbf{a}}_3=\mathbf{0}$, or

$$\alpha_1\tilde{A}_1+\alpha_2\tilde{A}_2+\alpha_3\tilde{A}_3=(0 \quad 0 \quad 0).$$

As a special case of this property, $\Delta(A)$ is zero if two rows, or alternately two columns of $A$, are identical.

(iii) The cyclic property (ii) of scalar triple products (p. 65) tells us that $\Delta(A)$ is unchanged if we cyclically permute either the rows or the columns of $A$.

(iv) The antisymmetric property (3.4) of vector products ensures that scalar triple products change sign if two of the vectors are interchanged. Hence $\Delta(A)$ changes sign if two rows of $A$ are interchanged, or if two columns of $A$ are interchanged.

(v) $\Delta(A)$ is linear in the components of the first column $A_1$, by (4.26); so if we form a second matrix by replacing the column $A_1$ by another column $A_0$, the determinants of the matrices are added by adding $A_0$ to $A_1$; that is,

$$\begin{aligned}\Delta(A_0 \quad A_2 \quad A_3)&+\Delta(A_1 \quad A_2 \quad A_3)\\ &=\Delta([A_0+A_1] \quad A_2 \quad A_3).\end{aligned} \tag{4.28}$$

The same rule applies to any row or any column of a square matrix.

(vi) Combining properties (ii) and (v), it follows that $\Delta(A)$ is unchanged if we add any multiples of columns $A_2$ and $A_3$ to the column $A_1$; for, if $\lambda$ and $\mu$ are any constants,

$$\begin{aligned}&\Delta([A_1+\lambda A_2+\mu A_3] \quad A_2 \quad A_3)\\ &\quad=\Delta(A_1 \quad A_2 \quad A_3)+\lambda\Delta(A_2 \quad A_2 \quad A_3)+\mu\Delta(A_3 \quad A_2 \quad A_3).\end{aligned}$$

Since the last two determinants are zero,

$$\Delta([A_1+\lambda A_2+\mu A_3] \quad A_2 \quad A_3)=\Delta(A). \tag{4.29}$$

(vii) If $A$ and $B$ are two $(3\times 3)$ matrices, the determinant of the matrix product $AB$ is the product of the determinants of $A$ and $B$:

$$\Delta(AB)=\Delta(A)\Delta(B). \tag{4.30}$$

This is a very simple and very important result, but there is no very obvious proof. For $(m \times m)$ matrices $A$ and $B$, one of the simplest proofs of (4.30) involves the use of $(2m \times 2m)$ matrices [Reference 4.3]. Since we are particularly concerned with relations between vectors and matrices, we give a proof of (4.30) for $(3 \times 3)$ matrices which depends upon vector properties that we have already established in Chapter 3. The proof for $(2 \times 2)$ matrices is left as a problem for the reader (Problem 4.1, Question 1).

First, let us write the matrix $A$ in the form (4.24b), in terms of rows, and $B$ in the form (4.24a), so that

$$B = (B_1 \quad B_2 \quad B_3)$$

in terms of its columns $B_1$, $B_2$, $B_3$. Recalling the discussion following (4.6), we see that the $(p, r)$ element of $AB$ is the 'scalar product' of the row $\tilde{A}_p$ with the column $B_r$. But $\tilde{\mathbf{a}}_p$ and $\mathbf{b}_r$ have been introduced as vectors whose component matrices are $\tilde{A}_p$ and $B_r$ respectively; so the $(p, r)$ element of $AB$ is equal to the scalar product $\tilde{\mathbf{a}}_p \cdot \mathbf{b}_r$. Thus

$$AB = \begin{pmatrix} \tilde{\mathbf{a}}_1 \cdot \mathbf{b}_1 & \tilde{\mathbf{a}}_1 \cdot \mathbf{b}_2 & \tilde{\mathbf{a}}_1 \cdot \mathbf{b}_3 \\ \tilde{\mathbf{a}}_2 \cdot \mathbf{b}_1 & \tilde{\mathbf{a}}_2 \cdot \mathbf{b}_2 & \tilde{\mathbf{a}}_2 \cdot \mathbf{b}_3 \\ \tilde{\mathbf{a}}_3 \cdot \mathbf{b}_1 & \tilde{\mathbf{a}}_3 \cdot \mathbf{b}_2 & \tilde{\mathbf{a}}_3 \cdot \mathbf{b}_3 \end{pmatrix}. \tag{4.31}$$

The determinant of this matrix is given by (4.26):

$$\begin{aligned} \Delta(AB) = (\tilde{\mathbf{a}}_1 \cdot \mathbf{b}_1)&[(\tilde{\mathbf{a}}_2 \cdot \mathbf{b}_2)(\tilde{\mathbf{a}}_3 \cdot \mathbf{b}_3) - (\tilde{\mathbf{a}}_3 \cdot \mathbf{b}_2)(\tilde{\mathbf{a}}_2 \cdot \mathbf{b}_3)] \\ + (\tilde{\mathbf{a}}_2 \cdot \mathbf{b}_1)&[(\tilde{\mathbf{a}}_3 \cdot \mathbf{b}_2)(\tilde{\mathbf{a}}_1 \cdot \mathbf{b}_3) - (\tilde{\mathbf{a}}_1 \cdot \mathbf{b}_2)(\tilde{\mathbf{a}}_3 \cdot \mathbf{b}_3)] \\ + (\tilde{\mathbf{a}}_3 \cdot \mathbf{b}_1)&[(\tilde{\mathbf{a}}_1 \cdot \mathbf{b}_2)(\tilde{\mathbf{a}}_2 \cdot \mathbf{b}_3) - (\tilde{\mathbf{a}}_2 \cdot \mathbf{b}_2)(\tilde{\mathbf{a}}_1 \cdot \mathbf{b}_3)]. \end{aligned}$$

Using the vector identity (3.33) for each of the square brackets, this equation becomes

$$\begin{aligned} \Delta(AB) = (\tilde{\mathbf{a}}_1 \cdot \mathbf{b}_1)&[(\tilde{\mathbf{a}}_2 \wedge \tilde{\mathbf{a}}_3) \cdot (\mathbf{b}_2 \wedge \mathbf{b}_3)] \\ + (\tilde{\mathbf{a}}_2 \cdot \mathbf{b}_1)&[(\tilde{\mathbf{a}}_3 \wedge \tilde{\mathbf{a}}_1) \cdot (\mathbf{b}_2 \wedge \mathbf{b}_3)] \\ + (\tilde{\mathbf{a}}_3 \cdot \mathbf{b}_1)&[(\tilde{\mathbf{a}}_1 \wedge \tilde{\mathbf{a}}_2) \cdot (\mathbf{b}_2 \wedge \mathbf{b}_3)]. \end{aligned} \tag{4.32}$$

Now, from the definitions (4.25a) and (4.25b),

$$\begin{aligned} \Delta(A)\Delta(B) &= [\tilde{\mathbf{a}}_1, \tilde{\mathbf{a}}_2, \tilde{\mathbf{a}}_3][\mathbf{b}_1, \mathbf{b}_2, \mathbf{b}_3] \\ &= [\tilde{\mathbf{a}}_1, \tilde{\mathbf{a}}_2, \tilde{\mathbf{a}}_3](\mathbf{b} \cdot \mathbf{b}_1), \end{aligned} \tag{4.33}$$

where

$$\mathbf{b} = \mathbf{b}_2 \wedge \mathbf{b}_3.$$

Equation (3.36b) allows us to write

$$[\tilde{\mathbf{a}}_1, \tilde{\mathbf{a}}_2, \tilde{\mathbf{a}}_3]\mathbf{b} = [\mathbf{b}, \tilde{\mathbf{a}}_2, \tilde{\mathbf{a}}_3]\tilde{\mathbf{a}}_1 + [\mathbf{b}, \tilde{\mathbf{a}}_3, \tilde{\mathbf{a}}_1]\tilde{\mathbf{a}}_2 + [\mathbf{b}, \tilde{\mathbf{a}}_1, \tilde{\mathbf{a}}_2]\tilde{\mathbf{a}}_3.$$

Substituting on the right of (4.33) and putting in the value of $\mathbf{b}$, we obtain

$$\begin{aligned}\Delta(A)\Delta(B) = &[\mathbf{b}_2 \wedge \mathbf{b}_3, \tilde{\mathbf{a}}_2, \tilde{\mathbf{a}}_3](\tilde{\mathbf{a}}_1 \cdot \mathbf{b}_1)\\ &+[\mathbf{b}_2 \wedge \mathbf{b}_3, \tilde{\mathbf{a}}_3, \tilde{\mathbf{a}}_1](\tilde{\mathbf{a}}_2 \cdot \mathbf{b}_1)\\ &+[\mathbf{b}_2 \wedge \mathbf{b}_3, \tilde{\mathbf{a}}_1, \tilde{\mathbf{a}}_2](\tilde{\mathbf{a}}_3 \cdot \mathbf{b}_1).\end{aligned}$$

Comparing this with (4.32), we have established the property (4.30).

The non-triviality of the result (4.30) is emphasised if we consider the numbers of terms on each side of the equation, when $A$ and $B$ are $(3 \times 3)$ matrices. In terms of the elements $\{a_{pq}\}$ of $A$ and $\{b_{qr}\}$ of $B$, each element of the matrix (4.31) consists of the sum of three terms, since it is a scalar product. $\Delta(AB)$ consists of 6 products, each of 3 such terms; so the direct expansion of $\Delta(AB)$ contains $6 \times 27 = 162$ terms. Each of $\Delta(A)$ and $\Delta(B)$, however, are the sum of 6 terms, so that $\Delta(A)\Delta(B)$ contains just 36 terms. Therefore 126 of the 162 terms of $\Delta(AB)$ must cancel, and the remaining 36 then factorise exactly to give $\Delta(A)\Delta(B)$. It is therefore not surprising that the proof of (4.30) involves some complication.

One of the fundamental concepts of matrix algebra is that of the **inverse** $A^{-1}$ of an $(m \times m)$ square matrix $A$, satisfying the matrix equations

$$A^{-1}A = AA^{-1} = I, \tag{4.34}$$

where $I$ is the unit $(m \times m)$ matrix. Once again, the general definition of $A^{-1}$ can be found in many textbooks [Reference 4.1]; we shall only consider $(3 \times 3)$ matrices, using vector methods and notation to establish our results. We write the matrix $A$ as

$$A = (A_1 \quad A_2 \quad A_3)$$

in terms of its columns (4.23a), and again regard $A_1$, $A_2$, $A_3$ as component matrices of three vectors $\mathbf{a}_1$, $\mathbf{a}_2$, $\mathbf{a}_3$. Now define the three vector products

$$\tilde{\mathbf{d}}_1 = \mathbf{a}_2 \wedge \mathbf{a}_3, \qquad \tilde{\mathbf{d}}_2 = \mathbf{a}_3 \wedge \mathbf{a}_1, \qquad \tilde{\mathbf{d}}_3 = \mathbf{a}_1 \wedge \mathbf{a}_2, \tag{4.35a}$$

and let $\tilde{D}_1$, $\tilde{D}_2$, $\tilde{D}_3$ be the component *row* matrices of these vectors. Then the **adjoint** $D$ of the matrix $A$ is defined as the matrix whose *rows* are $\tilde{D}_1, \tilde{D}_2, \tilde{D}_3$:

$$D = \begin{pmatrix} \tilde{D}_1 \\ \tilde{D}_2 \\ \tilde{D}_3 \end{pmatrix}. \tag{4.36a}$$

The matrix product $DA$ is similar to (4.31); each element is therefore a scalar triple product, the scalar product of a vector (4.35a) with one of $\mathbf{a}_1$, $\mathbf{a}_2$, $\mathbf{a}_3$. Thus

$$DA = \begin{pmatrix} [\mathbf{a}_2, \mathbf{a}_3, \mathbf{a}_1] & [\mathbf{a}_2, \mathbf{a}_3, \mathbf{a}_2] & [\mathbf{a}_2, \mathbf{a}_3, \mathbf{a}_3] \\ [\mathbf{a}_3, \mathbf{a}_1, \mathbf{a}_1] & [\mathbf{a}_3, \mathbf{a}_1, \mathbf{a}_2] & [\mathbf{a}_3, \mathbf{a}_1, \mathbf{a}_3] \\ [\mathbf{a}_1, \mathbf{a}_2, \mathbf{a}_1] & [\mathbf{a}_1, \mathbf{a}_2, \mathbf{a}_2] & [\mathbf{a}_1, \mathbf{a}_2, \mathbf{a}_3] \end{pmatrix}.$$

The off-diagonal elements of this matrix are all zero by (3.23), while the cyclic property of scalar triple products ensures that the diagonal elements are all equal to $[\mathbf{a}_1, \mathbf{a}_2, \mathbf{a}_3] = \Delta(A)$, using the definition (4.25a). Hence $D$ satisfies the matrix equation

$$DA = \Delta(A)I. \tag{4.37a}$$

Provided that $A$ is a **non-singular** matrix, meaning that

$$\Delta(A) \neq 0, \tag{4.38}$$

we can define the inverse of $A$ to be

$$A^{-1} = [\Delta(A)]^{-1}D; \tag{4.39}$$

then (4.37a) ensures that $A^{-1}$ satisfies

$$A^{-1}A = I,$$

one of the Equations (4.34).

We now establish a formula for the general element of $D$ in terms of $(2 \times 2)$ determinants. From (4.36a), it is clear that $d_{pq}$ is the $q$th element in the row $\tilde{D}_p$. For example, $d_{32}$ is the second element of $\tilde{D}_3$; but, by (4.35a), $\tilde{D}_3$ corresponds to the vector product $\mathbf{a}_1 \wedge \mathbf{a}_2$, whose components are derived from $A_1$ and $A_2$ in (4.23a) by using the formula (3.19). The second component of $\tilde{D}_3$ is thus

$$\begin{aligned} d_{32} &= a_{31}a_{12} - a_{11}a_{32} \\ &= -(a_{11}a_{32} - a_{31}a_{12}). \end{aligned}$$

Now consider the matrix formed by deleting the second row and the third column of the matrix (4.22); this process defines the $(2 \times 2)$ matrix

$$A[\text{row } 2, \text{col } 3] = \begin{pmatrix} a_{11} & a_{12} \\ a_{31} & a_{32} \end{pmatrix},$$

whose determinant, by (4.18), is equal to $(a_{11}a_{32}-a_{31}a_{12})$. Therefore the (3, 2) element of $D$ is

$$d_{32}=-\Delta(A[\text{row } 2, \text{col. } 3]).$$

This argument is quite general, and provides the formula

$$d_{pq}=(-1)^{p+q}\Delta(A[\text{row } q, \text{col. } p]) \tag{4.40}$$

for the general element of the adjoint matrix; these elements are known as the **minors** of $A$. The determinant in (4.40) is that of the matrix formed from $A$ by deleting the $q$th row and the $p$th column. Note that the deleted *row* in $A$ corresponds to the *column* suffix in $d_{pq}$, and vice versa; this switch of suffixes occurs because the *rows* of $D$ are multiplied into the *columns* of $A$ to form the product in (4.37a).

Equation (4.40) shows that the definition of the adjoint $D$ is symmetrical between the rows and columns of $A$. We could therefore have used the rows (4.23b) of $A$, rather than the columns (4.23a), to define $D$. If the rows (4.23b) are regarded as component matrices of vectors $\tilde{\mathbf{a}}_1$, $\tilde{\mathbf{a}}_2$, $\tilde{\mathbf{a}}_3$, we define, similarly to (4.35a),

$$\mathbf{d}_1=\tilde{\mathbf{a}}_2\wedge\tilde{\mathbf{a}}_3, \qquad \mathbf{d}_2=\tilde{\mathbf{a}}_3\wedge\tilde{\mathbf{a}}_1, \qquad \mathbf{d}_3=\tilde{\mathbf{a}}_1\wedge\tilde{\mathbf{a}}_2. \tag{4.35b}$$

Then if $D_r$ $(r=1, 2, 3)$ are the component matrices of $\mathbf{d}_r$, the symmetry of (4.40) ensures that they are the *columns* of $D$; thus

$$D=(D_1 \quad D_2 \quad D_3) \tag{4.36b}$$

is a definition equivalent to (4.36a).

Using the forms (4.24b) and (4.36b) for $A$ and $D$, the argument used to establish (4.37a) tells us that

$$\begin{aligned} AD&=[\tilde{\mathbf{a}}_1, \tilde{\mathbf{a}}_2, \tilde{\mathbf{a}}_3]I \\ &=\Delta(A)I. \end{aligned} \tag{4.37b}$$

So provided $A$ is non-singular, the inverse (4.39) of $A$ satisfies

$$AA^{-1}=I;$$

this establishes in full the equations (4.34), for the inverse $A^{-1}$ defined by (4.39) and (4.40).

It is important to show that the inverse $A^{-1}$ of a non-singular matrix $A$ is unique. Since we are assuming that $\Delta(A)\neq 0$, we have to show that the adjoint $D$ is uniquely determined by (4.37a). Suppose that $D^{(1)}$ and $D^{(2)}$ both satisfy (4.37a); then subtracting the two equations gives

$$EA = \varnothing,$$

where $E = D^{(1)} - D^{(2)}$ and $\varnothing$ is the zero matrix. Now write $E$ in terms of rows $\tilde{E}_1$, $\tilde{E}_2$, $\tilde{E}_3$, and $A$ in the form (4.24a); then

$$\begin{pmatrix} \tilde{E}_1 \\ \tilde{E}_2 \\ \tilde{E}_3 \end{pmatrix} (A_1 \quad A_2 \quad A_3) = \varnothing.$$

Consider the row $\tilde{E}_1$ for instance. If we regard it as the component matrix of a vector $\tilde{\mathbf{e}}_1$, then this vector satisfies

$$\tilde{\mathbf{e}}_1 \cdot \mathbf{a}_1 = \tilde{\mathbf{e}}_1 \cdot \mathbf{a}_2 = \tilde{\mathbf{e}}_1 \cdot \mathbf{a}_3 = 0.$$

But since $\Delta(A) \neq 0$, the vectors $\mathbf{a}_1$, $\mathbf{a}_2$, $\mathbf{a}_3$ are linearly independent; the only vector orthogonal to three linearly independent vectors is the zero vector, so $\tilde{\mathbf{e}}_1 = \mathbf{0}$. Likewise $\tilde{\mathbf{e}}_2 = \tilde{\mathbf{e}}_3 = \mathbf{0}$. Therefore the matrix $E$ is identically zero, and $D^{(1)} \equiv D^{(2)}$. So when $\Delta(A) \neq 0$, the adjoint $D$, and hence the inverse $A^{-1}$, is unique.

The existence and uniqueness of $A^{-1}$ enables us to solve the set of linear equations

$$AX = V, \tag{4.41}$$

where $V$ is a given column matrix, $X$ an unknown column matrix, and $A$ a non-singular square matrix. Multiplying (4.41) on the left by $A^{-1}$ gives $IX = A^{-1}V$, or simply

$$X = A^{-1}V. \tag{4.42}$$

The uniqueness of $A^{-1}$ ensures that (4.42) is the unique solution of (4.41). If $V = \varnothing$, with all elements zero, then (4.42) tells us that, when $\Delta(A) \neq 0$, $X = \varnothing$ is the only solution of

$$AX = \varnothing. \tag{4.43a}$$

When $\Delta(A) = 0$, (4.43a) has non-zero solutions. If we write $A$ in the form (4.24b) in terms of its rows $\tilde{A}_1$, $\tilde{A}_2$, $\tilde{A}_3$, then (4.43a) is equivalent to the vector equations

$$\tilde{\mathbf{a}}_1 \cdot \mathbf{x} = \tilde{\mathbf{a}}_2 \cdot \mathbf{x} = \tilde{\mathbf{a}}_3 \cdot \mathbf{x} = 0, \tag{4.43b}$$

where $\mathbf{x}$ is the vector with component matrix $X$. So $\mathbf{x}$ must be orthogonal to $\tilde{\mathbf{a}}_1$, $\tilde{\mathbf{a}}_2$, $\tilde{\mathbf{a}}_3$. The range of solutions of (4.43a) can then be classified in terms of the **rank** $r(A)$ of the matrix $A$, defined to be the *number of linearly independent rows* of $A$:

(i) If $\Delta(A) \neq 0$, the vectors $\tilde{\mathbf{a}}_r$ $(r = 1, 2, 3)$ are linearly independent and span the whole of 3-space; then $r(A) = 3$ and $\mathbf{x} = \mathbf{0}$ is the only vector satisfying (4.43b).

(ii) If $\Delta(A) = 0$, then $\tilde{\mathbf{a}}_r$ $(r = 1, 2, 3)$ are linear dependent. But if $r(A) = 2$, the set $\{\tilde{\mathbf{a}}_r\}$ will span a plane. If $\mathbf{n}$ is a unit normal to this plane (determined apart from sign), and $\lambda$ is any number, then any vector $\mathbf{x} = \lambda\mathbf{n}$ satisfies (4.43b), so that its component matrix satisfies (4.43a). The solutions therefore correspond to the ray of vectors $\{\lambda\,\mathbf{n}\}$.

This set of solutions can be obtained more precisely. Since $\Delta(A) = 0$, (4.37b) becomes

$$AD = \varnothing.$$

Therefore (4.43a) is satisfied if $X = D_1$, $D_2$ or $D_3$ (the columns of $D$). These columns correspond to the vector products (4.35b), which are all orthogonal to the plane of $\tilde{\mathbf{a}}_1$, $\tilde{\mathbf{a}}_2$, $\tilde{\mathbf{a}}_3$, and which are therefore all of the form $\lambda\mathbf{n}$. But since $r(A) = 2$, at least two of the vectors $\{\tilde{\mathbf{a}}_r\}$ are linearly independent, so that one or more of the vector products (4.35b) must be non-zero; thus (4.35b) defines the ray $\{\lambda\,\mathbf{n}\}$ explicitly.

(iii) If $\Delta(A) = 0$ and $r(A) = 1$, then the rows $\tilde{A}_1$, $\tilde{A}_2$, $\tilde{A}_3$ are multiples of a single row matrix; $\tilde{\mathbf{a}}_1$, $\tilde{\mathbf{a}}_2$, $\tilde{\mathbf{a}}_3$ are then parallel (or zero), and (4.43b) only restrict $\mathbf{x}$ to lie in the plane orthogonal to their common direction. There is no unique normal $\{\mathbf{n}\}$ to the three vectors $\{\tilde{\mathbf{a}}_r\}$, and every one of the vector products (4.35a) is zero. So the adjoint matrix is $D = \varnothing$.

(iv) If $r(A) = 0$, $A$ is the zero matrix. Then *any* column matrix $X$ satisfies (4.43a).

Generally, the set $\{X\}$ of solutions of (4.43a) correspond to the vectors $\{\mathbf{x}\}$ orthogonal to the space spanned by $\tilde{\mathbf{a}}_r$ $(r = 1, 2, 3)$. The dimension of the space $\{\mathbf{x}\}$ is therefore $3 - r(A)$.

## Problems 4.1

1 One $(2 \times 2)$ matrix $A$ is written in the form (4.21b), and a second $(2 \times 2)$ matrix $B$ in the form (4.21a). Use (4.20) to write $\Delta(A)\Delta(B)$ in terms of $\tilde{\mathbf{a}}_1$, $\tilde{\mathbf{a}}_2$, $\mathbf{b}_1$, $\mathbf{b}_2$.

Adapt (4.31) to $(2 \times 2)$ matrices, and express $\Delta(AB)$ in terms of $\tilde{\mathbf{a}}_1$, $\tilde{\mathbf{a}}_2$, $\mathbf{b}_1$, $\mathbf{b}_2$. Hence show that

$$\Delta(A)\Delta(B) = \Delta(AB).$$

2 If the product $AB$ of two matrices $A$ and $B$ exists, show that the transposed matrix is given by $(AB)^{\mathrm{T}} = B^{\mathrm{T}}A^{\mathrm{T}}$.

3 If $A$ is a $(3\times3)$ diagonal matrix with diagonal elements $a_r$ ($r = 1, 2, 3$), and $B = (b_{rs})$ is any $(3\times3)$ matrix, write down a formula for the general element of the matrix product $AB$. Use (4.26) to evaluate $\Delta(AB)$, and hence to show that $\Delta(AB) = \Delta(A)\Delta(B)$.

4 If

$$A = \begin{pmatrix} 1 & 2 & 3 \\ 1 & 1 & 2 \\ 2 & -1 & 2 \end{pmatrix},$$

evaluate $\Delta(A)$. Use (4.35a), (4.36a) and (4.39) to calculate the inverse $A^{-1}$.

Check that

$$A^{-1}A = AA^{-1} = I.$$

5 If $A$ is the matrix defined in Question 4, check that the inverse $A^{-1}$ given by (4.35b) and (4.36b) is the same as that given by (4.35a) and (4.36a).

6 Use the inverse $A^{-1}$ calculated in Questions 4 or 5 to solve the set of linear equations

$$x + 2y + 3z = 2,$$

$$x + y + 2z = 0,$$

$$2x - y + 2z = -3.$$

7 If $A$ and $B$ are two non-singular square matrices, show that the inverse of $AB$ is given by $(AB)^{-1} = B^{-1}A^{-1}$.

8 If $A$ and $B$ are $(3\times3)$ matrices, show that $r(AB) \leqslant r(A)$.

## 4.3 *Rotations and reflections in a plane*

Before studying rotations and reflections in 3-space, we investigate the simpler situation in a plane. This serves two purposes: first, we can discuss several general concepts in a particularly simple case; second, we are able to understand fully the property of addition of angles in a plane; this is one more familiar geometrical concept which needs to be derived through our axiomatic approach to vectors and Euclidean geometry.

Let $(\mathbf{i}, \mathbf{j})$ be an orthonormal basis in a plane, so that $\mathbf{i}$ and $\mathbf{j}$ are orthogonal unit vectors. As in §2.3, let $(\mathbf{i}, \mathbf{j})$ define a positive sense of rotation in the plane. Now let $(\mathbf{i}', \mathbf{j}')$ be a second orthonormal basis, chosen to satisfy (2.48):

$$\mathbf{i}' \cdot \mathbf{i} = \mathbf{j}' \cdot \mathbf{j} = \cos\alpha, \tag{2.48a}$$

$$\mathbf{i}' \cdot \mathbf{j} = -\mathbf{j}' \cdot \mathbf{i} = \sin\alpha; \tag{2.48b}$$

then the rotation from $\mathbf{i}'$ to $\mathbf{j}'$ is also $\frac{1}{2}\pi$ in the positive sense, as in Fig. 2.6. The relations between the two sets of basis vectors are

$$\mathbf{i}' = \cos\alpha\,\mathbf{i} + \sin\alpha\,\mathbf{j}, \tag{4.44a}$$

$$\mathbf{j}' = -\sin\alpha\,\mathbf{i} + \cos\alpha\,\mathbf{j}. \tag{4.44b}$$

The component matrices of $\mathbf{i}'$ and $\mathbf{j}'$ relative to the basis $(\mathbf{i}, \mathbf{j})$ are thus

$$\begin{pmatrix} \cos\alpha \\ \sin\alpha \end{pmatrix} \quad \text{and} \quad \begin{pmatrix} -\sin\alpha \\ \cos\alpha \end{pmatrix}. \tag{4.45}$$

By combining these two column matrices, we form the **transformation matrix**

$$R = \begin{pmatrix} \cos\alpha & -\sin\alpha \\ \sin\alpha & \cos\alpha \end{pmatrix}. \tag{4.46}$$

So, if the $(\mathbf{i}, \mathbf{j})$ frame is denoted by $F$, this definition ensures that *the component matrices of* $\mathbf{i}'$ *and* $\mathbf{j}'$ *in the frame $F$ are the columns of the transformation matrix $R$.*

The transpose of the matrix (4.46) is

$$R^{\mathrm{T}} = \begin{pmatrix} \cos\alpha & \sin\alpha \\ -\sin\alpha & \cos\alpha \end{pmatrix},$$

and it is easy to check that

$$R^{\mathrm{T}}R = RR^{\mathrm{T}} = I, \tag{4.47}$$

where $I$ is the $(2\times 2)$ unit matrix. Taking determinants on each side of (4.47) and using (4.30) and (4.27), we find that $[\Delta(R)]^2 = 1$ or

$$\Delta(R) = \pm 1. \tag{4.48}$$

A unique inverse $R^{-1}$ of $R$ therefore exists, and comparing (4.47) with (4.34), we see that $R^{-1}$ is given by

$$R^{-1} = R^{\mathrm{T}}. \tag{4.49}$$

Any real square matrix which satisfies (4.47) or (4.49) is called an **orthogonal matrix** and represents an **orthogonal transformation**. Clearly matrices of the form (4.46) satisfy $\Delta(R) = +1$; later on, we shall discuss orthogonal matrices $L$ for which $\Delta(L) = -1$.

Relations (4.44) can be written in matrix form, using row matrices with vectors as components:

$$(\mathbf{i}' \quad \mathbf{j}') = (\mathbf{i} \quad \mathbf{j})\begin{pmatrix} \cos\alpha & -\sin\alpha \\ \sin\alpha & \cos\alpha \end{pmatrix}$$

$$\equiv (\mathbf{i} \quad \mathbf{j})R. \tag{4.50}$$

Multiplying this equation on the right by $R^{-1}$, and using (4.34) and (4.49), we find

$$(\mathbf{i} \quad \mathbf{j}) = (\mathbf{i}' \quad \mathbf{j}')R^{-1} = (\mathbf{i}' \quad \mathbf{j}')\begin{pmatrix} \cos\alpha & \sin\alpha \\ -\sin\alpha & \cos\alpha \end{pmatrix}. \tag{4.51}$$

Comparing (4.51) with (4.50), we see that if the $(\mathbf{i}', \mathbf{j}')$ frame $F'$ is obtained from $F$ by a rotation $\alpha$, as in Fig. 2.6, then the frame $F$ is obtained from $F'$ by a rotation $-\alpha$. This is a 'geometrically obvious' fact.

Consider a given vector $\mathbf{v}$ in the plane of $\mathbf{i}$ and $\mathbf{j}$. It can be expressed in terms of the two bases in the form

$$\mathbf{v} = v_1\mathbf{i} + v_2\mathbf{j}, \tag{4.52a}$$

$$\mathbf{v} = v_1'\mathbf{i}' + v_2'\mathbf{j}'. \tag{4.52b}$$

We now find the relationship between the sets of components $(v_1, v_2)$ and $(v_1', v_2')$. Substituting from (4.44) into (4.52b),

$$\mathbf{v} = (v_1' \cos\alpha - v_2' \sin\alpha)\mathbf{i}$$
$$+ (v_1' \sin\alpha + v_2' \cos\alpha)\mathbf{j}.$$

Comparing with (4.52a) we find, using matrix notation,

$$\begin{pmatrix} v_1 \\ v_2 \end{pmatrix} = \begin{pmatrix} v_1' \cos\alpha - v_2' \sin\alpha \\ v_1' \sin\alpha + v_2' \cos\alpha \end{pmatrix}$$

$$= \begin{pmatrix} \cos\alpha & -\sin\alpha \\ \sin\alpha & \cos\alpha \end{pmatrix}\begin{pmatrix} v_1' \\ v_2' \end{pmatrix}. \tag{4.53a}$$

Using the notation (4.7) for component matrices, and the definition (4.46), this equation can be written

$$V = RV'. \tag{4.53b}$$

These equations can be solved for $V'$ in terms of $V$, as in (4.43); then

$$V' = R^{-1}V = R^{\mathrm{T}}V, \tag{4.54}$$

using (4.48); (4.53) and (4.54) give the relationship between the component matrices of the *same vector* $\mathbf{v}$ in the two frames of reference $F$ and $F'$. So (4.53) and (4.54) express the effect of a **rotation of the reference frame**, also termed a **passive transformation** of the component matrix.

We can also ask: what is the effect of **rotation of a vector v**, keeping the reference frame $F$ fixed? We must make clear what is meant by rotating a vector through an angle $\alpha$; this is shown in Fig. 4.1. As

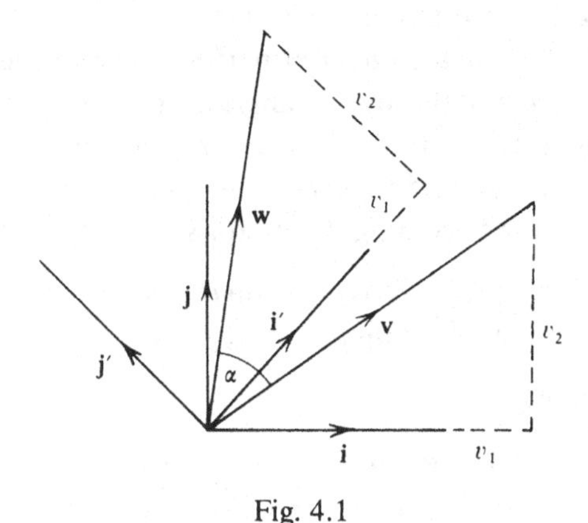


Fig. 4.1

above, let $(\mathbf{i}', \mathbf{j}')$ be the basis derived from $(\mathbf{i}, \mathbf{j})$ by a rotation $\alpha$, and defined by (4.44). If

$$\mathbf{v} = v_1\mathbf{i} + v_2\mathbf{j},$$

then the vector $\mathbf{w}$ given by rotating $\mathbf{v}$ through an angle $\alpha$ is

$$\mathbf{w} = v_1\mathbf{i}' + v_2\mathbf{j}'; \tag{4.55}$$

thus it has, by definition, the same components as $\mathbf{v}$, but relative to the rotated frame $F'$. To find the components of $\mathbf{w}$ in the frame $F$, substitute from (4.44) into (4.55); then

$$\begin{aligned}\mathbf{w} = (v_1 \cos\alpha - v_2 \sin\alpha)\mathbf{i}\\ +(v_1 \sin\alpha + v_2 \cos\alpha)\mathbf{j}.\end{aligned}$$

So if $\mathbf{w} = w_1\mathbf{i} + w_2\mathbf{j}$, the component matrix of $\mathbf{w}$ relative to $F$ is given by

$$\begin{pmatrix} w_1 \\ w_2 \end{pmatrix} = \begin{pmatrix} \cos\alpha & -\sin\alpha \\ \sin\alpha & \cos\alpha \end{pmatrix}\begin{pmatrix} v_1 \\ v_2 \end{pmatrix} \tag{4.56a}$$

or

$$W = RV. \tag{4.56b}$$

So *rotating the vector* **v** through an angle $\alpha$ in the plane is represented by the operation of $R$ on the component matrix $V$. This result should be carefully compared with (4.54), where the effect on $V$ of *rotating the frame* is the operation of the inverse $R^{-1}$. A little thought tells us that this is sensible: for if we rotate both the vector **v** and the frame of reference, there is no change in the components. In other words, the transformation of $V$ due to rotating the reference frame must be the inverse of that due to rotating the vector **v**.

Let us now consider the effect of performing two successive rotations on a vector **v**, first through an angle $\alpha$ and then through an angle $\beta$. The rotation through $\alpha$ changes **v** into the vector **w** with components given by (4.56). Using the same formula, the second rotation results in $w$ becoming the vector $\mathbf{t}=t_1\mathbf{i}+t_2\mathbf{j}$, with

$$\begin{pmatrix} t_1 \\ t_2 \end{pmatrix}=\begin{pmatrix} \cos\beta & -\sin\beta \\ \sin\beta & \cos\beta \end{pmatrix}\begin{pmatrix} w_1 \\ w_2 \end{pmatrix}.$$

Substituting from (4.56a),

$$\begin{aligned}\begin{pmatrix} t_1 \\ t_2 \end{pmatrix}&=\begin{pmatrix} \cos\beta & -\sin\beta \\ \sin\beta & \cos\beta \end{pmatrix}\begin{pmatrix} \cos\alpha & -\sin\alpha \\ \sin\alpha & \cos\alpha \end{pmatrix}\begin{pmatrix} v_1 \\ v_2 \end{pmatrix}\\ &=\begin{pmatrix} \cos\alpha\cos\beta-\sin\alpha\sin\beta & -\sin\alpha\cos\beta-\cos\alpha\sin\beta \\ \sin\alpha\cos\beta+\cos\alpha\sin\beta & \cos\alpha\cos\beta-\sin\alpha\sin\beta \end{pmatrix}\begin{pmatrix} v_1 \\ v_2 \end{pmatrix}\\ &=\begin{pmatrix} \cos(\alpha+\beta) & -\sin(\alpha+\beta) \\ \sin(\alpha+\beta) & \cos(\alpha+\beta) \end{pmatrix}\begin{pmatrix} v_1 \\ v_2 \end{pmatrix},\end{aligned} \tag{4.57}$$

using (2.50). So the effect of successive rotations through angles $\alpha$ and $\beta$ is the same as a rotation through the angle defined in Chapter 2 as the 'sum of angles' $\alpha+\beta$. This result establishes, in terms of vectors, the property of **addition of angles** in a plane: successive transformations of the form (4.56) are equivalent to a third transformation (4.57) of the same type. We note that these transformations are all expressed in terms of cosines and sines, and so through (2.5) and (2.43) are directly expressible in terms of scalar products. The properties of rotations in a plane have therefore been based on the axioms of vectors.

If we now denote the transformation matrix (4.46) by $R(\alpha)$, then the matrix multiplication performed in deriving (4.57) can be expressed as

$$R(\beta)R(\alpha)=R(\alpha+\beta). \tag{4.58a}$$

Since $\cos(\alpha+\beta)$ and $\sin(\alpha+\beta)$ are symmetrical between $\alpha$ and $\beta$, (4.58a) can also be written

$$R(\alpha)R(\beta)=R(\alpha+\beta), \tag{4.58b}$$

implying that the effect of performing two rotations in a plane is independent of the order in which they are performed.

The set of matrices $\{R(\alpha)\}$ form a **group**. The properties of a group, consisting of certain **group elements**, are:

(i) There is a rule for combining any two elements of the set to form another element, and this rule must satisfy the associative law. (4.57) tells us that rotation matrices combine by matrix multiplication. Matrices satisfy the associative law of multiplication $A(BC)=(AB)C$.

(ii) There is a unit element of the group which combines with any element of the group to give exactly the same element; the unit matrix, representing a zero rotation, is the unit element of the **rotation group**.

(iii) Every element of the group has an inverse which combines with the element to give the unit element. (4.46) and (4.47) tell us that the inverse of $R(\alpha)$ is

$$R^{-1}(\alpha)=R^{\mathrm{T}}(\alpha)=R(-\alpha),$$

expressing the obvious fact that the inverse of a rotation through an angle $\alpha$ is a rotation through $-\alpha$. The fact that $R(\alpha)R(\beta)=R(\beta)R(\alpha)$, for all $\alpha$ and $\beta$, tells us that the group is an **Abelian group**.

In establishing these group properties and the property of additivity of angles, we discussed the rotation of vectors. We could equally well have discussed the rotations of frames of reference; then $R(\alpha)$ is replaced in the above discussion by $R^{-1}(\alpha)=R(-\alpha)$, representing the inverse rotation. The whole discussion proceeds as above, but with the sign of every angle changed. So additivity of angles of rotation, and the group properties of rotations in a plane, apply equally to rotations of vectors and to rotations of frames of reference.

We took the rotation $\frac{1}{2}\pi$ from $\mathbf{i}$ to $\mathbf{j}$ as defining the positive sense of rotation in the plane, and also chose the rotation $\frac{1}{2}\pi$ from $\mathbf{i}'$ to $\mathbf{j}'$ to be in the positive sense. Now suppose that $\mathbf{j}'$ has been chosen with the opposite sign, so that the rotation $\frac{1}{2}\pi$ from $\mathbf{i}'$ to $\mathbf{j}'$ is in the negative sense, as shown in Fig. 4.2. Then (4.44a) remains unchanged, but

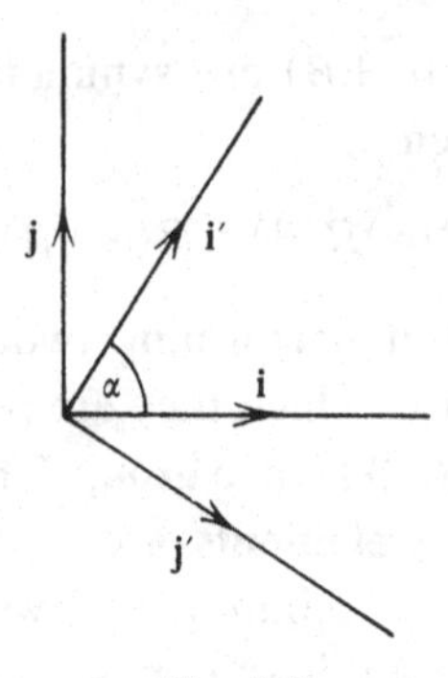


Fig. 4.2

(4.44b) is replaced by

$$\mathbf{j}' = \sin\alpha\ \mathbf{i} - \cos\alpha\ \mathbf{j}.$$

The transformation matrix (4.46) is replaced by

$$L(\alpha) = \begin{pmatrix} \cos\alpha & \sin\alpha \\ \sin\alpha & -\cos\alpha \end{pmatrix}. \tag{4.59}$$

There are several differences between transformations of type $L$ and type $R$. If we put $\alpha = 0$ in $R$, we obtain the unit matrix $I$, which represents no change in the component matrix $V$ in (4.54) and (4.56); $R(0)$ corresponds to zero rotation. But putting $\alpha = 0$ in $L(\alpha)$ gives

$$L(0) = \begin{pmatrix} 1 & 0 \\ 0 & -1 \end{pmatrix}; \tag{4.60}$$

clearly this matrix is its own inverse,

$$L^{-1}(0) = L(0); \tag{4.61}$$

replacing $R$ by $L(0)$ in (4.50) produces a change of basis from (**i**, **j**) to (**i′**, **j′**), where

$$\mathbf{i}' = \mathbf{i}, \quad \mathbf{j} = -\mathbf{j}'. \tag{4.62}$$

The components of a given vector **v** using the two bases are then related by

$$v_1' = v_1, \qquad v_2' = -v_2.$$

The change of basis (4.62) is a **reflection of the frame** in the **i**-axis.

We may also consider $L(0)$ as an operator acting on a vector **v**, with given basis (**i**, **j**); then the components of the transformed vector **w**, given by (4.56a), are

$$w_1 = v_1, \qquad w_2 = -v_2.$$

In this case, $L(0)$ represents the **reflection of the vector** in the **i**-axis.

Next, we note that the transformation matrix (4.59) can be written as the matrix product

$$L(\alpha)=R(\alpha)L(0)$$

of the rotation matrix $R(\alpha)$ and the **i**-axis reflection matrix (4.60). The transformation is therefore a combination of an **i**-axis reflection, followed by a rotation. It is geometrically obvious from Fig. 4.2 that $L(\alpha)$ cannot represent a pure rotation in the plane; algebraically, transformations $L(\alpha)$ can be distinguished from $R(\alpha)$ by evaluating their determinants. While $\Delta[R(\alpha)]=+1$, it is clear from (4.59) that $\Delta[L(\alpha)]=-1$, for all $\alpha$. So the sign of the determinant distinguishes between matrices representing a rotation and those representing a rotation plus a reflection.

The set of matrices $\{L(\alpha)\}$ does not form a group, since the unit matrix is not one of the set. However, the set of matrices $\{R(\alpha), L(\alpha)\}$ does form a group; in Problems 4.2, Question 4, the reader is asked to verify the group properties of this set.

**Example 4.1**

Show that the matrices $I$, $-I$, $L(0)$ and $-L(0)$ form an Abelian group. Explain the geometric significance of $-I$ and $-L(0)$.

The matrices are

$$I=\begin{pmatrix}1 & 0\\ 0 & 1\end{pmatrix},\qquad -I=\begin{pmatrix}-1 & 0\\ 0 & -1\end{pmatrix},$$

$$L(0)=\begin{pmatrix}1 & 0\\ 0 & -1\end{pmatrix},\qquad -L(0)=\begin{pmatrix}-1 & 0\\ 0 & 1\end{pmatrix}.$$

$I$ and $-I$ are their own inverses, and $[L(0)]^{-1}=L(0)$. Since $L(0)L(0)=I$, all products of two of the four matrices are members of the set. Since the matrices are all diagonal, the products are independent of the order of the matrices. So the group is Abelian.

The matrix $-I$ is given by putting $\alpha=\pi$ in (4.46), and so corresponds to a rotation in the plane through angle $\pi$; this is also called **reflection in the origin** or **central inversion.** Just as $L(0)$ represents reflection in the **i**-axis, $-L(0)$ represents reflection in the **j**-axis.

**Example 4.2**

Show that the rotation matrices (4.46) with $\alpha_p = 2\pi p/n$ ($p = 0, 1, \ldots, n-1$) form an Abelian group with $n$ members.

By (4.58) the products of $R(\alpha_p)$ and $R(\alpha_q)$ obey

$$R(\alpha_p)R(\alpha_q) = R(\alpha_q)R(\alpha_p) = R(\alpha_p + \alpha_q).$$

If $p+q \leqslant n-1$, $R(\alpha_p + \alpha_q)$ is a member of the set. If $n \leqslant p+q \leqslant 2n-2$, so that $2\pi \leqslant \alpha_p + \alpha_q < 4\pi$, $R(\alpha_p + \alpha_q) = R(\alpha_p + \alpha_q - 2\pi)$, again a member of the set. So the product of two members of the set is a member.

The unit matrix is the member of the set with $\alpha_p = 0$. For $p = 1, 2, \ldots, n-1$, the inverse of $R(\alpha_p)$ is $R(-\alpha_p) = R(2\pi - \alpha_p)$, corresponding to angle $2\pi(n-p)/n$; this is also a member of the set. So the set possesses all the properties of the group. [The group is a **cyclic group of order** $n$.]

## *Problems 4.2*

1. Write down the transformation matrices $R_1$ and $R_2$ corresponding to rotations through angles $\frac{1}{6}\pi$ and $\frac{1}{3}\pi$. Show by matrix multiplication that the matrix products $R_1R_2$ and $R_2R_1$ correspond to a rotation through $\frac{1}{2}\pi$.
2. Show that all $(2\times 2)$ matrices $A$ satisfying $AA^{\mathrm{T}} = I$ are either of the form (4.46) or (4.59).
3. The vector $\mathbf{v}$ is given as $\mathbf{v} = \mathbf{i} + 2\mathbf{j}$ relative to the frame $F$. A second frame $F'$ has orthonormal basis $(\mathbf{i}', \mathbf{j}')$, with $\mathbf{i}' = \frac{3}{5}\mathbf{i} - \frac{4}{5}\mathbf{j}$ and $\mathbf{i}' \cdot \mathbf{i} = \mathbf{j}' \cdot \mathbf{j}$. Find the components of $\mathbf{v}$ relative to the frame $F'$. If $\mathbf{w}$ is the vector formed from $\mathbf{v}$ by the rotation which takes $\mathbf{i}$ into $\mathbf{i}'$, find the components of $\mathbf{w}$ relative to both frames of reference.
4. Show that the set of matrices $\{R(\alpha), L(\alpha);\ 0 \leqslant \alpha < 2\pi\}$, defined by (4.46) and (4.59), form a group.
5. If $L$ is a $(2\times 2)$ orthogonal matrix, interpret the condition $\Delta(L) = \pm 1$ in terms of

   (*a*) the vector product of the columns of $L$,

   (*b*) areas in the plane.

   Explain carefully the effect of the choice of sign of $\Delta(L)$.

6 Show that the set of matrices $\{R(\alpha_p), L(\alpha_p)\}$ defined by (4.46) and (4.59), with $\alpha_p = 2\pi p/n$ $(p = 0, 1, \ldots, n-1)$ form a group. Draw diagrams to show how each member of the group transforms an arbitrary vector $\mathbf{v}$, as in (4.56), for the groups defined by $n = 3$ and $n = 4$.

## 4.4 *Rotations and reflections in 3-space*

Many of the properties of $(2\times 2)$ orthogonal transformations have a simple generalisation when we consider a change of orthonormal basis in 3-space. Let us consider two bases $(\mathbf{i}, \mathbf{j}, \mathbf{k})$ and $(\mathbf{i}', \mathbf{j}', \mathbf{k}')$, each consisting of three mutually orthogonal unit vectors. Any unit vector $\mathbf{u}$ can be expanded in the form (2.63),

$$\mathbf{u} = l_1\mathbf{i} + l_2\mathbf{j} + l_3\mathbf{k},$$

where, by (2.62), the direction cosines satisfy

$$\sum_{q=1}^{3} l_q^2 = 1.$$

Expressing each of the unit vectors $\mathbf{i}'$, $\mathbf{j}'$, $\mathbf{k}'$ in this form, we have

$$\left.\begin{aligned} \mathbf{i}' &= l_{11}\mathbf{i} + l_{21}\mathbf{j} + l_{31}\mathbf{k}, \\ \mathbf{j}' &= l_{12}\mathbf{i} + l_{22}\mathbf{j} + l_{32}\mathbf{k}, \\ \mathbf{k}' &= l_{13}\mathbf{i} + l_{23}\mathbf{j} + l_{33}\mathbf{k}, \end{aligned}\right\} \tag{4.63a}$$

with

$$\sum_{q=1}^{3} l_{q1}^2 = \sum_{q=1}^{3} l_{q2}^2 = \sum_{q=1}^{3} l_{q3}^2 = 1. \tag{4.64}$$

Since $\mathbf{i}'$, $\mathbf{j}'$, $\mathbf{k}'$ are mutually orthogonal, their scalar products, given in terms of components by (2.36), are zero; thus

$$\sum_{q=1}^{3} l_{q1}l_{q2} = \sum_{q=1}^{3} l_{q1}l_{q3} = \sum_{q=1}^{3} l_{q2}l_{q3} = 0. \tag{4.65}$$

Using the Kronecker delta defined by (4.15), the six equations (4.64) and (4.65) can be written compactly as

$$\sum_{q} l_{qp}l_{qr} = \delta_{pr} \qquad (p, r = 1, 2, 3), \tag{4.66}$$

where the summation is over the values $q = 1, 2, 3$; (4.66) constitute the **orthonormality conditions** for $(\mathbf{i}', \mathbf{j}', \mathbf{k}')$. We now define the matrix

$$L = \begin{pmatrix} l_{11} & l_{12} & l_{13} \\ l_{21} & l_{22} & l_{23} \\ l_{31} & l_{32} & l_{33} \end{pmatrix}, \tag{4.67}$$

which is called an **orthogonal matrix**; we see from (4.63a) that *the columns of L are the component matrices of* $\mathbf{i}', \mathbf{j}', \mathbf{k}'$ *relative to the basis* $(\mathbf{i}, \mathbf{j}, \mathbf{k})$. If $L^{\mathrm{T}}$ is the transpose of $L$, with elements given by (4.8), then the expression $\Sigma_q \, l_{qp} l_{qr}$ in (4.66) is just the $(p, r)$ component of the matrix product $L^{\mathrm{T}}L$, for $p, r = 1, 2, 3$. Since $(\delta_{pr})$ is the unit matrix, (4.66) can be written in matrix form as

$$L^{\mathrm{T}}L = I; \tag{4.68a}$$

this is of exactly the same form as one of the equations (4.47) for $(2 \times 2)$ orthogonal matrices. As for $(2 \times 2)$ matrices, taking determinants in (4.68a) gives

$$\Delta(L) = \pm 1, \tag{4.69}$$

and multiplying on the right by the unique inverse $L^{-1}$ gives

$$L^{-1} = L^{\mathrm{T}}; \tag{4.70}$$

multiplication on the left by $L$ then gives

$$LL^{\mathrm{T}} = I, \tag{4.68b}$$

analogous to the second equation (4.47). In general, an orthogonal matrix $L$ is a square matrix satisfying (4.68) or (4.70), analogous to (4.47) or (4.49) for $(2 \times 2)$ matrices. Orthogonal $(3 \times 3)$ matrices are again of two types, dependent upon the sign of $\Delta(L)$ in (4.69). We shall see later that, as in 2-space, $L$ represents a pure rotation when $\Delta(L) = +1$; but when $\Delta(L) = -1$, $L$ represents a rotation plus a reflection. Writing the basis vectors as row matrices, the relations (4.63a) connecting two frames $F$ and $F'$ can be written, as in (4.50),

$$(\mathbf{i}' \quad \mathbf{j}' \quad \mathbf{k}') = (\mathbf{i} \quad \mathbf{j} \quad \mathbf{k})L; \tag{4.63b}$$

multiplying on the right by $L^{-1}$ and using (4.68) and (4.70),

$$(\mathbf{i} \quad \mathbf{j} \quad \mathbf{k}) = (\mathbf{i}' \quad \mathbf{j}' \quad \mathbf{k}')L^{\mathrm{T}}. \tag{4.71}$$

So, just as $L$ in (4.63) transforms the basis vectors of $F$ to those of $F'$, $L^{\mathrm{T}}$ in (4.71) effects the inverse transformation from $F'$ to $F$. Therefore the columns of $L^{\mathrm{T}}$ are the component matrices of $\mathbf{i}$, $\mathbf{j}$, $\mathbf{k}$ relative to the basis $(\mathbf{i}', \mathbf{j}', \mathbf{k}')$. Remembering that transposition interchanges the rows and columns of $L$, we see that *the rows of $L$ are the component matrices of* $\mathbf{i}$, $\mathbf{j}$, $\mathbf{k}$ *relative to the basis* $(\mathbf{i}', \mathbf{j}', \mathbf{k}')$. So both the rows and columns of an orthogonal matrix $L$ have a direct interpretation in terms of the two bases.

**Example 4.3**

The basis vectors of a frame $F'$, relative to frame $F$, are the unit vectors associated with $\mathbf{a}_1$, $\mathbf{a}_2$ and $\mathbf{a}_1 \wedge \mathbf{a}_2$, where

$$\mathbf{a}_1 = \mathbf{i} + 2\mathbf{j} + 2\mathbf{k},$$
$$\mathbf{a}_2 = \mathbf{j} - \mathbf{k}.$$

Find the component matrices of each set of basis vectors in terms of the other set.

First note that $\mathbf{a}_1$ and $\mathbf{a}_2$ are orthogonal, as required; the unit vectors associated with them are

$$\mathbf{i}' = \tfrac{1}{3}(\mathbf{i} + 2\mathbf{j} + 2\mathbf{k}),$$
$$\mathbf{j}' = \frac{1}{\sqrt{2}}(\mathbf{j} - \mathbf{k}).$$

The third member of the triad is thus

$$\mathbf{k}' = \mathbf{i}' \wedge \mathbf{j}' = \frac{1}{3\sqrt{2}}(-4\mathbf{i} + \mathbf{j} + \mathbf{k}).$$

The components of $\mathbf{i}'$, $\mathbf{j}'$, $\mathbf{k}'$ in frame $F$ form the columns of the transformation matrix (4.67), which is thus

$$L = \begin{pmatrix} \frac{1}{3} & 0 & -\frac{4}{3\sqrt{2}} \\ \frac{2}{3} & \frac{1}{\sqrt{2}} & \frac{1}{3\sqrt{2}} \\ \frac{2}{3} & -\frac{1}{\sqrt{2}} & \frac{1}{3\sqrt{2}} \end{pmatrix}.$$

The component matrices of **i**, **j**, **k** relative to frame $F'$ are the rows of $L$, namely

$$\begin{pmatrix} \frac{1}{3} \\ 0 \\ -\frac{4}{3\sqrt{2}} \end{pmatrix}, \quad \begin{pmatrix} \frac{2}{3} \\ \frac{1}{\sqrt{2}} \\ \frac{1}{3\sqrt{2}} \end{pmatrix}, \quad \begin{pmatrix} \frac{2}{3} \\ -\frac{1}{\sqrt{2}} \\ \frac{1}{3\sqrt{2}} \end{pmatrix}.$$

Now consider a given vector **v**, expanded in terms of the two triads as

$$\mathbf{v} = v_1\mathbf{i} + v_2\mathbf{j} + v_3\mathbf{k}, \tag{4.72a}$$

$$\mathbf{v} = v_1'\mathbf{i}' + v_2'\mathbf{j}' + v_3'\mathbf{k}'. \tag{4.72b}$$

Substituting from (4.63a) into (4.72b) gives

$$\mathbf{v} = \Big(\sum_q l_{1q}v_q'\Big)\mathbf{i} + \Big(\sum_q l_{2q}v_q'\Big)\mathbf{j} + \Big(\sum_q l_{3q}v_q'\Big)\mathbf{k},$$

$\Sigma_q$ denoting the sum over the values 1, 2, 3 of $q$. Comparing with (4.72a), we find the relations

$$v_p = \sum_q l_{pq}v_q' \qquad (p = 1, 2, 3) \tag{4.73a}$$

between the components in the two frames. Introducing, as in (4.7b), the component matrices $V$ and $V'$ of the vector **v**, (4.73a) is then of the form (4.53b):

$$V = LV'. \tag{4.73b}$$

Just as in 2-space, it follows that the inverse relation is

$$V' = L^{-1}V = L^{\mathrm{T}}V. \tag{4.74}$$

Once again, (4.73b) and (4.74) express the effect of a **change of the reference frame** on the component matrix of a given vector **v**.

When we first defined a triad (**i**, **j**, **k**) in Chapter 2, we arbitrarily chose to call this particular triad 'right-handed', while the triad (**i**, **j**, −**k**) was 'left-handed'. With this choice for the one particular triad (**i**, **j**, **k**), we are now able to define 'handedness' for any other triad (**i**′, **j**′, **k**′). Given **i**′ and **j**′, the third unit vector **k**′ must equal

$\pm\mathbf{i}' \wedge \mathbf{j}'$. If $\mathbf{k}' = \mathbf{i}' \wedge \mathbf{j}'$, so that

$$[\mathbf{i}', \mathbf{j}', \mathbf{k}'] = +1,$$

the triad is defined to be **right-handed**. If $L$ is the orthogonal matrix defined by (4.63), with columns equal to the component matrices of $\mathbf{i}'$, $\mathbf{j}'$, $\mathbf{k}'$, then the definition (4.25a) gives

$$\Delta(L) = +1$$

as the condition for a right-handed triad. Equally, if $\mathbf{k}' = -\mathbf{i}' \wedge \mathbf{j}'$, then

$$\Delta(L) = [\mathbf{i}', \mathbf{j}', \mathbf{k}'] = -1,$$

and the triad is defined to be **left-handed**. Since all orthogonal matrices satisfy (4.69), all triads are either right-handed or left-handed.

If we picture a reference frame $F$ as a set of unit vectors along three Cartesian axes, we can consider rotating $F$ continuously until it coincides with a second frame $F'$. This cannot happen if $F$ and $F'$ have opposite 'handedness'; there can be no *continuous* change from a right-handed to a left-handed frame, since at some stage the determinant associated with the frame must change abruptly from $+1$ to $-1$. We shall now see, by explicit construction of the transformation matrix, that all transformations with determinant $+1$ can be carried out continuously as a series of rotations. When we are dealing specifically with matrices representing rotations, we shall denote them by $R$, rather than $L$, as in §4.3; then $\Delta(R) = +1$ always.

In Fig. 4.3, we show how the transformation from the frame $(\mathbf{i}, \mathbf{j}, \mathbf{k})$ to any other right-handed frame $(\mathbf{i}', \mathbf{j}', \mathbf{k}')$ can be effected by three successive rotations through angles $\alpha$, $\beta$, $\gamma$. The unit vectors $\mathbf{i}, \mathbf{j}, \mathbf{k}$ are represented by three mutually orthogonal displacements $\mathbf{OX}$, $\mathbf{OY}$, $\mathbf{OZ}$. The first rotation is through the angle $\alpha$ about the axis $OZ$, so

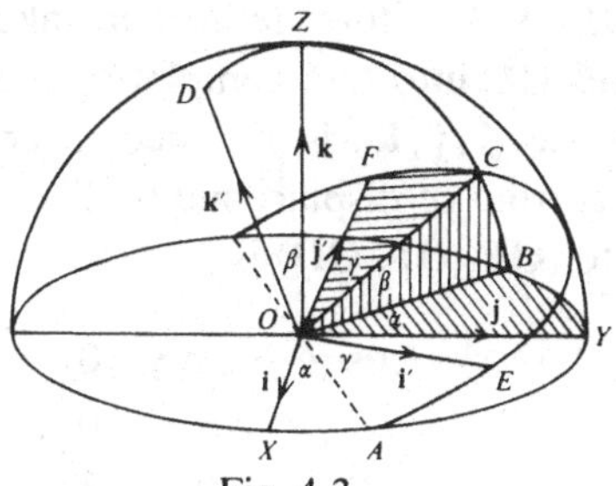


Fig. 4.3

that the rotated triad is represented by (**OA**, **OB**, **OZ**). The transformation matrix representing the rotation is

$$R_1 = \begin{pmatrix} \cos\alpha & -\sin\alpha & 0 \\ \sin\alpha & \cos\alpha & 0 \\ 0 & 0 & 1 \end{pmatrix}; \qquad (4.75a)$$

this transforms (**i**, **j**) as in (4.50) and (4.46), leaving **k** unchanged. The next rotation is through an angle $\beta$ about the axis $OA$, the first axis of the displaced triad; after this second rotation, the position of the triad is represented by (**OA**, **OC**, **OD**). The transformation matrix corresponding to the second rotation is

$$R_2 = \begin{pmatrix} 1 & 0 & 0 \\ 0 & \cos\beta & -\sin\beta \\ 0 & \sin\beta & \cos\beta \end{pmatrix}; \qquad (4.75b)$$

this matrix corresponds to the transformation of type (4.63) from the frame (**OA**, **OB**, **OZ**) to the frame (**OA**, **OC**, **OD**), and operates on unit vectors parallel to **OA**, **OC**, **OD**. So, representing unit vectors by displacements, two applications of (4.63b), with $L = R_2$ and $L = R_1$ respectively, give

$$\begin{aligned}(\mathbf{OA} \quad \mathbf{OC} \quad \mathbf{OD}) &= (\mathbf{OA} \quad \mathbf{OB} \quad \mathbf{OZ})R_2 \\ &= (\mathbf{OX} \quad \mathbf{OY} \quad \mathbf{OZ})R_1R_2. \qquad (4.76)\end{aligned}$$

We now identify the unit vector $\mathbf{k}'$ as that represented by **OD**; if $\mathbf{k}'$ is given it is clear from Fig. 4.3 that its direction defines the angles $\alpha$ and $\beta$, and that any unit vector $\mathbf{k}'$ corresponds to angles in the ranges $0 \leqslant \alpha \leqslant 2\pi$, $0 \leqslant \beta \leqslant \pi$. This identification of $\mathbf{k}'$ tells us that $\mathbf{i}'$ and $\mathbf{j}'$ must lie in the plane defined by $O$, $A$ and $C$; so $\mathbf{i}'$ is represented by some unit displacement **OE** in this plane. The third rotation $\gamma$ ($0 \leqslant \gamma < 2\pi$) about axis $OD$ is therefore defined to take **OA** into **OE**; this same rotation takes **OC** into **OF**, completing the right-handed triad (**OE**, **OF**, **OD**); since ($\mathbf{i}'$, $\mathbf{j}'$, $\mathbf{k}'$) is assumed to be right-handed, it is represented by this triad of displacements. The matrix transforming (**OA**, **OC**, **OD**) into (**OE**, **OF**, **OD**) is

$$R_3 = \begin{pmatrix} \cos\gamma & -\sin\gamma & 0 \\ \sin\gamma & \cos\gamma & 0 \\ 0 & 0 & 1 \end{pmatrix}; \qquad (4.75c)$$

so applying (4.63b) again, with $L = R_3$, and using (4.76),

$$\begin{aligned}(\mathbf{OE} \quad \mathbf{OF} \quad \mathbf{OD}) &= (\mathbf{OA} \quad \mathbf{OC} \quad \mathbf{OD})R_3 \\ &= (\mathbf{OX} \quad \mathbf{OY} \quad \mathbf{OZ})R_1R_2R_3. \end{aligned} \tag{4.77}$$

Thus the complete transformation between the bases is of the form (4.63b), with transformation matrix

$$R = R_1R_2R_3. \tag{4.78}$$

The three rotation matrices $R_r$ $(r = 1, 2, 3)$ are given by (4.75a, b, c); forming the matrix product (4.78) gives

$$R = \begin{pmatrix} \cos\alpha\cos\gamma - \sin\alpha\cos\beta\sin\gamma & -\cos\alpha\sin\gamma - \sin\alpha\cos\beta\cos\gamma & \sin\alpha\sin\beta \\ \sin\alpha\cos\gamma + \cos\alpha\cos\beta\sin\gamma & -\sin\alpha\sin\gamma + \cos\alpha\cos\beta\cos\gamma & -\cos\alpha\sin\beta \\ \sin\beta\sin\gamma & \sin\beta\cos\gamma & \cos\beta \end{pmatrix}. \tag{4.79}$$

This transformation matrix corresponds to the three rotations $\alpha, \beta, \gamma$ performed in succession; the angles $\alpha, \beta, \gamma$ defining the transformation are called **Euler's angles**. It is not hard to check directly that $R$ *is* an orthogonal matrix with $\Delta R = +1$ (see Problems 4.3, Question 2), and that $R^{-1} = R^{\mathrm{T}}$. Since the matrix elements of $R$ are all continuous functions [Reference 4.4] of $\alpha$, $\beta$, $\gamma$ and $R = I$ when $\alpha = \beta = \gamma = 0$, the transformation from frame $F$ to $F'$ can be carried out continuously. Later in this section we shall show that the transformation $R$ represents a rotation about an axis that can be determined.

Let us now consider transformations with determinant $-1$. One of the simplest is the transformation $\mathbf{i}' = -\mathbf{i}$, $\mathbf{j}' = -\mathbf{j}$, $\mathbf{k}' = -\mathbf{k}$, with transformation matrix

$$P \equiv \begin{pmatrix} -1 & 0 & 0 \\ 0 & -1 & 0 \\ 0 & 0 & -1 \end{pmatrix}. \tag{4.80}$$

This transformation corresponds to **reflection in the origin** (**central inversion**) of the frame $(\mathbf{i}, \mathbf{j}, \mathbf{k})$.

Now consider any transformation matrix $L$ with $\Delta(L) = -1$; it can be expressed as the matrix product

$$L = PR, \tag{4.81}$$

where $P$ is the reflection (4.80) and $R = -L$. Since the columns of $L$

represent a triad, so do those of $R$, so that $R$ is an orthogonal matrix; and since $\Delta(L)=\Delta(P)=-1$, (4.81) and (4.30) ensure that $\Delta(R)=+1$. So $R$ is a matrix representing a pure rotation; hence $L=PR$ is a combination of a rotation and a reflection in the origin.

We have so far discussed transformations of reference frames. As in 2-space, we can also consider transformations of vectors relative to a given frame $F$, taken as right-handed. First we consider rotations of a vector $\mathbf{v}$. If $(\mathbf{i}', \mathbf{j}', \mathbf{k}')$ is another right-handed frame, and $\mathbf{v}$ is given by (4.72a), the rotated vector $\mathbf{w}$ is defined to be

$$\mathbf{w}=v_1\mathbf{i}'+v_2\mathbf{j}'+v_3\mathbf{k}'.$$

Substituting from (4.63a), with $L$ replaced by a rotation matrix $R=(r_{pq})$,

$$\mathbf{w}=\left(\sum_q r_{1q}v_q\right)\mathbf{i}+\left(\sum_q r_{2q}v_q\right)\mathbf{j}+\left(\sum_q r_{3q}v_q\right)\mathbf{k},$$

so that the components of $\mathbf{w}$ in frame $F$ are

$$w_p=\sum_q r_{pq}v_1. \tag{4.82a}$$

In matrix form, this equation is the same as (4.56b) in 2-space:

$$W=RV. \tag{4.82b}$$

Once again, we can compare the effect of the transformations (4.74) and (4.82) on the component matrix $V$. The effect of **rotating the vector** is the operation (4.82) of $R$ on $V$, while the effect of **rotating the reference frame** is the operation of $R^{-1}$ on $V$.

The set of orthogonal matrices $R$ representing rotations, with $\Delta(R)=+1$, form a group since

(i) The matrix $R_1R_2$ formed by matrix multiplication of two rotation matrices is also a rotation matrix, satisfying (4.68),

(ii) the unit matrix $I$ is a member of the set, and

(iii) each matrix $R$ has an inverse $R^{-1}=R^T$ which is also a rotation matrix.

Since any member $R$ of this group is of form (4.79) and can be derived from $I$ by continuous variation of the values of $\alpha$, $\beta$, $\gamma$, the group is called a **continuous group**; it is known as the **3-dimensional rotation group**. It is not difficult (see Problems 4.3, Question 4) to find rotation matrices $R_1$, $R_2$ with $R_1R_2\neq R_2R_1$; the group is therefore *not* Abelian.

The set of all $(3\times 3)$ orthogonal matrices also forms a non-Abelian group, since it satisfies the conditions (i), (ii) and (iii) above. It is not, however, a continuous group, since matrices $L$ with $\Delta(L)=-1$ cannot be obtained from $I$ by continuous variation. The set of matrices $L$ with $\Delta(L)=-1$, as in 2-space, is not a group, since it does not contain the unit matrix $I$.

We are intuitively familiar with the concept of an **axis of rotation**; it is a straight line which is unchanged by a particular rotation. We shall now show that to every non-zero rotation there corresponds a unique axis of rotation. If the rotation is represented by an orthogonal matrix $R$ with $\Delta(R)=+1$, then we look for a vector $\mathbf{v}$ whose component matrix $V$ satisfies

$$RV=V$$

or

$$(R-I)V=0, \tag{4.83a}$$

so that it is unchanged by the operation of $R$. Written out in full, this matrix equation is of the form (4.43a):

$$\begin{pmatrix} r_{11}-1 & r_{12} & r_{13} \\ r_{21} & r_{22}-1 & r_{23} \\ r_{31} & r_{32} & r_{33}-1 \end{pmatrix}\begin{pmatrix} v_1 \\ v_2 \\ v_3 \end{pmatrix}=\begin{pmatrix} 0 \\ 0 \\ 0 \end{pmatrix}. \tag{4.83b}$$

At the end of §4.2, we showed that this set of equations would have a solution $V$ if the matrix of coefficients $R-I$ had zero determinant. We shall now show that $\Delta(R-I)=0$.

Since $R^{\mathrm{T}}=R^{-1}$,

$$I-R=R(R^{\mathrm{T}}-I).$$

Taking determinants on both sides, and using (4.30),

$$\Delta(I-R)=\Delta(R)\Delta(R^{\mathrm{T}}-I).$$

Since $\Delta(R)=+1$ and the determinant of $R^{\mathrm{T}}-I$ equals that of its transpose $R-I$, this gives

$$\Delta(I-R)=\Delta(R-I)=-\Delta(I-R).$$

Therefore $\Delta(I-R)=0$, showing that (4.83) has a solution $V$; as in §4.2, there is therefore a unique ray of vectors invariant under the rotation, defining an axis of rotation, provided that the rank $r(I-R)$ is 2. We have already shown that $r(I-R)<3$; $r(I-R)=0$ only if $R=I$, representing zero rotation. It can also be shown (Problems 4.3,

Question 6) that $r(I-R)=1$ also implies that $R=I$. So $r(I-R)=2$ for all non-zero rotations, and a unique axis of rotation is defined.

**Example 4.4**

Show that the eight diagonal $(3\times 3)$ matrices, with each diagonal element taking values $+1$ or $-1$, form an Abelian group. Discuss the geometric interpretation of the matrices.

The eight matrices are of form

$$\begin{pmatrix} \pm 1 & 0 & 0 \\ 0 & \pm 1 & 0 \\ 0 & 0 & \pm 1 \end{pmatrix}.$$

The set contains the unit matrix, and each matrix is its own inverse. If we multiply two of the matrices (in either order), we simply multiply corresponding diagonal elements, giving another diagonal matrix of the set. So the eight matrices form an Abelian group.

The matrix $P=-I$ denotes reflection in the origin or central inversion. The matrix with diagonal elements $\{1, 1, -1\}$ changes the sign of the third component, and so denotes reflection in the $(\mathbf{i}, \mathbf{j})$ plane. The matrix with diagonal elements $\{-1, -1, 1\}$, as in Example 4.1, denotes a rotation through angle $\pi$ about the $\mathbf{k}$-axis, which is unchanged by the transformation. The other matrices have similar interpretations.

**Example 4.5**

Relative to a frame $F$, a vector $\mathbf{w}$ is obtained from a vector $\mathbf{v}$ by a rotation represented by the matrix $R$. The transformation from the frame $F$ to a second frame $F'$ is represented by the matrix $L$. Show that, relative to the frame $F'$, the rotation from $\mathbf{v}$ to $\mathbf{w}$ is represented by the matrix

$$L^{-1}RL=L^{\mathrm{T}}RL.$$

If $V$ and $W$ are the component matrices of $\mathbf{v}$ and $\mathbf{w}$ relative to $F$, then by (4.56),

$$W=RV.$$

But by (4.53), the component matrices $V'$ and $W'$ relative to $F'$ are related to $V$ and $W$ by

$$V=LV', \qquad W=LW'.$$

Substituting in the above equation gives

$$LW' = RLV'$$

or

$$W' = L^{-1}RLV'.$$

Thus the matrix representing the rotation, in frame $F'$, is

$$L^{-1}RL = L^{\mathrm{T}}RL. \tag{4.84}$$

**Example 4.6**

Relative to the frame $F$, the rotation from $F$ to a second frame $F'$ is represented by the matrix $R_1$. Relative to $F'$, a second rotation from $F'$ to frame $F''$ is represented by $R_2$. Show that, relative to $F$, the rotation from $F$ to $F''$ is represented by the matrix $R_1R_2$.

In frame $F$, let the rotation from $F'$ to $F''$ be represented by $R$. Now use the result of Example 4.5 above, with $L$ replaced by $R_1$. Then the rotation from $F'$ to $F''$, in frame $F'$, is represented by $R_1^{-1}RR_1$. But this is equal to $R_2$: so $R_2 = R_1^{-1}RR_1$, or

$$R = R_1R_2R_1^{-1}.$$

In frame $F$, the rotation from $F$ to $F''$ is represented by the matrix product $R_2R$ ($R$ acts first, then $R_2$, with fixed frame $F$). Substituting for $R$, this matrix product becomes

$$(R_1R_2R_1^{-1})R_1 = R_1R_2,$$

as required.

## *Problems 4.3*

1 The vectors $\mathbf{j}'$, $\mathbf{k}'$ of a left-handed frame $F'$ are the unit vectors corresponding to

$$\mathbf{a}_1 = \mathbf{i} + 2\mathbf{j} - 2\mathbf{k}$$

and

$$\mathbf{a}_2 = 2\mathbf{i} - 2\mathbf{j} - \mathbf{k}$$

respectively. Find the component matrix of

$$\mathbf{v} = 3\mathbf{i} - \mathbf{j} + 2\mathbf{k}$$

relative to the frame $F'$.

2 Check by direct matrix multiplication that the matrix $R$ given by (4.79) satisfies $R^{\mathrm{T}}R = RR^{\mathrm{T}} = I$. Show by direct calculation that $\Delta(R) = +1$.

3 Show that the matrix

$$\frac{1}{\sqrt{6}}\begin{pmatrix} \sqrt{2} & \sqrt{2} & \sqrt{2} \\ \sqrt{3} & 0 & -\sqrt{3} \\ -1 & 2 & -1 \end{pmatrix}$$

corresponds to a rotation. Find a unit vector $\mathbf{u}$ in the direction of the axis of rotation.

4 Write down the set of nine matrices which correspond to rotations through angles $\frac{1}{2}\pi$, $\pi$ and $\frac{3}{2}\pi$ about axes in directions $\mathbf{i}$, $\mathbf{j}$ and $\mathbf{k}$. Find a further set of rotation matrices which, together with $I$ and these nine matrices, form a group. To what rotations do these other matrices correspond?

5 Relative to frame $F$, the rotation to frame $F^{(1)}$ corresponds to matrix $R_1$; and for $k = 2, \ldots, n$, the rotation from frame $F^{(k-1)}$ to $F^{(k)}$, relative to frame $F^{(k-1)}$, corresponds to matrix $R_k$. Show that the rotation from $F$ to $F^{(n)}$, relative to frame $F$, corresponds to the matrix

$$R = R_1R_2 \ldots R_{n-1}R_n.$$

[This is an extension of the result in Example 4.6.]

6 If $R$ is an orthogonal matrix with $\Delta(R) = +1$, and $R \neq I$, show that $r(I - R) \neq 1$.

[*Hint*: The columns of $I - R$ correspond to vectors $\mathbf{i} - \mathbf{i}'$, $\mathbf{j} - \mathbf{j}'$, $\mathbf{k} - \mathbf{k}'$; if $r(I - R) = 1$, these three vectors are parallel or zero.]

## 4.5 *Vector products and axial vectors*

If we reflect the vector $\mathbf{v} = v_1\mathbf{i} + v_2\mathbf{j} + v_3\mathbf{k}$ in the origin, its components are transformed by the matrix $P$, given by (4.80). The components of the reflected vectors $\mathbf{v}'$ are given by

$$v_p' = -v_p \qquad (p = 1, 2, 3). \tag{4.85}$$

A second vector $\mathbf{w} = w_1\mathbf{i} + w_2\mathbf{j} + w_3\mathbf{k}$ has a reflected vector with components

$$w_p' = -w_p.$$

The vector product $\mathbf{v} \wedge \mathbf{w}$ is defined by (3.19), so that the vector product of the reflected vectors $\mathbf{v}'$ and $\mathbf{w}'$ is given by

$$\mathbf{v}' \wedge \mathbf{w}' = \mathbf{v} \wedge \mathbf{w}. \tag{4.86}$$

There is no change of sign of the components of $\mathbf{v} \wedge \mathbf{w}$ due to reflection; that this is to be expected can be seen from Fig. 4.4. The

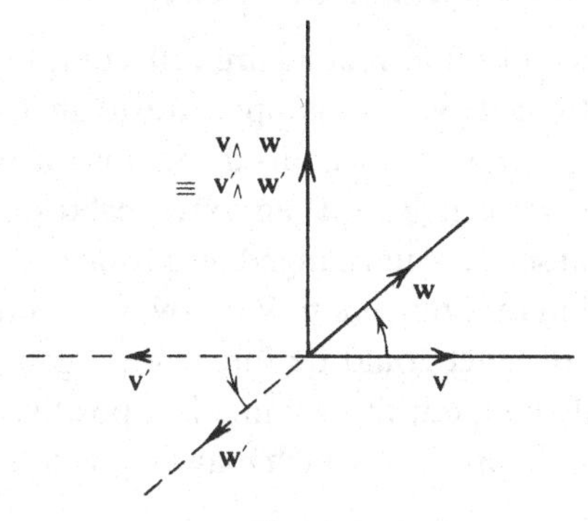


Fig. 4.4

sense of rotation from $\mathbf{v}$ to $\mathbf{w}$ is the same as that from $\mathbf{v}' = -\mathbf{v}$ to $\mathbf{w}' = -\mathbf{w}$, so that the vector products $\mathbf{v} \wedge \mathbf{w}$ and $\mathbf{v}' \wedge \mathbf{w}'$ are in the same sense. Thus $\mathbf{v} \wedge \mathbf{w}$ and $\mathbf{v}$ behave differently under the reflection operation, and $\mathbf{v} \wedge \mathbf{w}$ is not in this respect a vector in the original vector space. The different nature of $\mathbf{v} \wedge \mathbf{w}$ stems from the introduction of a 'sense of rotation' in its definition in §3.1; this 'sense of rotation' is unchanged when $\mathbf{v}$ and $\mathbf{w}$ are replaced by $-\mathbf{v}$ and $-\mathbf{w}$. We distinguish these two types of vectors by referring to a vector $\mathbf{v}$ with reflection property (4.85) as a **polar vector**; $\mathbf{v}$ is a vector in the original vector space. The vector product $\mathbf{v} \wedge \mathbf{w}$, with reflection property (4.86), is called an **axial vector** or a **pseudovector**. Provided that we do not wish to discuss reflection properties, axial vectors behave in every sense like vectors; for example, under a rotation of vectors defined by an orthogonal matrix $R$ with $\Delta(R) = +1$, the vector $\mathbf{n}$ in the definition (3.1) of a vector product will rotate with the vectors $\mathbf{a}$ and $\mathbf{b}$ in order to remain orthogonal to them, while $a$, $b$ and $\sin\theta$ (defined in terms of scalar products) remain unchanged; so a vector product behaves exactly like a vector under the operation of a rotation.

Axial vectors are not necessarily of the form of a vector product, although they are frequently associated with vector products. In mechanics, the moment of a force is a vector product, and has the reflection properties of an axial vector. The vorticity of a fluid, measuring the rotation of a fluid about an axis, is an axial vector; it is defined as an integral of a vector product.

Suppose that **v**, **w** and **t** are three polar vectors, each changing sign under reflection. Then the scalar triple product

$$[\mathbf{v}, \mathbf{w}, \mathbf{t}] = \mathbf{v} \cdot (\mathbf{w} \wedge \mathbf{t})$$

changes sign when all three vectors are reflected; this is because the 'handedness' of the set $(\mathbf{v}, \mathbf{w}, \mathbf{t})$ is opposite to that of $(-\mathbf{v}, -\mathbf{w}, -\mathbf{t})$. Under rotations, $[\mathbf{v}, \mathbf{w}, \mathbf{t}]$ is unchanged, because it was defined as a scalar product; so we call $[\mathbf{v}, \mathbf{w}, \mathbf{t}]$ an **axial scalar** or a **pseudoscalar**. Generally, an axial scalar is unchanged by a rotation, but changes sign under the operation of a reflection. We saw in §3.3 that the modulus of a scalar triple product could be interpreted geometrically as the volume of a parallelepiped; the volume is a positive scalar quantity, while the scalar triple product, which may be positive or negative, is a pseudoscalar.

## *4.6 Tensors in 3-space*

Given a basis $(\mathbf{i}, \mathbf{j}, \mathbf{k})$ in 3-space, a vector **v** is uniquely determined by its three components $v_p$; these components obey transformation laws (4.74) and (4.82). A second-rank tensor **T** is a mathematical entity described by nine quantities $t_{pq}$ $(p, q = 1, 2, 3)$, relative to a given basis; each suffix $p$, $q$ of $t_{pq}$ transforms like a vector suffix under a rotation. So if the nine components of **T** relative to a basis $(\mathbf{i}', \mathbf{j}', \mathbf{k}')$ are $t'_{rs}$ $(r, s = 1, 2, 3)$, then the transformation law (4.73a) is generalised to

$$t_{pq} = \sum_r \sum_s l_{pr} l_{qs} t'_{rs} \qquad (p, q = 1, 2, 3), \tag{4.87}$$

where $L$ is the transformation matrix and the summations are over values 1, 2, 3. The inverse transformation, as in (4.74), involves the inverse transformation matrix $L^{-1} = L^{\mathrm{T}}$, and is therefore

$$\begin{aligned} t'_{rs} &= \sum_p \sum_q l^{\mathrm{T}}_{rp} l^{\mathrm{T}}_{sq} t_{pq} \\ &= \sum_p \sum_q t_{pq} l_{pr} l_{qs}, \end{aligned} \tag{4.88}$$

using the definition (4.8) of a transposed matrix.

It is sometimes convenient to regard $(t_{pq})$ as a $(3\times 3)$ matrix $T$, with $p$ labelling the rows and $q$ the columns. Again using the transposed matrix $L^{\mathrm{T}}$, (4.87) can be written

$$t_{pq}=\sum_r\sum_s l_{pr}t'_{rs}l^{\mathrm{T}}_{sq}; \qquad (4.89a)$$

since the sums are now over *adjacent* suffixes, (4.89a) can be written as the matrix equation

$$T=LT'L^{\mathrm{T}}, \qquad (4.89b)$$

relating the two $(3\times 3)$ matrices $T=(t_{pq})$ and $T'=(t'_{rs})$. Multiplying (4.89b) on the left by $L^{\mathrm{T}}=L^{-1}$, and on the right by $L$, we obtain the matrix equation

$$T'=L^{\mathrm{T}}TL; \qquad (4.90)$$

this is just the matrix form of (4.88).

**Example 4.7**

Comparing (4.90) with result (4.84) of Example 4.5, we see that the matrix $R$ representing the rotation of a vector transforms like a second-rank tensor when the frame of reference is changed.

Tensors of rank $n$ $(n>2)$ can be defined similarly; relative to a frame $F$, a **tensor T of rank** $n$ has $3^n$ components $t_{pq\ldots r}$ $(p,q,\ldots,r=1,2,3)$, where there are $n$ suffixes $p,q,\ldots,r$. If $\{t'_{ij\ldots k}\}$ are the components of **T** relative to frame $F'$, the transformation laws (4.87) and (4.88) are generalised to

$$t_{pq\ldots r}=\sum_i\sum_j\ldots\sum_k l_{pi}l_{qj}\ldots l_{rk}t'_{ij\ldots k} \qquad (4.91)$$

and

$$t'_{ij\ldots k}=\sum_p\sum_q\ldots\sum_r t_{pq\ldots r}l_{pi}l_{qj}\ldots l_{rk}. \qquad (4.92)$$

These equations define tensor transformations when the *frame* is changed. As in (4.82), the transformation matrix $L$ in (4.91) or (4.92) is replaced by $L^{\mathrm{T}}=L^{-1}$ to define the transformation when the *tensor*, rather than the frame of reference, is subjected to a transformation; we note that each tensor suffix is then being subjected to the vector-type transformation.

Tensors are used in a wide variety of physical contexts, such as the study of stress and strain in solids, flow of electricity in crystals, heat flow in fluids, and in the special and general theories of relativity. We shall not consider any particular application, but we now study two tensors which merit special attention.

The first important tensor is the **unit tensor**, a tensor of second rank, which has the same definition as the unit matrix, through (4.15). We must remember, however, that the tensor is defined relative to a given frame $F$, and transforms according to (4.88) or (4.90) under change of reference frame. We shall now show, however, that the unit tensor is unchanged by a change of reference frame: putting $t_{pq} = \delta_{pq}$ in (4.88), the transformed tensor has components

$$\begin{aligned} t'_{rs} &= \sum_p \sum_q \delta_{pq} l_{pr} l_{qs} \\ &= \sum_p l_{pr} l_{ps} \\ &= \delta_{rs}, \end{aligned}$$

using (4.66). This proof may be written more simply using the matrix form (4.90): putting $T = I$ gives $T' = L^{\mathrm{T}}L = I$. So we have shown that *the unit tensor is invariant under orthogonal transformations, including reflections.*

The second important tensor is a tensor of the third rank which arises out of the coordinate expression of a vector product. Suppose that $\mathbf{t} = \mathbf{v} \wedge \mathbf{w}$; then, relative to a given right-handed frame $F$, the components of $t$ are given by

$$t_p = \sum_q \sum_r \varepsilon_{pqr} v_q w_r, \tag{4.93}$$

where the 27 numbers $\varepsilon_{pqr}$ $(p, q, r = 1, 2, 3)$ are defined as follows:

(i) If $(p, q, r)$ is an even permutation of the numbers $(1, 2, 3)$, for example $(2, 3, 1)$, then

$$\varepsilon_{pqr} = +1.$$

(ii) If $(p, q, r)$ is an odd permutation of the numbers $(1, 2, 3)$, for example $(2, 1, 3)$, then

$$\varepsilon_{pqr} = -1.$$

(iii) If any two, or all three, of the numbers $p, q, r$ are equal, then

$$\varepsilon_{pqr} = 0.$$

This means that 21 of the 27 numbers are zero. The reader should check that (4.93) is equivalent to (3.19).

The tensor $\varepsilon_{pqr}$ has been defined relative to a given right-handed frame $F$. We now show that, subject to a rotation of the frame with matrix $R$, the set $\{\varepsilon_{pqr}\}$ transforms into itself by the tensor transformation law (4.91) or (4.92). In terms of components relative to a second right-handed frame $F'$, the relation $\mathbf{t}=\mathbf{v}\wedge\mathbf{w}$ is

$$t'_i=\sum_j\sum_k \varepsilon_{ijk}v'_jw'_k, \tag{4.94a}$$

with $\varepsilon_{ijk}$ defined exactly as in (4.93). The components $\{v'_j\}$ and $\{v_q\}$ of $\mathbf{v}$ are related by (4.73), and the same holds for the vectors $\mathbf{w}$ and $\mathbf{t}$. Substituting these values into (4.93), we find

$$\sum l_{ps}t'_s=\sum_q\sum_r \varepsilon_{pqr}\sum_j l_{qj}v'_j\sum_k l_{rk}w'_k.$$

This is a set of three equations, given by $p=1,2,3$. If we multiply these equations by $l_{pi}$ and sum over $p$, the left-hand side becomes, for each value of $i$,

$$\sum_p\sum_s l_{pi}l_{ps}t'_s=\sum_s \delta_{is}t'_s=t'_i,$$

using (4.66) and (4.16a). So we obtain

$$t'_i=\sum_j\sum_k\left[\sum_p\sum_q\sum_r \varepsilon_{pqr}l_{pi}l_{qj}l_{rk}\right]v'_jw'_k,$$

for $i=1,2,3$. Comparing with (4.94a), we see that

$$\varepsilon_{ijk}=\sum_p\sum_q\sum_r \varepsilon_{pqr}l_{pi}l_{qj}l_{rk}. \tag{4.95a}$$

This shows that, under rotations, the set $\{\varepsilon_{pqr}\}$ transforms into itself according to the tensor transformation law (4.92), as a third-rank tensor. It therefore also transforms into itself under the inverse transformation (4.91).

Equations (4.93) and (4.94) follow from $\mathbf{t}=\mathbf{v}\wedge\mathbf{w}$ only if $F$ and $F'$ are each right-handed frames, so that the transformation law (4.95) has been established only when $L$ is a rotation matrix. If $F'$ is a left-handed frame, then the signs of $\mathbf{i}'$, $\mathbf{j}'$, $\mathbf{k}'$ will be opposite to those of $\mathbf{i}$, $\mathbf{j}$, $\mathbf{k}$ in (3.17), and hence those of the components in (3.19) will be changed; so (4.94a) will become

$$t'_i=-\sum_j\sum_k \varepsilon_{ijk}v'_jw'_k. \tag{4.94b}$$

The change of sign carries through to (4.95a), which becomes

$$\varepsilon_{ijk} = -\sum_p \sum_q \sum_r \varepsilon_{pqr} l_{pi} l_{qj} l_{rk}. \tag{4.95b}$$

So the transformation law (4.95a) is modified when $L$ represents a rotation plus reflection, by a change of sign. For a reflection in the origin, $L = P = -I$, as in (4.80), so that $l_{pi} = -\delta_{pi}$. Then (4.95b) becomes

$$\varepsilon_{ijk} = +\sum_p \sum_q \sum_r \varepsilon_{pqr} \delta_{pi} \delta_{qj} \delta_{rk} = +\varepsilon_{ijk}.$$

The sign change in (4.95b) is therefore necessary to preserve mathematical consistency. Since the transformation law (4.95a) holds only for rotation matrices $L$, and has to be modified to (4.95b) if the handedness of the frame changes, $\varepsilon_{ijk}$ is often termed a **pseudo-tensor** or **axial tensor**.

## *Problems 4.4*

1 If $T_{pq}$ $(p, q = 1, 2)$ are the components of a second-rank tensor in a plane, relative to a frame $F$, find the explicit values of the components relative to frame $F'$

(*a*) when the transformation from $F$ to $F'$ is represented by the rotation matrix (4.46),

(*b*) when the transformation is represented by the matrix (4.59).

Use these results to show directly that the unit tensor in a plane transforms into itself under all orthogonal transformations.

2 From the definitions of $\delta_{pq}$ and $\varepsilon_{pqr}$, show that

$$\sum_p \varepsilon_{pqr} \varepsilon_{pst} = \delta_{qs} \delta_{rt} - \delta_{qt} \delta_{rs}$$

for all $q$, $r$, $s$, $t$ taking the values 1, 2, 3.
Use this identity to establish in component form the identity

$$(\mathbf{a} \wedge \mathbf{b}) \cdot (\mathbf{c} \wedge \mathbf{d}) = (\mathbf{a} \cdot \mathbf{c})(\mathbf{b} \cdot \mathbf{d}) - (\mathbf{a} \cdot \mathbf{d})(\mathbf{b} \cdot \mathbf{c}).$$

3 Use the tensor identity of Question 2 above to establish the vector equality

$$\mathbf{a} \wedge (\mathbf{b} \wedge \mathbf{c}) = (\mathbf{a} \cdot \mathbf{c})\mathbf{b} - (\mathbf{a} \cdot \mathbf{b})\mathbf{c}.$$

## *4.7 General linear transformations*

In §4.3 and §4.4 we studied the operations of rotation and reflection, corresponding to orthogonal matrices. This group of transformations is only a subset of all linear transformations of a column matrix, defined by (4.13); each $(3\times 3)$ matrix $A$ defines a linear transformation in 3-space. As for orthogonal transformations, we may consider either transformations of the reference frame or transformations of the vector.

Let us first consider transformations of the basis vectors $\mathbf{i}$, $\mathbf{j}$, $\mathbf{k}$ of a frame $F$. The transformation analogous to (4.63) will define the three vectors $\mathbf{a}_1$, $\mathbf{a}_2$, $\mathbf{a}_3$ by

$$\left.\begin{aligned}\mathbf{a}_1&=a_{11}\mathbf{i}+a_{21}\mathbf{j}+a_{31}\mathbf{k},\\ \mathbf{a}_2&=a_{12}\mathbf{i}+a_{22}\mathbf{j}+a_{32}\mathbf{k},\\ \mathbf{a}_3&=a_{13}\mathbf{i}+a_{23}\mathbf{j}+a_{33}\mathbf{k},\end{aligned}\right\}\qquad (4.96a)$$

or

$$(\mathbf{a}_1 \quad \mathbf{a}_2 \quad \mathbf{a}_3)=(\mathbf{i} \quad \mathbf{j} \quad \mathbf{k})A, \qquad (4.96b)$$

where $A$ is the matrix $(a_{rs})$ whose components are the three columns (4.23a). Since $A$ is not generally an orthogonal matrix, we do not know that its determinant $\Delta(A)$ is non-zero. If $\Delta(A)=0$, we have shown in §4.3 that the columns of $A$ are linearly dependent. So, using (4.96a), there are constants $\alpha_1$, $\alpha_2$, $\alpha_3$ (not all zero) such that

$$\alpha_1\mathbf{a}_1+\alpha_2\mathbf{a}_2+\alpha_3\mathbf{a}_3=\mathbf{0};$$

that is to say, the three vectors $\mathbf{a}_r$ $(r=1, 2, 3)$ are linearly dependent, and so do not span the 3-space. But if $\Delta(A)\neq 0$, the columns of $A$ are linearly independent, and by (4.96a), the vectors $\mathbf{a}_r$ $(r=1, 2, 3)$ span the 3-space. A unique inverse $A^{-1}$ of $A$ then exists, and we can multiply (4.96b) on the right by $A^{-1}$ to give

$$(\mathbf{i} \quad \mathbf{j} \quad \mathbf{k})=(\mathbf{a}_1 \quad \mathbf{a}_2 \quad \mathbf{a}_3)A^{-1}, \qquad (4.97a)$$

analogous to (4.71), except that $A^{-1}$ is not in general equal to the transpose $A^{\mathrm{T}}$. Then (4.97a) expresses the basis vectors $\mathbf{i}$, $\mathbf{j}$, $\mathbf{k}$ in terms of $\mathbf{a}_1$, $\mathbf{a}_2$, $\mathbf{a}_3$, and enables us to expand any vector $\mathbf{v}$ as a linear combination of $\{\mathbf{a}_r\}$. So if $\Delta(A)\neq 0$, the linearly independent set $\{\mathbf{a}_r\}$ form a basis in 3-space. The vectors $\mathbf{a}_r$ are not in general unit vectors, and they are not mutually orthogonal. A basis whose vectors are not mutually orthogonal is called an **oblique frame of reference**. For the present, we assume that $\Delta(A)\neq 0$.

If a vector $\mathbf{v}$ is expressible in terms of the two bases as

$$\mathbf{v}=v_1\mathbf{i}+v_2\mathbf{j}+v_3\mathbf{k} \qquad (4.98a)$$

and

$$\mathbf{v}=v_1'\mathbf{a}_1+v_2'\mathbf{a}_2+v_3'\mathbf{a}_3, \qquad (4.98b)$$

then, just as in (4.73b), the component (column) matrices $V$ and $V'$ in the two frames are related by

$$V = AV'; \tag{4.99a}$$

multiplying by $A^{-1}$ on the left gives

$$V' = A^{-1}V. \tag{4.99b}$$

We may also consider the transformation of a vector $\mathbf{v}$ into a different vector $\mathbf{w}$ due to replacement of the basis vectors $\mathbf{i}, \mathbf{j}, \mathbf{k}$ by the vectors $\mathbf{a}_1$, $\mathbf{a}_2$, $\mathbf{a}_3$. Such transformations can be used to describe the change of position vectors of points in an elastic solid, when it is deformed by the application of given stresses. Provided that the deformation or **strain** of the body can be assumed to be linear, the transform of the vector $\mathbf{v}$ will be of the form

$$\mathbf{w} = v_1\mathbf{a}_1 + v_2\mathbf{a}_2 + v_3\mathbf{a}_3.$$

Just as in (4.82), the components of $\mathbf{w}$ are given by

$$w_p = \sum_q a_{pq}v_q \tag{4.100a}$$

or

$$W = AV. \tag{4.100b}$$

The set of all $(3\times3)$ transformation matrices $A$ with $\Delta(A)\neq0$ form a group. The operation of successive transformations $A$ and $B$ is represented by the matrix product $BA$; also, (4.30) gives $\Delta(BA) = \Delta(B)\Delta(A)\neq0$, so that $BA$ belongs to the set. Since each matrix $A$ has a unique inverse, and the unit matrix belongs to the set, the set is a group. Just as for the group of orthogonal matrices, the subset of matrices with $\Delta(A)>0$ themselves form a group [Problems 4.5, Question 1]. Any $(3\times3)$ matrix $B$ with $\Delta(B)<0$ can be written in the form

$$B = PA, \tag{4.101}$$

where $P$ is the reflection matrix defined by (4.80), and $\Delta(A)>0$. The set $\{B\}$ is not a group, as it does not contain the unit matrix.

If $(\mathbf{a}_1, \mathbf{a}_2, \mathbf{a}_3)$ and $(\mathbf{b}_1, \mathbf{b}_2, \mathbf{b}_3)$ are two oblique frames of reference, then by (4.96) and (4.97), they are related to $(\mathbf{i}, \mathbf{j}, \mathbf{k})$ by transformations of the form

$$(\mathbf{a}_1 \quad \mathbf{a}_2 \quad \mathbf{a}_3) = (\mathbf{i} \quad \mathbf{j} \quad \mathbf{k})A_1,$$
$$(\mathbf{i} \quad \mathbf{j} \quad \mathbf{k}) = (\mathbf{b}_1 \quad \mathbf{b}_2 \quad \mathbf{b}_3)A_2^{-1},$$

with $\Delta(A_1)\neq 0$ and $\Delta(A_2)\neq 0$. So if we form the matrix product $M=A_2^{-1}A_1$, with $\Delta(M)\neq 0$, the bases are related by the transformation

$$(\mathbf{a}_1 \quad \mathbf{a}_2 \quad \mathbf{a}_3)=(\mathbf{b}_1 \quad \mathbf{b}_2 \quad \mathbf{b}_3)M, \tag{4.96c}$$

a simple generalisation of (4.98b). The inverse relation, corresponding to (4.97a), is

$$(\mathbf{b}_1 \quad \mathbf{b}_2 \quad \mathbf{b}_3)=(\mathbf{a}_1 \quad \mathbf{a}_2 \quad \mathbf{a}_3)M^{-1}. \tag{4.97b}$$

If the vector $\mathbf{v}$ has expansion

$$\mathbf{v}=v_1''\mathbf{b}_1+v_2''\mathbf{b}_2+v_3''\mathbf{b}_3 \tag{4.98c}$$

then the component (column) matrices in the two oblique frames are related by

$$V''=MV' \tag{4.99c}$$

and

$$V'=M^{-1}V'', \tag{4.99d}$$

analogous to (4.73) and (4.74), or to (4.99a) and (4.99b).

If a matrix $A$ defines an oblique frame of reference $F_1$ through (4.96), it is sometimes convenient to consider simultaneously a second frame of reference $F_2$, with basis vectors $\tilde{\mathbf{c}}_r$ $(r=1,2,3)$ defined by

$$(\tilde{\mathbf{c}}_1 \quad \tilde{\mathbf{c}}_2 \quad \tilde{\mathbf{c}}_3)=(\mathbf{i} \quad \mathbf{j} \quad \mathbf{k})(A^{\mathrm{T}})^{-1}, \tag{4.102}$$

where $(A^{\mathrm{T}})^{-1}$ is the inverse of the transpose of $A$. Now suppose a vector $\mathbf{v}$ is defined by (4.98b), with column matrix $V'$ satisfying (4.99). Let another vector $\mathbf{w}$ be expressed as

$$\mathbf{w}=w_1\mathbf{i}+w_2\mathbf{j}+w_3\mathbf{k}$$

and

$$\mathbf{w}=w_1'\tilde{\mathbf{c}}_1+w_2'\tilde{\mathbf{c}}_2+w_3'\tilde{\mathbf{c}}_3, \tag{4.103}$$

with respect to frames $F$ and $F_2$. Then the column component matrices $W$ and $W'$ are related, as in (4.99a), by

$$W=(A^{\mathrm{T}})^{-1}W',$$

so that the transposed row matrices satisfy

$$W^{\mathrm{T}}=W'^{\mathrm{T}}A^{-1}. \tag{4.104}$$

The scalar product of the vectors $\mathbf{v}$ and $\mathbf{w}$ is, by (4.12b), the single component of $W^{\mathrm{T}}V$. But (4.100b) and (4.104) give

$$W^{\mathrm{T}}V=W'^{\mathrm{T}}A^{-1}AV'=W'^{\mathrm{T}}V'; \tag{4.105}$$

so the familiar form (2.36) for the scalar product is retained if one vector $\mathbf{v}$ is referred to an oblique frame $F_1$, while the second vector is referred to the associated oblique frame $F_2$. The frames $F_1$ and $F_2$, used in this way, form a bi-orthogonal coordinate system.

The basis $(\tilde{\mathbf{c}}_1, \tilde{\mathbf{c}}_2, \tilde{\mathbf{c}}_3)$ is called the **reciprocal basis** of the basis $(\mathbf{a}_1, \mathbf{a}_2, \mathbf{a}_3)$, and is of considerable importance in the study of crystals, where non-orthogonal frames arise naturally. From (4.102), the component matrices of $\tilde{\mathbf{c}}_r$ $(r = 1, 2, 3)$ are just the columns of $(A^{\mathrm{T}})^{-1}$, *equal to the rows of* $A^{-1}$. The rows of $A^{-1}$ are given by (4.39) and (4.36a) to be $[\Delta(A)]^{-1}\tilde{\mathbf{D}}_r$. So the reciprocal basis vectors $\{\tilde{\mathbf{c}}_r\}$ are given by

$$\tilde{\mathbf{c}}_r = [\Delta(A)]^{-1}\tilde{\mathbf{d}}_r, \tag{4.106a}$$

or using (4.35a) and the $\varepsilon$-tensor,

$$\tilde{\mathbf{c}}_r = \tfrac{1}{2}[\Delta(A)]^{-1}\varepsilon_{rst}\mathbf{a}_s \wedge \mathbf{a}_t. \tag{4.106b}$$

The close association of the basis $\{\tilde{\mathbf{c}}_r\}$ with the 'reciprocal' $A^{-1}$ is the reason for the name 'reciprocal basis'. It is not difficult to see that the basis $(\mathbf{a}_1, \mathbf{a}_2, \mathbf{a}_3)$ is the reciprocal basis of $(\tilde{\mathbf{c}}_1, \tilde{\mathbf{c}}_2, \tilde{\mathbf{c}}_3)$ [see Problems 4.5, Question 2].

We have already pointed out that when $\Delta(A) = 0$, the vectors $\mathbf{a}_r$ $(r = 1, 2, 3)$ defined by (4.96) are linearly dependent, and do not span the 3-space; in Problems 4.5, Question 3, the reader is asked to prove that the dimension of the space spanned by $\{\mathbf{a}_r\}$ is equal to the rank of the matrix $A$.

Certain types of $(n \times n)$ matrices represent transformations known as **elementary operations**; these matrices are called **elementary matrices**. We shall exemplify them by using $(3 \times 3)$ matrices. An example of the first type of elementary operation is represented by the diagonal matrix

$$E_3(c) = \begin{pmatrix} 1 & 0 & 0 \\ 0 & 1 & 0 \\ 0 & 0 & c \end{pmatrix}. \tag{4.107}$$

When $E_3(c)$ acts on the column matrix $V$, given by (4.7b), it transforms $V$ to

$$E_3(c)V = \begin{pmatrix} v_1 \\ v_2 \\ cv_3 \end{pmatrix}. \tag{4.108}$$

If $c>0$, this transformation is known as a **dilatation** along the third axis with factor $c$, and represents a change of scale in this direction. If $c<0$, the transformation represents a dilatation with factor $|c|$, together with a reflection in the (1, 2) plane. Provided $c\neq\pm1$, this transformation is not an orthogonal transformation. If $c>0$, the product of dilatations with factor $c$ along each of the three axes gives a **uniform dilatation**, represented by

$$E(c)=\begin{pmatrix} c & 0 & 0 \\ 0 & c & 0 \\ 0 & 0 & c \end{pmatrix}. \tag{4.109}$$

Dilatations may represent either a change of unit of measurement along an axis (a passive transformation), or a change in the physical dimension of some system, such as the expansion of a solid when it is heated (an active transformation).

The second type of elementary operation is represented by a matrix such as

$$E_{21}(k)=\begin{pmatrix} 1 & k & 0 \\ 0 & 1 & 0 \\ 0 & 0 & 1 \end{pmatrix}, \tag{4.110}$$

which transforms a column matrix $V$ into

$$E_{21}(k)V=\begin{pmatrix} v_1+kv_2 \\ v_2 \\ v_3 \end{pmatrix}. \tag{4.111}$$

If the transformation represents the (active) transformation of a vector, $E_{21}(k)$ represents a **shear** parallel to the first axis; the first component $v_1$ is increased by an amount proportional to the second component. If the transformation is passive, it represents a change of frame in which the vector $j$ of the triad $(\mathbf{i}, \mathbf{j}, \mathbf{k})$ is replaced by the vector $\mathbf{j}-k\mathbf{i}$. Provided $k\neq0$, the matrix $E_{21}(k)$ is not orthogonal, and the new frame is oblique.

The third type of elementary operation is simply an interchange of two components of $V$, and is therefore represented by an orthogonal transformation. The interchange of $V_1$ and $V_3$, for example, is effected by the matrix

$$E_{13}=\begin{pmatrix} 0 & 0 & 1 \\ 0 & 1 & 0 \\ 1 & 0 & 0 \end{pmatrix}. \tag{4.112}$$

All matrices of this type have determinant equal to $-1$. The operation $E_{13}$ can be regarded either as an interchange of basis vectors $\mathbf{i}$ and $\mathbf{k}$, or of the components $v_1$ and $v_3$ of a vector relative to a given triad.

Elementary operations in $n$-space, represented by $(n\times n)$ matrices, can be defined [Reference 4.5] as straightforward generalisations of matrices such as (4.107), (4.110) and (4.112). The most important property of these elementary matrices, for any value of $n$, is that *any non-singular* $(n\times n)$ *matrix* $A$ *can be written as a product of elementary matrices.* This theorem provides a basis for the study of matrices and transformations in $n$-space. We shall not, however, prove the theorem or develop this theory, since we have taken an alternative approach, based on vectors, to the study of matrices and transformations.

## Problems 4.5

1 If $A$ is a non-singular matrix, and $\{\mathbf{a}_r\}$ and $\{\tilde{\mathbf{c}}_r\}$ are defined by (4.96) and (4.102), show that

$$\tilde{\mathbf{c}}_r\cdot\mathbf{a}_s=\delta_{rs}\qquad(r,s=1,2,3).$$

Hence show that the components $\{v_r'\}$ of $\mathbf{v}$ in (4.98b) are given by

$$v_r'=\tilde{\mathbf{c}}_r\cdot\mathbf{v}.$$

2 If $\Delta(A)\neq 0$ and $\{\mathbf{a}_r\}$ and $\{\tilde{\mathbf{c}}_r\}$ are defined by (4.96) and (4.102), show that $\{\mathbf{a}_r\}$ is the reciprocal basis to $\{\tilde{\mathbf{c}}_r\}$.

3 If $r(A)$ is the rank of a $(3\times 3)$ matrix $A$, show that the vectors $\{\mathbf{a}_r\}$ defined by (4.96) span a space of dimension $r(A)$.

4 Find the inverses of the elementary matrices defined by (4.107), (4.110) and (4.112), and express them as elementary matrices.

5 Show that numbers $b$, $c$, $k$, $l$ can be chosen so that the general $(2\times 2)$ matrix $\{a_{rs}\}$ can be written as the product

$$\begin{pmatrix}1&k\\0&1\end{pmatrix}\begin{pmatrix}0&1\\1&0\end{pmatrix}\begin{pmatrix}b&0\\0&1\end{pmatrix}\begin{pmatrix}1&0\\0&c\end{pmatrix}\begin{pmatrix}1&l\\0&1\end{pmatrix}$$

of elementary $(2\times 2)$ matrices.

# 5

# Curves and surfaces: vector calculus

## 5.1 *Definition of curves and surfaces*

When the value of a scalar $f(u)$ depends upon, and is uniquely determined by, a variable $u$, we say that $f$ is a function of $u$. In just the same way, we can consider a vector $\mathbf{v}(u)$ which is uniquely determined when $u$ is given, lying in some definite range of values; we then say that $\mathbf{v}(u)$ is a **vector function** of the variable $u$. Now suppose that $O$ is a fixed point in Euclidean space, and that the position vector $\mathbf{r}(u)$ relative to $O$ is a function of $u$; then we have defined a point $R$ whose position is determined by the value of the variable $u$. Let us assume that $(\mathbf{i}, \mathbf{j}, \mathbf{k})$ is a **fixed frame of reference** with origin at $O$, meaning that the three basis vectors do not depend on the variable $u$. Then the position vector $\mathbf{r}$ can be expressed as

$$\mathbf{r}(u) = X(u)\mathbf{i} + Y(u)\mathbf{j} + Z(u)\mathbf{k}, \tag{5.1}$$

where $X(u)$, $Y(u)$, $Z(u)$ are functions of $u$, determining the rectangular coordinates through

$$x = X(u), \qquad y = Y(u), \qquad z = Z(u). \tag{5.2}$$

**Example 5.1**

The position vector

$$\mathbf{r} = \cos\theta\,\mathbf{i} + \sin\theta\,\mathbf{j},$$

with $\theta$ varying in the range $0 \leqslant \theta < 2\pi$, corresponds to a point $P$ which traverses the 'unit circle' in the $(x, y)$ plane,

$$r^2 = x^2 + y^2 = 1,$$

as $\theta$ varies from 0 to $2\pi$. The rectangular coordinates, as functions of $\theta$, are

$$x = \cos\theta, \qquad y = \sin\theta, \qquad z = 0.$$

If the functions $X(u)$, $Y(u)$, $Z(u)$ are defined over the range

$$u_0 \leqslant u \leqslant u_1$$

and are *continuous* functions [Reference 5.1] in the range, then the set of points corresponding to the position vectors

$$\{\mathbf{r}(u);\ u_0 \leqslant u \leqslant u_1\} \tag{5.3a}$$

is defined to be a finite curve. So that no section of the curve is represented more than once in (5.3a), we assume that no two open sub-intervals [Reference 5.2] of the range $u_0 \leqslant u \leqslant u_1$ correspond to the same set of points; it may happen, however, that the position vectors of a *finite* number of points correspond to two (or more) values of $u$. Provided that $\mathbf{r}(u_0) \neq \mathbf{r}(u_1)$, these position vectors define, respectively, the **initial point** and the **final point** of the curve; these two points are known as the **end-points** of the curve. The expression (5.1) is called the **parametric equation** of the curve, with $u$ as the parameter. A curve may be specified in many ways: suppose that $u$ is defined as a continuous strictly increasing function $u = U(t)$ of a variable $t$ [Reference 5.3], and that the range $t_0 \leqslant t \leqslant t_1$ corresponds to the range $u_0 \leqslant u \leqslant u_1$, with $u_0 = U(t_0)$ and $u_1 = U(t_1)$. The position vector $\mathbf{r}(u)$ can then be regarded as the function $\mathbf{r}[U(t)]$ of the parameter $t$, with rectangular components

$$x = X[U(t)], \qquad y = Y[U(t)], \qquad z = Z[U(t)]. \tag{5.4}$$

Since $U(t)$ is continuous, $x$, $y$, $z$ are given as continuous functions of $t$ in the range $t_0 \leqslant t \leqslant t_1$ [Reference 5.1]. So $t$ can parametrise the curve.

Continuity of the function $X(u)$ for a given value of $u$ means that, if $\varepsilon$ is any positive number, there is a second positive number $\delta_1$ such that

$$|X(u+\delta u) - X(u)| < \varepsilon \tag{5.5a}$$

whenever $|\delta u| < \delta_1$. Since $Y(u)$, $Z(u)$ are also continuous, positive numbers $\delta_2$, $\delta_3$ also exist such that

$$|Y(u+\delta u) - y(u)| < \varepsilon \tag{5.5b}$$

and

$$|Z(u+\delta u) - Z(u)| < \varepsilon \tag{5.5c}$$

provided that $|\delta u| < \delta_2$ and $|\delta u| < \delta_3$.

If $\delta$ is the least of the numbers $\delta_r$ $(r=1, 2, 3)$ it follows from (5.5) that

$$\begin{aligned}|\mathbf{r}(u+\delta u)-\mathbf{r}(u)|^2 \\ &= |X(u+\delta u)-X(u)|^2+|Y(u+\delta u)-Y(u)|^2 \\ &\quad +|Z(u+\delta u)-Z(u)|^2\end{aligned}$$

provided that $|\delta u|<\delta$, so that

$$|\mathbf{r}(u+\delta u)-\mathbf{r}(u)|<\varepsilon\sqrt{3}; \tag{5.6}$$

since $\varepsilon\sqrt{3}$ can be chosen as small as we please, (5.6) expresses **continuity** of $\mathbf{r}(u)$ as a vector function of $u$, in the same form as continuity of scalar functions.

A curve $\Gamma$ will **intersect** itself if there are two (or perhaps more) different values $u=u_2$, $u_3$, in the range $u_0<u<u_1$, such that $\mathbf{r}(u_2)=\mathbf{r}(u_3)$. If $\mathbf{r}(u_0)=\mathbf{r}(u_1)$, the two ends of the range correspond to the same point, and $\Gamma$ is said to be **closed**. A curve which has no point corresponding to two different parameters is called a **simple open curve**, or more simply, an **arc**. A closed curve which does not intersect itself is called a **simple closed curve**. The circle in Example 5.1 is a simple closed curve; as in this example, we often omit one end-point in order to ensure a one-to-one correspondence between values of the parameter $u$ and points of a simple closed curve.

If the parameter range $u_0\leqslant u\leqslant u_1$ of an open curve is extended by allowing $u_1$ to increase it may happen that $|\mathbf{r}(u)|\to\infty$ as $u_1$ approaches a particular value $u_\infty$; provided that $X(u)$, $Y(u)$ and $Z(u)$ are defined to be continuous in the range $u_0\leqslant u<u_\infty$, the set of position vectors

$$\{\mathbf{r}(u);\ u_0\leqslant u<u_\infty\} \tag{5.3b}$$

will define a **semi-infinite curve**, with one end-point given by $\mathbf{r}(u_0)$. It may be that the limiting value $u_\infty$ is $+\infty$, but this is not necessarily so. In the same way, if $|\mathbf{r}(u_0)|\to\infty$ as $u_0$ decreases to the value $u_{-\infty}$, the set of position vectors

$$\{\mathbf{r}(u);\ u_{-\infty}<u<u_\infty\} \tag{5.3c}$$

defines an **infinite curve**, provided that $\mathbf{r}(u)$ is continuous in the range; again, $u_{-\infty}$ may equal $-\infty$.

**Example 5.2**

If $a$, $b$ are constants, the equations

$$x=a\cos\theta, \qquad y=a\sin\theta, \qquad z=b\theta,$$

with $\theta_0 \leq \theta \leq \theta_1$, define an arc of a **circular helix**, with axis along the $z$-axis; the helix is shown in Fig. 5.1. As $\theta$ increases, the $(x, y)$ coordinates repeatedly describe a circle of radius $a$, while the $z$-coordinate increases in proportion to $\theta$. Since each $z$-value corresponds to only one value of $\theta$, the curve does not intersect itself. If $\theta_1 \to +\infty$ the curve becomes semi-infinite; if $\theta_0 \to -\infty$ also, the curve becomes infinite, and is then the complete circular helix.

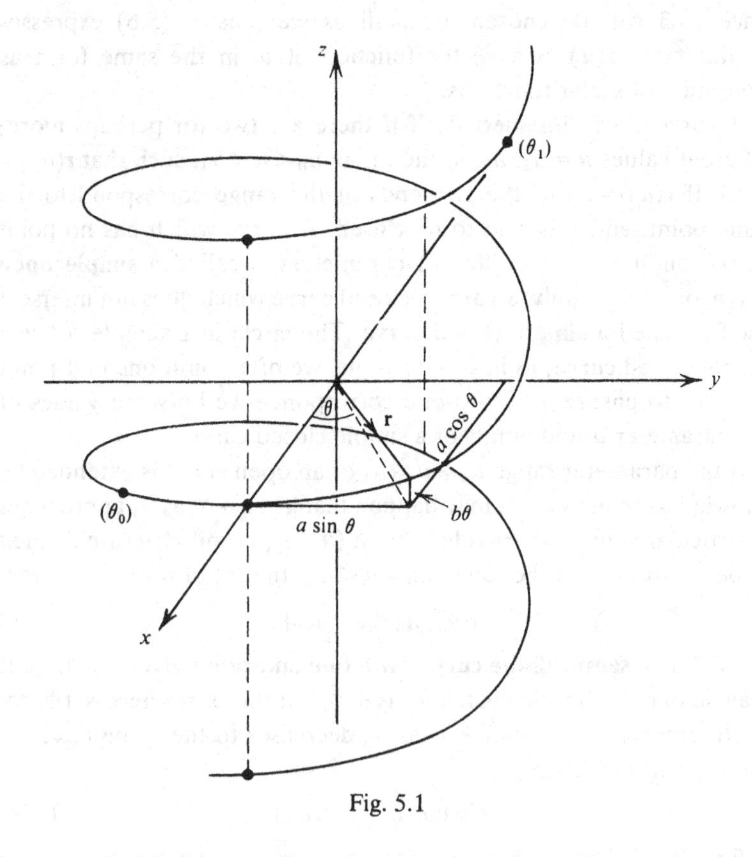


Fig. 5.1

If the parameter $\theta$ is replaced by a new parameter

$$u = \tan^{-1}\theta,$$

the range $-\infty < \theta < \infty$ is in one-to-one correspondence with the range $-\frac{1}{2}\pi < u < \frac{1}{2}\pi$. So with $u$ as parameter, the limits of the range are $u_{-\infty} = -\frac{1}{2}\pi$ and $u_{\infty} = \frac{1}{2}\pi$.

Quite frequently, it is convenient to divide a curve $\Gamma$ into several pieces $\Gamma_1, \Gamma_2, \ldots, \Gamma_n$, with the final point of each piece $\Gamma_r$ ($r = 1, 2, \ldots, n-1$) coinciding with the initial point of the next piece $\Gamma_{r+1}$. If the final point of $\Gamma_n$ coincides with the initial point of $\Gamma_1$, the curve is closed. Each piece can then be defined by a different parametrisation and range, through (5.1) and (5.3). A curve consisting of a semi-circle $\Gamma_1$ and the diameter $\Gamma_2$ joining its end-points, or a square consisting of four sides $\Gamma_1$, $\Gamma_2$, $\Gamma_3$, $\Gamma_4$, exemplifies the division of a closed curve into several pieces.

**Example 5.3**

Show that a closed curve is defined by the three arcs

$$\Gamma_1: \{\mathbf{r}(y) = \mathbf{i} + y\mathbf{j};\ -1 \leqslant y \leqslant 1\},$$
$$\Gamma_2: \{\mathbf{r}(z) = (1-z)\mathbf{i} + (1-z)\mathbf{j} + z\mathbf{k};\ 0 \leqslant z \leqslant 1\},$$
$$\Gamma_3: \{\mathbf{r}(z) = (1-z)\mathbf{i} - (1-z)\mathbf{j} + z\mathbf{k};\ 1 \leqslant z \leqslant 0\}.$$

The arc $\Gamma_1$ is the section of line from $\mathbf{r}_1 = \mathbf{i} - \mathbf{j}$ (with $y = -1$) to $\mathbf{r}_2 = \mathbf{i} + \mathbf{j}$ (with $y = 1$); $\Gamma_2$ is the line section from $\mathbf{r}_2 = \mathbf{i} + \mathbf{j}$ (with $z = 0$) to $\mathbf{r}_3 = \mathbf{k}$ (with $z = 1$); $\Gamma_3$ is the line section from $\mathbf{r}_3 = \mathbf{k}$ to $\mathbf{r}_1 = \mathbf{i} - \mathbf{j}$. So the arcs define a triangle whose vertices have position vectors $\mathbf{r}_1$, $\mathbf{r}_2$, $\mathbf{r}_3$.

We note that the range of the parameter $z$ for $\Gamma_3$ is not an increasing range. We can choose a parameter $u$ which increases continuously along the complete curve if we define

$$\Gamma_1: u = y;\ -1 \leqslant u \leqslant 1;$$
$$\Gamma_2: u = 1 + z;\ 1 \leqslant u \leqslant 2;$$
$$\Gamma_3: u = 3 - z;\ 2 \leqslant u \leqslant 3.$$

Then the curve is defined in the form (5.3a) with $u_0 = -1$, $u_1 = 3$. Since $\mathbf{r}(-1) = \mathbf{r}(3)$, the curve is closed.

It is sometimes convenient to eliminate the parameter $u$ from (5.2) to obtain two relations between the coordinates $x$, $y$, $z$. The circle of Example 5.1, for instance, can be defined by the two equations

$$x^2 + y^2 = 1, \qquad z = 0.$$

In Example 5.2, the parameter $\theta$ can be eliminated by writing $\theta = z/b$, giving the equations

$$x = a\cos(z/b), \qquad y = a\sin(z/b)$$

for the circular helix.

Curves have been defined by considering continuous vector functions of a single parameter $u$. Let us now suppose that a position vector $\mathbf{r}$ is a function

$$\mathbf{r}(u_s) = X(u_s)\mathbf{i} + Y(u_s)\mathbf{j} + Z(u_s)\mathbf{k} \tag{5.7}$$

of *two* real variables $u_s$ $(s = 1, 2)$. Then the points corresponding to $\mathbf{r}(u_s)$, for a range of values of $(u_1, u_2)$, will form a **finite surface** $\sigma$ provided that several conditions are satisfied; these are:

(i) The variables $u_1$, $u_2$ can be chosen so that their ranges are of the form

$$a \leqslant u_1 \leqslant b, \tag{5.8a}$$

$$\alpha(u_1) \leqslant u_2 \leqslant \beta(u_1), \tag{5.8b}$$

where $a$ and $b$ $(>a)$ are constants, and $\alpha(u_1)$ and $\beta(u_1)$ are continuous functions of $u_1$ in the range $a \leqslant u_1 \leqslant b$. The functions $\alpha(u_1)$ and $\beta(u_1)$ may, of course, also be constant, but $\beta(u_1) > \alpha(u_1)$ for all values of $u_1$, except possibly $u_1 = a$ and $u_1 = b$.

(ii) The functions $X(u_s)$, $Y(u_s)$, $Z(u_s)$ are continuous functions [Reference 5.4] of $u_1$ and $u_2$ throughout the range defined by (5.8).

(iii) Except at a finite number of points, and along a finite number of curves, there is only one set of parameter values $\{u_s\}$ corresponding to each position vector $\mathbf{r}$.

It is important to distinguish between **open surfaces** (such as a circular disc), which have a boundary, and **closed surfaces** (such as the surface of a sphere), which do not. But before we attempt to define the boundary of a surface, we shall study several examples of finite surfaces.

**Example 5.4**

If polar coordinates $\rho$, $\phi$ are chosen as parameters of points in the plane $z = 0$, so that

$$\mathbf{r} = \rho\cos\phi\,\mathbf{i} + \rho\sin\phi\,\mathbf{j},$$

the 'unit disc' corresponds to the parameter range

$$0 \leqslant \rho \leqslant 1, \qquad 0 \leqslant \phi \leqslant 2\pi.$$

This range is of the form (5.8) with $\rho = u_1$ and $\phi = u_2$, and with $\alpha(u_1) \equiv 0$ and $\beta(u_1) \equiv 2\pi$, both constant. The limiting values of $\rho$ and $\phi$ define three different sets of points, and it is important to distinguish them:

(*a*) $\phi = 0, 2\pi$.

For a given value of $\rho$, these values of $\phi$ represent the same point; this duplication can be avoided by changing the range of $\phi$ to $0 \leqslant \phi < 2\pi$. The line $\phi = 0$ is a radius of the disc, and *any* radius of the disc could be chosen to define the limits of $\phi$. For example, the range of $\phi$ could be chosen as $-\frac{1}{2}\pi \leqslant \phi < \frac{3}{2}\pi$; then the value $\phi = 0$ would lie inside the range of $\phi$. The fact that $\phi = 0$ does not *necessarily* define the limiting values of $\phi$ means that this radius is not part of the boundary of the disc.

(*b*) $\rho = 1$.

There is no other set of parameter values to define the 'unit circle', which is the boundary of the disc.

(*c*) $\rho = 0$.

For every value of $\phi$ in the range $0 \leqslant \phi < 2\pi$, $\rho = 0$ corresponds to the origin. This is an isolated point, not on the boundary, and corresponds to a whole range of values of the parameters. By using a different coordinate system, we can avoid including the origin in defining the limits of the parameters; using $(x, y)$ as parameters, for instance, the unit disc corresponds to the ranges

$$-1 \leqslant x \leqslant 1, \tag{5.9a}$$

$$-(1-x^2)^{\frac{1}{2}} \leqslant y \leqslant (1-x^2)^{\frac{1}{2}}. \tag{5.9b}$$

The end-points of the ranges define *only* the unit circle $y^2 = 1 - x^2$, which is the boundary. So the use of rectangular coordinates shows that neither the point $\rho = 0$ nor the line $\phi = 0$ *need* be used to define the limits of the parameter range.

**Example 5.5**

Points on the unit sphere

$$x^2 + y^2 + z^2 = 1$$

can be expressed parametrically in the form

$$\mathbf{r}(\theta, \phi) = \sin\theta\cos\phi\,\mathbf{i} + \sin\theta\sin\phi\,\mathbf{j} + \cos\theta\,\mathbf{k},$$

using (2.70) with $r = 1$.

If the ranges of $\theta$, $\phi$ are

$$\theta_1 \leqslant \theta \leqslant \theta_2, \qquad 0 \leqslant \phi \leqslant 2\pi,$$

where $0 < \theta_1 < \theta_2 < \pi$, then $\mathbf{r}(\theta, \phi)$ defines that part $\sigma$ of the surface of the unit sphere shown in Fig. 5.2, lying between the simple closed curves

$$\Gamma_1 \colon \{\theta = \theta_1; \qquad 0 \leqslant \phi < 2\pi\},$$
$$\Gamma_2 \colon \{\theta = \theta_2; \qquad 0 \leqslant \phi < 2\pi\}.$$

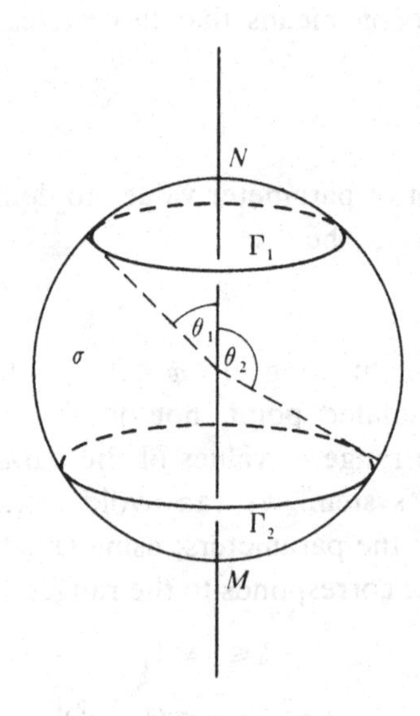


Fig. 5.2

These closed curves form the boundary of $\sigma$. The curves $\phi = 0$ and $\phi = 2\pi$ coincide, and do not form part of the boundary; just as in Example 5.4, the limits of the range of $\phi$ could be chosen differently. To avoid duplicating the curve $\phi = 0$, we have taken the range of $\phi$ to be $0 \leqslant \phi < 2\pi$.

If the value of $\theta_1$ tended to zero, the boundary curve $\Gamma_1$ would shrink to the point $N$. In the limit $\theta_1 = 0$, $\Gamma_1$ would cease to be part of the boundary; the point $N$ would then be an 'interior point' of the surface, and would correspond to the whole set of parameter values $\{\theta=0, 0\leqslant\phi<2\pi\}$. This is analogous to the description of the origin by polar coordinates in Example 5.4.

If $\theta_1\to 0$ and $\theta_2\to 2\pi$, $\Gamma_1$ and $\Gamma_2$ shrink to the points $N$ and $M$ respectively. In this limit, the surface has no boundary, and has become a 'closed' surface.

**Example 5.6**

Define ranges of two parameters which correspond to the plane triangle bounded by the closed curve of Example 5.3.

The three line sections $\Gamma_1$, $\Gamma_2$, $\Gamma_3$ lie in the plane

$$x+z=1.$$

We choose $y$, $z$ as the parameters to describe the triangle. All points on the boundary have $z$-values in the range $0\leqslant z\leqslant 1$; for any value of $z$ in this range, the values of $y$ on the boundary curves $\Gamma_2$, $\Gamma_3$ are $\pm(1-z)$; so the range of $y$ is $-(1-z)\leqslant y\leqslant 1-z$, and is dependent on the value of $z$. Thus the range of values of the parameters $y$, $z$ is

$$0\leqslant z\leqslant 1, \qquad z-1\leqslant y\leqslant 1-z.$$

In examples 5.4, 5.5 and 5.6, we have introduced the terms 'boundary', 'interior point' and 'closed surface' without giving them precise definitions. These examples, however, indicate some of the problems that arise in defining these terms, and suggest how they may be overcome. First, we define from (5.8) a **limit set** of points; this is the set of points on a surface $\sigma$ corresponding to an end of one of the ranges (5.8), namely

$$\{u_1, u_2; u_1=a \text{ or } u_1=b, \alpha(u_1)\leqslant u_2\leqslant\beta(u_1)\} \tag{5.10a}$$

and

$$\{u_1, u_2; a<u_1<b, u_2=\alpha(u_1) \text{ or } u_2=\beta(u_1)\}. \tag{5.10b}$$

The examples above indicate that the limit set contains all the points of the boundary of $\sigma$, but that it may contain other points also. We define a **boundary point** to be one whose parameters belong to the limit set, for *every* parametric system satisfying conditions (i), (ii) and (iii) on p. 140; the **boundary** of $\sigma$ is the set of all boundary points. Then in Example 5.4, the unit circle is the boundary of the unit disc; in Example 5.5, $\Gamma_1$ and $\Gamma_2$ constitute the boundary of the region $\sigma$ on the unit sphere; and in Example 5.6 the line sections $\Gamma_1$, $\Gamma_2$, $\Gamma_3$ form the boundary of the triangle.

Points of a surface $\sigma$ which are not boundary points are **interior points**. In order to show that this definition of boundary points and interior points is reasonable, we now study an example of a surface which has a 'fold' in it.

**Example 5.7**

A surface $\sigma$ consists of the two sets of points $\sigma_1$ and $\sigma_2$, defined in terms of rectangular coordinates by

$$\sigma_1: \{z = 0, 0 \leqslant x \leqslant 1, 0 \leqslant y \leqslant 1\}$$

$$\sigma_2: \{y = 0, 0 \leqslant x \leqslant 1, 0 < z \leqslant 1\},$$

and corresponding to the shaded region in Fig. 5.3. If $0 < x_1 < 1$, show that the point $B$ with coordinates $(x_1, 0, 0)$ is an interior point of $\sigma$.

The point $B$ is in the set $\sigma_1$ defined above, but lies on the boundary of that set. We must show that there is a parametric system for $\sigma$ which satisfies conditions (i), (ii) and (iii), but with $B$ *not* belonging to the limit set. We choose $x$ as one parameter, and define a second parameter $u$ by

$$u = y \text{ for points in } \sigma_1,$$

$$u = -z \text{ for points in } \sigma_2.$$

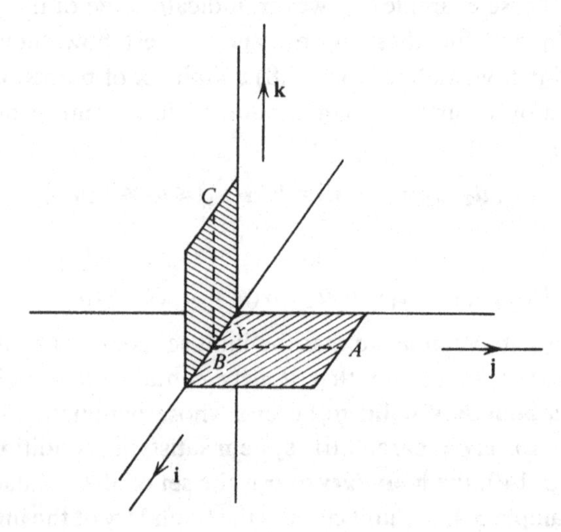


Fig. 5.3

Consider points in $\sigma$ with a fixed value of $x$, corresponding to the two lines $AB$ and $BC$. Then if $|u|$ is the absolute value of $u$, all the points on these two lines have position vectors

$$\mathbf{r}(x_1, u) = x_1\mathbf{i} + \tfrac{1}{2}(|u| + u)\mathbf{j} + \tfrac{1}{2}(|u| - u)\mathbf{k},$$

with $u$ taking values in the range $-1 \leq u \leq 1$. This is true for all values of $x_1$, so that $\sigma$ is defined by

$$\{\mathbf{r}(x, u);\, 0 \leq x \leq 1,\, -1 \leq u \leq 1\},$$

with range of the form (5.8). The coordinates $x$, $\frac{1}{2}(|u|+u)$, $\frac{1}{2}(|u|-u)$ are continuous functions of $x$, $u$ in this range, in particular at $u=0$, and there is only one set of parameters corresponding to each point; so the parametric system obeys conditions (i), (ii) and (iii). But if $0<x_1<1$ and $u=0$, giving $\mathbf{r}=x_1\mathbf{i}$, the parameters $(x_1, 0)$ do not belong to the limit set (5.10); since $\mathbf{r}=x_1\mathbf{i}$ is the position vector of $B$, $B$ is not on the boundary of $\sigma$.

The essence of this example is that, for any interior point, we can always find a set of parameters $(u_1, u_2)$ which go continuously *through* the values corresponding to the point, even when the surface has a 'fold'.

One complication we have not considered is the possibility of a surface $\sigma$ intersecting itself. A true intersection will define a curve on $\sigma$ whose points will correspond to two (or possibly more) distinct sets of parameter values. In Examples 5.4 and 5.5, however, we have noted curves whose points *do* correspond to two sets of parameters on a surface, but which do not intersect themselves. But by choosing a different set of parameters, this 'double representation' disappears. We therefore define a **self-intersection** of $\sigma$ to be a curve on $\sigma$ whose points correspond to two (or more) distinct sets of parameters, for *all* sets of parameters satisfying conditions (i), (ii) and (iii). We shall only be interested in surfaces which have no self-intersections, which are called **simple surfaces**. When we use the word 'surface' in future, however, we shall take it to mean 'simple surface', since we are not interested in self-intersecting surfaces.

So far, we have only considered finite surfaces, whose points correspond to finite position vectors $\mathbf{r}$. If $\Gamma$ is the boundary of a surface $\sigma$, then by increasing the range (5.8) of parameters, the position vectors of all or part of $\Gamma$ may tend to infinity in magnitude; this defines an

**infinite surface**. Extension of the range of parameters may mean that some of the limits of the parameter values become infinite, but this need not necessarily be so.

**Example 5.8**

If the spherical polar coordinate $\theta$ takes a fixed value $\alpha\,(0<\alpha<\frac{1}{2}\pi)$, the position vector

$$\mathbf{r}(r,\phi)=r[\sin\alpha\cos\phi\,\mathbf{i}+\sin\alpha\sin\phi\,\mathbf{j}+\cos\alpha\,\mathbf{k}],$$

with parameters $r$ and $\phi$, lies on the right circular cone

$$x^2+y^2=z^2\tan^2\alpha, \tag{5.11}$$

shown in Fig. 5.4. The origin $O$, the vertex of the cone, corresponds to the range of parameter values $\{r=0, 0\leqslant\phi<2\pi\}$. The range of parameters

$$0\leqslant r\leqslant r_1, \qquad 0\leqslant\phi<2\pi,$$

defines the shaded section of the cone, with boundary curve

$$\Gamma\colon\{r=r_1, \theta=\alpha, 0\leqslant\phi<2\pi\}.$$

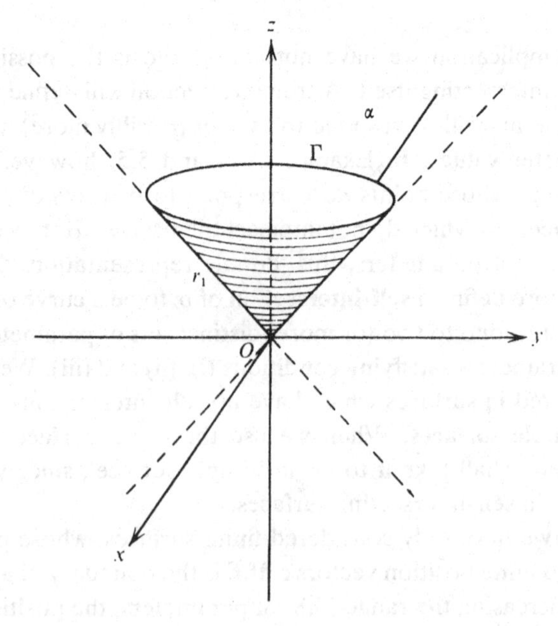


Fig. 5.4

If $r_1 \to \infty$, all points on $\Gamma$ recede to infinity, so that the range

$$0 \leqslant r < \infty, \qquad 0 \leqslant \phi < 2\pi,$$

defines the half of the right circular cone which has $z \geqslant 0$. If we defined a new variable $u$ by

$$u = \frac{r}{1+r},$$

and used $(u, \phi)$ as parameters, the infinite half-cone would correspond to the *finite* range

$$0 \leqslant u < 1, \qquad 0 \leqslant \phi < 2\pi.$$

The half-cone with $z \leqslant 0$ corresponds to the fixed value $\theta = \pi - \alpha$; points on this half-cone can also be represented by leaving $\theta = \alpha$, but allowing $r$ to take negative values in the expression for $\mathbf{r}(r, \phi)$.

In Examples 5.5 and 5.8, we have defined surfaces by fixing the value of one of the three spherical polar coordinates $(r, \theta, \phi)$, and using the other two as parameters of the surface. In this way, we can parametrise spherical surfaces (fixing $r$), cones (fixing $\theta$) and planes containing the $z$-axis (fixing $\phi$). A variety of surfaces can be parametrised by fixing the value of one of three coordinates. Coordinates other than $(x, y)$ in the plane $z = 0$ are called **curvilinear coordinates** in the plane.

Our definition of a surface depends upon the existence of a parametric system $(u_1, u_2)$ for which the range is given by (5.8). Although this parametric system exists, it may not be the most convenient for a given surface. In Example 5.7, for instance, it is probably easier to treat the regions $\sigma_1$ and $\sigma_2$ of the surface separately, rather than to introduce the parametric system $(x, u)$ with range of the form (5.8). More generally, it may be simpler to treat a surface $\sigma$ as the union of a finite number of surfaces $\sigma_1, \sigma_2, \ldots, \sigma_n$, each of which shares part of its boundary with other members of the set $\{\sigma_r\}$. The surface of a cube, for example, is most simply looked upon as the union of its six square faces $\sigma_1, \sigma_2, \ldots, \sigma_6$.

## ■ *Problems 5.1*

1 Draw a diagram of the curve

$$\mathbf{r}(\theta) = \theta \cos \theta\, \mathbf{i} + \theta \sin \theta\, \mathbf{j} + \lambda\theta\, \mathbf{k}\ (\lambda > 0),$$

with $-\pi \leqslant \theta \leqslant \frac{3}{2}\pi$.

2 If a curve is defined by one vector function $\mathbf{r}_1(u)$ in the range $u_1 \leq u \leq u_2$ and by a second vector function $\mathbf{r}_2(u)$ in the range $u_3 \leq u \leq u_4$, with $\mathbf{r}_1(u_2) = \mathbf{r}_2(u_3)$, give a single formula for the vector function describing the curve over a continuous range of a parameter.

3 A surface $\sigma$ is defined by the relation $z = \lambda\phi$ ($\lambda$ constant) between two of the cylindrical polar coordinates $\rho$, $\phi$, $z$; the ranges are $0 \leq \rho \leq \rho_0$ and $\phi_0 \leq \phi \leq \phi_1$. Sketch the surface when $\phi_0 = 0$ and $\phi_1 = \pi$. What is the boundary of $\sigma$? Discuss ways in which the surface can become infinite. Show in your figure the curves defined by adding a second relation $\rho = \mu\phi$ ($\mu$ constant).

4 The coordinates $(x, y)$ in the plane $z = 0$ are defined in terms of two parameters $\xi$, $\eta$ by

$$x = c \cosh \xi \cos \eta,$$
$$y = c \sinh \xi \sin \eta.$$

Find the equations relating $x$ and $y$ of the curves given by putting (i) $\xi$ constant, (ii) $\eta$ constant. Draw a diagram showing these curves for two values each of $\xi$ and $\eta$. Define ranges of $\xi$ and $\eta$ which correspond to the ranges $-\infty < x < \infty$, $-\infty < y < \infty$.

5 **Oblate spheroidal coordinates** $(\xi, \eta, \phi)$ are related to rectangular coordinates $(x, y, z)$ by

$$x = c \cosh \xi \cos \eta \cos \phi,$$
$$y = c \cosh \xi \cos \eta \sin \phi,$$
$$z = c \sinh \xi \sin \eta.$$

Find the relations between cylindrical polar coordinates and oblate spheroidal coordinates. Find the equations relating $x$, $y$, $z$ for the surfaces defined by putting (i) $\xi = \text{constant}$, (ii) $\eta = \text{constant}$, (iii) $\phi = \text{constant}$.

[*Hint*: use results obtained in Question 4 above.]

6 **Oblique coordinates** $(\xi, \eta, \zeta)$ in space are related to rectangular coordinates by

$$\begin{pmatrix} x \\ y \\ z \end{pmatrix} = A \begin{pmatrix} \xi \\ \eta \\ \zeta \end{pmatrix},$$

where $A = (a_{rs})$ is a $(3 \times 3)$ matrix with $\Delta(A) \neq 0$. Find the equations of the oblique axes $\xi = 0$, $\eta = 0$ and $\zeta = 0$ in terms of $x$, $y$, $z$. Explain the importance of the condition $\Delta(A) \neq 0$.

7 Coordinates $(\xi, \eta)$ in the plane $z=0$ are related to $(x, y)$ by

$$x=\xi^2-\eta^2,$$
$$y=2\xi\eta.$$

Draw a diagram to show how a point in the plane is determined by $(\xi, \eta)$, and give ranges of $\xi$, $\eta$ which correspond to the ranges $0<x<\infty$, $-\infty<y<\infty$.

## *5.2 Differentiation of vectors; moving axes*

Given a suitable vector function $\mathbf{v}(u)$ of a single parameter $u$, we can define the derivative $\mathrm{d}\mathbf{v}/\mathrm{d}u$ exactly as we do for a scalar function of $u$; we shall assume that the reader is familiar with the differential and integral calculus of scalar functions [Reference 5.5]. Let $\mathbf{v}(u+\delta u)$ be the value of the vector $\mathbf{v}$ when the parameter takes the value $u+\delta u$. Then the **derivative** of $\mathbf{v}(u)$ with respect to $u$ is defined as

$$\dot{\mathbf{v}}(u) \equiv \frac{\mathrm{d}\mathbf{v}}{\mathrm{d}u} = \lim_{\delta u \to 0} \frac{\mathbf{v}(u+\delta u)-\mathbf{v}(u)}{\delta u}, \tag{5.12}$$

whenever this limit exists. The geometrical significance of the derivative can be seen by considering position vectors $\mathbf{r}(u)=\mathbf{OP}$ and $\mathbf{r}(u+\delta u)=\mathbf{OQ}$ on a curve $\Gamma$, as shown in Fig. 5.5; then the displacement $\mathbf{PQ}$ corresponds to the vector

$$\delta\mathbf{r}=\mathbf{r}(u+\delta u)-\mathbf{r}(u), \tag{5.13}$$

giving the change or 'increment' in $\mathbf{r}$ when the parameter increases

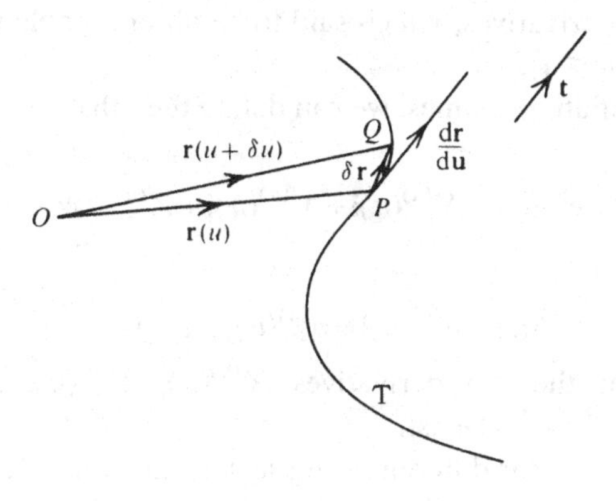


Fig. 5.5

from $u$ to $u+\delta u$. The derivative of $\mathbf{r}(u)$ is then

$$\dot{\mathbf{r}}(u)=\frac{\mathrm{d}\mathbf{r}}{\mathrm{d}u}=\lim_{\delta u\to 0}\frac{\delta\mathbf{r}}{\delta u}. \tag{5.14}$$

When $\mathrm{d}\mathbf{r}/\mathrm{d}u$ exists, this vector is **tangential** to the curve $\Gamma$ at $P$. We shall give a fuller definition of the tangent later on. The increment $\delta\mathbf{r}$ defined by (5.13) is unchanged if the origin is changed; in other words $\delta\mathbf{r}$, and hence $\dot{\mathbf{r}}(u)$ is an *intrinsic* property of the curve defined by $\mathbf{r}(u)$.

If a curve is defined in parametric form by (5.1), where $(\mathbf{i}, \mathbf{j}, \mathbf{k})$ is a fixed frame of reference, then the derivative of $\mathbf{r}(u)$ is given by

$$\dot{\mathbf{r}}(u)=\dot{X}(u)\mathbf{i}+\dot{Y}(u)\mathbf{j}+\dot{Z}(u)\mathbf{k}, \tag{5.15}$$

where $\dot{X}, \dot{Y}, \dot{Z}$ are the derivatives of the functions $X(u), Y(u), Z(u)$; this result follows directly from the definition (5.14), since $\mathbf{i}, \mathbf{j}, \mathbf{k}$ do not depend on $u$. The existence of the derivative $\dot{\mathbf{r}}(u)$ is equivalent to the existence of the three derivatives $\dot{X}(u)$, $\dot{Y}(u)$, $\dot{Z}(u)$. More generally, the derivative of a vector function

$$\mathbf{v}(u)=v_1(u)\mathbf{i}+v_2(u)\mathbf{j}+v_3(u)\mathbf{k} \tag{5.16}$$

with respect to $u$ is

$$\dot{\mathbf{v}}(u)=\dot{v}_1(u)\mathbf{i}+\dot{v}_2(u)\mathbf{j}+\dot{v}_3(u)\mathbf{k}, \tag{5.17}$$

provided $\dot{v}_1$, $\dot{v}_2$, $\dot{v}_3$ exist. When $\dot{\mathbf{v}}(u)$ exists for a given value of $u$, we say that $\mathbf{v}(u)$ is **differentiable** at $u$. If $\mathbf{v}(u)$ is defined in the range $u_0\leqslant u\leqslant u_1$, and the limit (5.12) exists at $u=u_0$ or $u=u_1$, with $u+\delta u$ restricted to the range $(u_0, u_1)$, we say that $\dot{\mathbf{v}}(u_0)$ or $\dot{\mathbf{v}}(u_1)$ exists as a **one-sided derivative** of $\mathbf{v}$. If $\dot{\mathbf{v}}(u)$ exists for $u_0<u<u_1$, $\mathbf{v}(u)$ is differentiable in this *open* interval; if, in addition, $\dot{\mathbf{v}}(u_0)$ and $\dot{\mathbf{v}}(u_1)$ exist as one-sided derivatives, $\mathbf{v}(u)$ is said to be differentiable in the *closed* interval $u_0\leqslant u\leqslant u_1$.

By differentiating $n$ times, we can define the $n$th derivatives of $\mathbf{r}(u)$ and $\mathbf{v}(u)$,

$$\mathbf{r}^{(n)}(u)=X^{(n)}(u)\mathbf{i}+Y^{(n)}(u)\mathbf{j}+Z^{(n)}(u)\mathbf{k}, \tag{5.18}$$

and

$$\mathbf{v}^{(n)}(u)=v_1^{(n)}(u)\mathbf{i}+v_2^{(n)}(u)\mathbf{j}+v_3^{(n)}(u)\mathbf{k}, \tag{5.19}$$

provided that the $n$th derivatives $X^{(n)}(u)$, $Y^{(n)}(u)$, $Z^{(n)}(u)$ and $v_1^{(n)}(u)$, $v_2^{(n)}(u)$, $v_3^{(n)}(u)$ exist.

The usual rules for differentiating scalar functions [Reference 5.6] can be immediately extended to vector functions. If $\mathbf{v}(u)$ and $\mathbf{w}(u)$ are differentiable vector functions and $f(u)$ is a differentiable scalar function, then

$$\frac{d}{du}[f(u)\mathbf{v}(u)]=\frac{df(u)}{du}\mathbf{v}(u)+f(u)\frac{d\mathbf{v}(u)}{du}, \tag{5.20}$$

$$\frac{d}{du}[\mathbf{v}(u)\cdot\mathbf{w}(u)]=\frac{d\mathbf{v}(u)}{du}\cdot\mathbf{w}(u)+\mathbf{v}(u)\cdot\frac{d\mathbf{w}(u)}{du}, \tag{5.21}$$

$$\frac{d}{du}[\mathbf{v}(u)\wedge\mathbf{w}(u)]=\frac{d\mathbf{v}(u)}{du}\wedge\mathbf{w}(u)+\mathbf{v}(u)\wedge\frac{d\mathbf{w}(u)}{du}, \tag{5.22}$$

$$\frac{d}{du}\left[\frac{\mathbf{v}(u)}{f(u)}\right]=\frac{1}{[f(u)]^2}\left[f(u)\frac{d\mathbf{v}(u)}{du}-\frac{df(u)}{du}\mathbf{v}(u)\right]. \tag{5.23}$$

In (5.22), the order of the vectors must be preserved. The formulae (5.20)–(5.23) can be established directly from the definition (5.14) of derivatives, or by expressing all the functions in component form through (5.16), (2.36) and (3.19), and using the rules for differentiating scalar functions.

The **chain rule** governing changes of variables also applies to vector functions: if $\mathbf{v}(u)$ is a differentiable function of $u$ and $u=g(s)$ is a differentiable function of $s$, so that

$$\mathbf{v}(u)=\mathbf{v}[g(s)],$$

then

$$\frac{d\mathbf{v}}{ds}=\left[\frac{d\mathbf{v}(u)}{du}\right]_{u=g(s)}\frac{dg(s)}{ds}; \tag{5.24}$$

the subscript '$u=g(s)$' indicates that this substitution is made *after* $\mathbf{v}(u)$ has been differentiated.

If $\mathbf{v}(u_1, u_2, \ldots, u_n)$ is a vector function of $n$ variables $u_1, u_2, \ldots, u_n$ we can define the **first partial derivatives** of $\mathbf{v}$ as

$$\frac{\partial\mathbf{v}}{\partial u_r}=\lim_{\delta u_r\to 0}\frac{\mathbf{v}(u_1,\ldots,u_r+\delta u_r,\ldots,u_n)-\mathbf{v}(u_1,\ldots,u_r,\ldots,u_n)}{\delta u_r}, \tag{5.25}$$

whenever these limits exist. By repeated differentiation one can form second and third partial derivatives of $\mathbf{v}$, such as

$$\frac{\partial^2\mathbf{v}}{\partial u_1\partial u_2}, \frac{\partial^2\mathbf{v}}{\partial u_2^2}, \frac{\partial^3\mathbf{v}}{\partial u_1\partial u_2^2}, \frac{\partial^3\mathbf{v}}{\partial u_1\partial u_2\partial u_3}.$$

Just as for scalar functions, a partial derivative is independent of the order in which differentiations are performed, provided the derivative is continuous in all the variables $u_1, u_2, \ldots, u_n$. This theorem, for second derivatives of scalar functions of two variables, is Theorem A1 of Appendix A.

**Example 5.9**

Show that if $v(u) = |\mathbf{v}(u)|$, then in general

$$\frac{dv}{du} \neq \left|\frac{d\mathbf{v}(u)}{du}\right|.$$

If $\mathbf{v}(u)$ is given by (5.16),

$$v = (v_1^2 + v_2^2 + v_3^2)^{\frac{1}{2}},$$

so that

$$\frac{dv}{du} = \frac{v_1\, dv_1/du + v_2\, dv_2/du + v_3\, dv_3/du}{(v_1^2 + v_2^2 + v_3^2)^{\frac{1}{2}}}.$$

But

$$\frac{d\mathbf{v}}{du} = \frac{dv_1}{du}\mathbf{i} + \frac{dv_2}{du}\mathbf{j} + \frac{dv_3}{du}\mathbf{k},$$

so that

$$\left|\frac{d\mathbf{v}}{du}\right| = \left[\left(\frac{dv_1}{du}\right)^2 + \left(\frac{dv_2}{du}\right)^2 + \left(\frac{dv_3}{du}\right)^2\right]^{\frac{1}{2}},$$

which is not the same as $dv/du$.

**Example 5.10**

If $\mathbf{r}(u)$ is a vector function of a parameter $u$, if $\mathbf{a}$, $\mathbf{b}$ and a unit vector $\mathbf{n}$ are constant vectors, and if $k$ is a constant, differentiate

$$\frac{k\mathbf{r} \wedge \mathbf{a} - (\mathbf{r} \cdot \mathbf{b})\mathbf{r}}{r^2 - (\mathbf{r} \cdot \mathbf{n})^2}$$

with respect to $u$.

Let

$$\mathbf{v}(u) = k\mathbf{r} \wedge \mathbf{a} - (\mathbf{r} \cdot \mathbf{b})\mathbf{r}$$

and

$$f(u) = r^2 - (\mathbf{r} \cdot \mathbf{n})^2.$$

Then, using (5.20)–(5.22),

$$\frac{d\mathbf{v}}{du} = k\frac{d\mathbf{r}}{du} \wedge \mathbf{a} - (\mathbf{r} \cdot \mathbf{b})\frac{d\mathbf{r}}{du} - \left(\frac{d\mathbf{r}}{du} \cdot \mathbf{b}\right)\mathbf{r}$$

and

$$\frac{df}{du}=\frac{d}{du}[\mathbf{r}\cdot\mathbf{r}-(\mathbf{r}\cdot\mathbf{n})^2]$$

$$=2\frac{d\mathbf{r}}{du}\cdot[\mathbf{r}-\mathbf{n}(\mathbf{r}\cdot\mathbf{n})]$$

$$=2\frac{d\mathbf{r}}{du}\cdot\mathbf{r}_\perp,$$

where $\mathbf{r}_\perp$ is the component of $\mathbf{r}$ orthogonal to $\mathbf{n}$. The result is then given by substituting these expressions for $dv/du$ and $df/du$ into (5.23).

**Example 5.11**

The position vector $\mathbf{r}$ in spherical polar coordinates $(r, \theta, \phi)$ is given by

$$\mathbf{r}=r\sin\theta\cos\phi\,\mathbf{i}+r\sin\theta\sin\phi\,\mathbf{j}+r\cos\theta\,\mathbf{k}.$$

The three first partial derivatives of $\mathbf{r}$, defined by (5.25), are

$$\frac{\partial\mathbf{r}}{\partial r}=\sin\theta\cos\phi\,\mathbf{i}+\sin\theta\sin\phi\,\mathbf{j}+\cos\theta\,\mathbf{k}, \tag{5.26a}$$

$$\frac{\partial\mathbf{r}}{\partial\theta}=r\cos\theta\cos\phi\,\mathbf{i}+r\cos\theta\sin\phi\,\mathbf{j}-r\sin\theta\,\mathbf{k}, \tag{5.26b}$$

$$\frac{\partial\mathbf{r}}{\partial\phi}=-r\sin\theta\sin\phi\,\mathbf{i}+r\sin\theta\cos\phi\,\mathbf{j}. \tag{5.26c}$$

The second partial derivatives $\partial^2\mathbf{r}/\partial r^2$, $\partial^2\mathbf{r}/\partial r\partial\theta$ and $\partial^2\mathbf{r}/\partial\theta\partial\phi$ are given by

$$\frac{\partial^2\mathbf{r}}{\partial r^2}=0,$$

$$\frac{\partial}{\partial r}\left(\frac{\partial\mathbf{r}}{\partial\theta}\right)=\cos\theta\cos\phi\,\mathbf{i}+\cos\theta\sin\phi\,\mathbf{j}-\sin\theta\,\mathbf{k},$$

and

$$\frac{\partial}{\partial\theta}\left(\frac{\partial\mathbf{r}}{\partial\phi}\right)=-r\cos\theta\sin\phi\,\mathbf{i}+r\cos\theta\cos\phi\,\mathbf{j}.$$

The last two derivatives are also equal to

$$\frac{\partial}{\partial\theta}\left(\frac{\partial\mathbf{r}}{\partial r}\right)\quad\text{and}\quad\frac{\partial}{\partial\phi}\left(\frac{\partial\mathbf{r}}{\partial\theta}\right)\text{ respectively.}$$

Formulae (5.15) and (5.17) for the derivative of a vector $\mathbf{r}$ or $\mathbf{v}$ were established on the assumption that the triad $\mathbf{i}$, $\mathbf{j}$, $\mathbf{k}$ was fixed, and did not vary when $u$ varied. It is, however, sometimes convenient to use a triad $(\mathbf{i}(u), \mathbf{j}(u), \mathbf{k}(u))$, which varies with the parameter $u$; the derivative $d\mathbf{v}/du$ of a vector $\mathbf{v}(u)$ will then depend partly on the rates of change $d\mathbf{i}/du$, $d\mathbf{j}/du$, $d\mathbf{k}/du$. We shall now establish a formula for these derivatives, from which we can derive a more general formula for $d\mathbf{v}/du$.

Let us suppose that $\mathbf{i}(u+\delta u)$, $\mathbf{j}(u+\delta u)$, $\mathbf{k}(u+\delta u)$ are the basis vectors when the parameter value is $u+\delta u$. Since we assume that the triad is always right-handed, the triads for parameter values $u$ and $u+\delta u$ must be related through (4.63b), with $L=R$ representing a rotation:

$$(\mathbf{i}(u+\delta u) \quad \mathbf{j}(u+\delta u) \quad \mathbf{k}(u+\delta u)) = (\mathbf{i}(u) \quad \mathbf{j}(u) \quad \mathbf{k}(u))R. \tag{5.27}$$

If we write

$$\delta\mathbf{i}(u) = \mathbf{i}(u+\delta u) - \mathbf{i}(u),$$

and so on, for the incremental changes in $\mathbf{i}(u)$, $\mathbf{j}(u)$, $\mathbf{k}(u)$, then subtracting $(\mathbf{i}(u) \quad \mathbf{j}(u) \quad \mathbf{k}(u))$ from each side of (5.27), and dividing by $\delta u$, we find

$$\left(\frac{\delta\mathbf{i}(u)}{\delta u} \quad \frac{\delta\mathbf{j}(u)}{\delta u} \quad \frac{\delta\mathbf{k}(u)}{\delta u}\right) = (\mathbf{i}(u) \quad \mathbf{j}(u) \quad \mathbf{k}(u))\frac{R-I}{\delta u},$$

where $I$ is the unit $(3\times 3)$ matrix. Assuming that the derivatives of $\mathbf{i}$, $\mathbf{j}$, $\mathbf{k}$ exist, we can take the limit of this equation as $\delta u \to 0$, giving

$$\left(\frac{d\mathbf{i}(u)}{du} \quad \frac{d\mathbf{j}(u)}{du} \quad \frac{d\mathbf{k}(u)}{du}\right) = (\mathbf{i}(u) \quad \mathbf{j}(u) \quad \mathbf{k}(u)) \lim_{\delta u \to 0} \frac{\delta R}{\delta u}, \tag{5.28}$$

where

$$\delta R = R - I \tag{5.29}$$

is a $(3\times 3)$ matrix. Since the limit of $\delta R/\delta u$ exists, every element of the matrix $\delta R$ must tend to zero as $\delta u \to 0$, and the same is true of the transposed matrix $\delta R^{\mathrm{T}} = R^{\mathrm{T}} - I$. Therefore, forming the matrix product $\delta R^{\mathrm{T}}\delta R$ and dividing by $\delta u$ gives

$$\lim_{\delta u \to 0} \frac{\delta R^{\mathrm{T}}\delta R}{\delta u} \to \varnothing. \tag{5.30}$$

But, using (5.29) and the orthogonality property $R^{\mathrm{T}}R = I$,

$$\begin{aligned}\delta R^{\mathrm{T}}\,\delta R &= (R^{\mathrm{T}} - I)(R - I)\\ &= R^{\mathrm{T}}R + I - R - R^{\mathrm{T}}\\ &= -(\delta R + \delta R^{\mathrm{T}}).\end{aligned} \tag{5.31}$$

Dividing by $\delta u$ and letting $\delta u \to 0$, we have from (5.30),

$$\lim_{\delta u \to 0}\left(\frac{\delta R}{\delta u} + \frac{\delta R^{\mathrm{T}}}{\delta u}\right) = \varnothing.$$

Now defining the limit of the matrix $\delta R/\delta u$ to be

$$\lim_{\delta u \to 0}\frac{\delta R}{\delta u} = \frac{\mathrm{d}R}{\mathrm{d}u} = (\omega_{rs}) \qquad (r, s = 1, 2, 3) \tag{5.32}$$

the equation becomes

$$\omega_{rs} + \omega_{sr} = 0 \qquad (r, s = 1, 2, 3) \tag{5.33}$$

so that $(\omega_{rs})$ is an anti-symmetric matrix. It is convenient to re-label three of the elements of the matrix $(\omega_{rs})$ by writing

$$\omega_1 = \omega_{32}, \qquad \omega_2 = \omega_{13}, \qquad \omega_3 = \omega_{21}; \tag{5.34}$$

then (5.33) implies that

$$\omega_{11} = \omega_{22} = \omega_{33} = 0$$

and

$$\omega_{23} = -\omega_1, \qquad \omega_{31} = -\omega_2, \qquad \omega_{12} = -\omega_3.$$

Therefore the matrix (5.32) can be written

$$\frac{\mathrm{d}R}{\mathrm{d}u} = \begin{pmatrix} 0 & -\omega_3 & \omega_2 \\ \omega_3 & 0 & -\omega_1 \\ -\omega_2 & \omega_1 & 0 \end{pmatrix}; \tag{5.35}$$

substituting into (5.28), the rates of change of the unit vectors are of the form

$$\frac{\mathrm{d}\mathbf{i}}{\mathrm{d}u} = \omega_3\mathbf{j} - \omega_2\mathbf{k}, \tag{5.36a}$$

$$\frac{\mathrm{d}\mathbf{j}}{\mathrm{d}u} = \omega_1\mathbf{k} - \omega_3\mathbf{i}, \tag{5.36b}$$

$$\frac{\mathrm{d}\mathbf{k}}{\mathrm{d}u} = \omega_2\mathbf{i} - \omega_1\mathbf{j}. \tag{5.36c}$$

The geometrical interpretation of $\omega_r$ $(r=1, 2, 3)$ can be seen if we consider the matrix

$$R_3 = \begin{pmatrix} \cos\delta\theta_3 & -\sin\delta\theta_3 & 0 \\ \sin\delta\theta_3 & \cos\delta\theta_3 & 0 \\ 0 & 0 & 1 \end{pmatrix},$$

representing, as in (4.77c), a rotation through angle $\delta\theta_3$ about the $z$-axis, corresponding to the parameter change $\delta u$. Putting $R = R_3$ in (5.29) and using (5.32),

$$\frac{dR_3}{du} = \lim_{\delta u \to 0} \frac{1}{\delta u} \begin{pmatrix} \cos\delta\theta_3 - 1 & -\sin\delta\theta_3 & 0 \\ \sin\delta\theta_3 & \cos\delta\theta_3 - 1 & 0 \\ 0 & 0 & 0 \end{pmatrix}.$$

If $\sin\delta\theta_3/\delta u$ has a limit as $\delta u \to 0$ [Reference 5.7], it is equal to

$$\lim_{\delta u \to 0} \frac{\delta\theta_3}{\delta u} = \frac{d\theta_3}{du},$$

and then

$$\lim_{\delta u \to 0} \frac{\cos\delta\theta_3 - 1}{\delta u} = 0.$$

Thus

$$\frac{dR_3}{du} = \begin{pmatrix} 0 & -\dfrac{d\theta_3}{du} & 0 \\ \dfrac{d\theta_3}{du} & 0 & 0 \\ 0 & 0 & 0 \end{pmatrix}.$$

Comparing this with the general result (5.35), we see that for a rotation about the $z$-axis, $\omega_3$ is equal to the **rate of rotation** $d\theta_3/du$. In general, the matrix $dR/du$ can be looked upon as a combination of rotations about the three axes $\mathbf{i}$, $\mathbf{j}$, $\mathbf{k}$, at rates $\omega_1$, $\omega_2$, $\omega_3$.

Now that the rates of change of the basis vectors are given by (5.36), we can find an expression for the rate of change of the vector

$$\mathbf{v}(u) = v_1(u)\,\mathbf{i}(u) + v_2(u)\,\mathbf{j}(u) + v_3(u)\,\mathbf{k}(u), \tag{5.37}$$

when both the components $\{v_r(u)\}$ and the basis vectors vary with $u$. Assuming that $\{v_r(u)\}$ and $\mathbf{i}(u)$, $\mathbf{j}(u)$, $\mathbf{k}(u)$ are differentiable,

$$\frac{d\mathbf{v}}{du} = \frac{dv_1}{du}\mathbf{i} + \frac{dv_2}{du}\mathbf{j} + \frac{dv_3}{du}\mathbf{k} + v_1\frac{d\mathbf{i}}{du} + v_2\frac{d\mathbf{j}}{du} + v_3\frac{d\mathbf{k}}{du}. \tag{5.38}$$

The first three terms give the rate of change of the vector $\mathbf{v}$ *relative* to the variable triad $(\mathbf{i}, \mathbf{j}, \mathbf{k})$, so we define

$$\left(\frac{d\mathbf{v}}{du}\right)_{\text{rel}} = \frac{dv_1}{du}\mathbf{i} + \frac{dv_2}{du}\mathbf{j} + \frac{dv_3}{du}\mathbf{k}. \tag{5.39}$$

Using (5.36), the last three terms in (5.38) become

$$(\omega_2 v_3 - \omega_3 v_2)\mathbf{i} + (\omega_3 v_1 - \omega_1 v_3)\mathbf{j} + (\omega_1 v_2 - \omega_2 v_1)\mathbf{k};$$

if we now define a vector

$$\boldsymbol{\omega} = \omega_1\mathbf{i} + \omega_2\mathbf{j} + \omega_3\mathbf{k}, \tag{5.40}$$

then, by (3.19), the three terms are just $\boldsymbol{\omega} \wedge \mathbf{v}$. So equation (5.38) can be written

$$\frac{d\mathbf{v}}{du} = \left(\frac{d\mathbf{v}}{du}\right)_{\text{rel}} + \boldsymbol{\omega} \wedge \mathbf{v}. \tag{5.41}$$

In dynamical problems, when $u$ represents time, so that $\boldsymbol{\omega}$ is the angular velocity vector, (5.41) is known as the **moving axes formula**.

Since $\mathbf{v}$ is a vector, given by (5.37), $d\mathbf{v}/du$ and $(d\mathbf{v}/du)_{\text{rel}}$ are also vectors. The remaining term $\boldsymbol{\omega} \wedge \mathbf{v}$ in (5.41), however, has the appearance of an axial vector; if this were so, and the formula (5.41) were referred to left-handed axes, this term would change sign. But since $\boldsymbol{\omega}$ represents a rotation, it is natural to define it to be an axial vector itself, with components retaining their sign under reflection; then $\boldsymbol{\omega} \wedge \mathbf{v}$ behaves like a vector under all transformations of axes, and (5.41) is valid in both right-handed and left-handed frames.

## *Problems 5.2*

1 The vectors $\mathbf{v}(u)$ and $\mathbf{w}(u)$ are differentiable functions of a parameter $u$; $\mathbf{a}$ and $\mathbf{b}$ are constant vectors and $k$ is constant. Find the derivatives with respect to $u$ of

(i) $[\mathbf{a}, \mathbf{v}, \mathbf{w}]$,

(ii) $|\mathbf{v}+\mathbf{w}|$,

(iii) $\dfrac{(\mathbf{a}+\mathbf{w}) \wedge \mathbf{v}}{|\mathbf{v}+\mathbf{w}|}$,

(iv) $\dfrac{\mathbf{a}\cdot\mathbf{v}+\mathbf{b}\cdot\mathbf{w}+k\mathbf{v}\cdot\mathbf{w}}{|\mathbf{v}\wedge\mathbf{w}|^2}$.

2 The position vector on a surface is given in terms of parameters $u$, $\alpha$ by $\mathbf{r}(u, \alpha) = u^2 \sec\alpha\, \mathbf{i} + u^2 \tan\alpha\, \mathbf{j} + u^4\mathbf{k}$. Find the first and second

partial derivatives of $\mathbf{r}(u, \alpha)$. What limitations must be imposed on $u$, $\alpha$ in order that $\mathbf{r}(u, \alpha)$ defines a surface? Explain the reasons for these limitations.

3 A vector $\mathbf{v}(u)$ has fixed modulus $|\mathbf{v}|$. If the derivative $d\mathbf{v}/du$ is continuous, show that it is of the form

$$\frac{d\mathbf{v}}{du} = \boldsymbol{\omega} \wedge \mathbf{v},$$

where $\boldsymbol{\omega}$ may vary with $u$. If $u$ represents time, and $\boldsymbol{\omega}$ is independent of $u$, show that $\mathbf{v}$ makes a fixed angle with $\boldsymbol{\omega}$, and that it rotates with uniform angular velocity about an axis parallel to $\boldsymbol{\omega}$.

## *5.3 Differential geometry of curves*

Suppose that a curve $\Gamma$ is defined by a position vector $\mathbf{r}(u)$, as in (5.3a), in terms of a particular parameter $u$. Suppose also that the derivative $d\mathbf{r}/du$ exists in the range $u_0 \leq u \leq u_1$; at any point of $\Gamma$, the vector $d\mathbf{r}/du$ is in the tangential direction. The magnitude $|\delta\mathbf{r}|$ of the increment (5.13) is the increment of distance between points of $\Gamma$, equal to the length of the chord $PQ$ in Fig. 5.5. In defining distance along $\Gamma$, we wish to ensure that the distance increases (or decreases) as $u$ increases (or decreases); the increment of distance must therefore have the same sign as $\delta u$. So, denoting the sign of $\delta u$ by $\operatorname{sgn}(\delta u)$, we take

$$\delta s = |\delta\mathbf{r}| \operatorname{sgn}(\delta u) \tag{5.42}$$

to be the increment of distance $PQ$. Note that a different choice of parameter, for instance the parameter $(-u)$, could lead to a change in sign of $\delta s$, and hence to distance increasing in the opposite direction along $\Gamma$.

Dividing (5.42) by $\delta u$ gives

$$\frac{\delta s}{\delta u} = \frac{|\delta\mathbf{r}|}{\delta u \operatorname{sgn}(\delta u)} = \left|\frac{\delta\mathbf{r}}{\delta u}\right|;$$

since we are assuming that $d\mathbf{r}/du$ exists for $u_0 \leq u \leq u_1$, we can let $\delta u \to 0$ in this equation, giving

$$\frac{ds}{du} = \lim_{\delta u \to 0} \frac{\delta s}{\delta u} = \lim_{\delta u \to 0} \left|\frac{\delta\mathbf{r}}{\delta u}\right| = \left|\frac{d\mathbf{r}}{du}\right|. \tag{5.43}$$

This defines the rate of increase of distance along $\Gamma$ as a function of $u$, and is non-negative for all values of $u$. We now make the first of a further series of assumptions about the differentiability of the function $\mathbf{r}(u)$ defining $\Gamma$; in practice the functions (5.2) can usually be differentiated any number of times except perhaps at a few isolated values of $u$, and the assumptions we make ensure that we can carry out the processes of differentiation and integration without analytic difficulties arising. For the present, we assume that the derivative $\mathrm{d}\mathbf{r}/\mathrm{d}u$ is non-zero and continuous for $u_0 \leqslant u \leqslant u_1$; then the modulus $\mathrm{d}s/\mathrm{d}u$ of this vector will be a positive and continuous function of $u$ in the range. When this condition is satisfied, we say that $\Gamma$ is a **smooth curve**. Since a function which is continuous in a closed interval is integrable [Reference 5.8], we can define the **arc length** as the integral along $\Gamma$ of $\mathrm{d}s/\mathrm{d}u$ with respect to $u$:

$$s(u_0, u_1) = \int_{u=u_0}^{u_1} \frac{\mathrm{d}s}{\mathrm{d}u}\,\mathrm{d}u. \tag{5.44}$$

For a **finite curve,** $s(u_0, u_1)$ is finite. If we regard $\mathbf{r}(u_0)$ as the position vector of a fixed initial point $P_0$ on $\Gamma$, then

$$s(u) \equiv s(u_0, u) \tag{5.45}$$

defines the **distance along the curve** from $P_0$ to the point with parameter $u$. Further, (5.44) ensures that $\mathrm{d}s/\mathrm{d}u$ is the derivative of $s(u)$, as the notation implies [Reference 5.9]. Since $\mathrm{d}s/\mathrm{d}u$ is positive, $s(u)$ is a strictly increasing function of $u$ [Reference 5.3], and there is therefore a one-to-one correspondence between $s$ and $u$. This enables us to change variable from $u$ to $s$ without any ambiguity arising; it is often useful to describe a curve as a function of the variable $s$. By imposing the condition that $\mathrm{d}\mathbf{r}/\mathrm{d}u$ is continuous, we are restricting the class of parameters used to describe the curve; if $u$ and $t$ are two suitable parameters, then $\mathrm{d}u/\mathrm{d}t$ must exist, be continuous and non-zero; $\mathrm{d}u/\mathrm{d}t$ must therefore be of fixed sign. When we impose further conditions on the differentiability of $\mathbf{r}(u)$, these restrict the choice of parameter in a similar way; in future, though, we shall not state these further restrictions.

If increments $\delta u$, $\delta\mathbf{r}$ and $\delta s$ are related by (5.42), then

$$\frac{\delta\mathbf{r}}{\delta s} = \frac{\delta\mathbf{r}/\delta u}{\delta s/\delta u};$$

as $\delta u \to 0$, both numerator and denominator here have non-zero limits $\dot{\mathbf{r}}$ and $\dot{s}$; therefore the limit

$$\mathbf{r}' \equiv \frac{d\mathbf{r}}{ds} = \lim_{\delta s \to 0} \frac{\delta \mathbf{r}}{\delta s} \tag{5.46}$$

exists, and is given by

$$\mathbf{r}' = \frac{d\mathbf{r}/du}{ds/du} = \frac{\dot{\mathbf{r}}}{\dot{s}}. \tag{5.47}$$

Here we have used a 'prime' to denote differentiation with respect to the distance $s$, in contrast to the 'dot' denoting differentiation with respect to the general parameter $u$.

From (5.42), $|\delta \mathbf{r}/\delta s| = 1$, so by (5.46) the vector

$$\mathbf{t} = \mathbf{r}' \tag{5.48}$$

is a unit vector; since $\delta \mathbf{r}$ is tangential to $\Gamma$ in the limit $\delta s \to 0$, $\mathbf{t}$ is called the **unit tangent vector** to $\Gamma$; it is shown in Fig. 5.5. The line through a point $P$ of the curve, parallel to $\mathbf{t}$, is the **tangent** to the curve at $P$.

When $u = u_0$, the distance $s(u_0)$ given by (5.45) and (5.44) is zero; denote $s(u_1)$, the distance between points with parameters $u_0$ and $u_1$, by $s_1$. We can then change variables [Reference 5.10] from $u$ to $s$ in (5.44), giving simply

$$s_1 = \int_0^{s_1} ds.$$

This is the simplest example of an **integral along a curve**. Generally, if $f(s)$ is an integrable function of $s$, then its integral along $\Gamma$ from the points with $s = s_1$ and $s = s_2$ is

$$\int_{s_1}^{s_2} f(s)\, ds. \tag{5.49a}$$

If the integrand is given as a function $g(u)$ of the parameter $u$, we can change variable in (5.49a) to give [Reference 5.10]

$$\int_{u_1}^{u_2} g(u) \frac{ds}{du}\, du = \int_{u_1}^{u_2} g(u) \left| \frac{d\mathbf{r}}{du} \right| du \tag{5.49b}$$

as the integral. We often encounter an integrand of the form $f(s) = \mathbf{v}(s) \cdot \mathbf{t}(s)$, the component of a vector function in the tangential direction; the notation

$$d\mathbf{s} = \mathbf{t}\, ds$$

can then be used to express the integral (5.49a) as

$$\int_{s_1}^{s_2} \mathbf{v}\cdot\mathbf{t}\,\mathrm{d}s = \int_{s_1}^{s_2} \mathbf{v}\cdot\mathrm{d}\mathbf{s}. \tag{5.50}$$

If $s_2 \to \infty$ as $u_2 \to u_\infty$, the upper limit in the integral (5.49a) becomes infinite; likewise, the lower limit $s_1$ may become $-\infty$. When such a limit of the integral (5.49) is required, we shall assume that it exists. Conditions for the existence of integrals with infinite limits are established in standard works on analysis [Reference 5.11].

In practice, we may wish to integrate along a **piecewise smooth curve**, which consists of a finite number of smooth pieces $\Gamma_1, \Gamma_2, \ldots, \Gamma_n$, as detailed on p. 139; the integral is defined to be the sum of the integrals along the $n$ separate smooth curves.

**Example 5.12**

In the plane $z = 0$, the **Archimedean spiral** is defined in terms of polar coordinates $(\rho, \phi)$ by $\rho = a\phi$ ($a$ = constant, $\phi \geqslant 0$). Find the arc length along the curve from the origin to $\phi = \phi_1$, and find the tangent vector $\mathbf{t}$ for any positive value of $\phi$. Integrate the function $\rho$ along the curve from the origin to $\phi = \phi_1$.

The position vector $\mathbf{r}$, as a function of $\phi$, is given by

$$\begin{aligned}\mathbf{r} &= \rho\cos\phi\,\mathbf{i} + \rho\sin\phi\,\mathbf{j}\\ &= a\phi(\cos\phi\,\mathbf{i} + \sin\phi\,\mathbf{j}).\end{aligned}$$

So

$$\frac{\mathrm{d}\mathbf{r}}{\mathrm{d}\phi} = a[(\cos\phi - \phi\sin\phi)\mathbf{i} + (\sin\phi + \phi\cos\phi)\mathbf{j}],$$

and from (5.43) with $u = \phi$,

$$\begin{aligned}\frac{\mathrm{d}s}{\mathrm{d}\phi} &= a[(\cos\phi - \phi\sin\phi)^2 + (\sin\phi + \phi\cos\phi)^2]^{\frac{1}{2}}\\ &= a(1+\phi^2)^{\frac{1}{2}}.\end{aligned}$$

So the arc length $s_1$ from $\phi = 0$ to $\phi = \phi_1$ is, by (5.44),

$$\begin{aligned}s_1 &= a\int_{\phi=0}^{\phi_1} (1+\phi^2)^{\frac{1}{2}}\,\mathrm{d}\phi\\ &= \tfrac{1}{2}a[\phi_1(1+\phi_1^2)^{\frac{1}{2}} + \sinh^{-1}\phi_1].\end{aligned}$$

The unit tangent vector for any $\phi > 0$ is, from (5.47),

$$\begin{aligned}\mathbf{t} &= \frac{\mathrm{d}\mathbf{r}/\mathrm{d}\phi}{\mathrm{d}s/\mathrm{d}\phi} \\ &= (1+\phi^2)^{-\frac{1}{2}}[(\cos\phi - \phi\sin\phi)\mathbf{i} + (\sin\phi + \phi\cos\phi)\mathbf{j}].\end{aligned}$$

The integral of $\rho = a\phi$ along the curve is

$$\int_0^{\phi_1} a^2\phi(1+\phi^2)^{\frac{1}{2}}\,\mathrm{d}\phi = \tfrac{1}{3}a^2[(1+\phi_1^2)^{\frac{3}{2}} - 1].$$

**Example 5.13**

Find the arc length from $\theta = 0$ to $\theta = \theta_1$ along the circular helix of Example 5.2. Show that the tangent vector $\mathbf{t}$ makes a constant angle with the plane $z = 0$.

The position vector on the helix is given by

$$\mathbf{r} = a\cos\theta\,\mathbf{i} + a\sin\theta\,\mathbf{j} + b\theta\mathbf{k}.$$

So

$$\frac{\mathrm{d}\mathbf{r}}{\mathrm{d}\theta} = -a\sin\theta\,\mathbf{i} + a\cos\theta\,\mathbf{j} + b\mathbf{k}$$

and hence, from (5.43),

$$\frac{\mathrm{d}s}{\mathrm{d}\theta} = (a^2 + b^2)^{\frac{1}{2}} = c,$$

say, defining the constant $c$. Thus the arc length from $\theta = 0$ to $\theta = \theta_1$ is

$$s_1 = \int_{\theta=0}^{\theta_1} \frac{\mathrm{d}s}{\mathrm{d}\theta}\,\mathrm{d}\theta = c\theta_1.$$

The tangent vector is

$$\mathbf{t} = \frac{\mathrm{d}\mathbf{r}}{\mathrm{d}s} = -\frac{a}{c}\sin\theta\,\mathbf{i} + \frac{a}{c}\cos\theta\,\mathbf{j} + \frac{b}{c}\mathbf{k}.$$

This vector makes an angle $\beta \equiv \arctan(b/a)$ with the plane $z = 0$, for all values of $\theta$.

Since $\mathbf{t}$ is a unit vector at every point on the curve,

$$\mathbf{t}(s)\cdot\mathbf{t}(s) = 1$$

for all values of $s$. If we now make the assumption that the vector $\mathbf{t}(s)$ is a differentiable function of $s$, we can differentiate this equation with respect to $s$, using (5.21); this gives

$$\mathbf{t}\cdot\mathbf{t}'=0,$$

so that the derivative $\mathbf{t}'=\mathrm{d}\mathbf{t}/\mathrm{d}s$ is orthogonal to $\mathbf{t}$ at all points, unless it is zero. If $\mathbf{t}'=\mathbf{0}$ over a range of values of $s$, $\mathbf{t}$ is a constant vector, and this part of the curve is just a straight line. Otherwise $\mathbf{t}'\neq\mathbf{0}$, and we can define the **unit normal vector n** as the unit vector corresponding to $\mathbf{t}'$, at every point of the curve. So differentiating (5.48) gives

$$\mathbf{r}''=\mathbf{t}'=\kappa\mathbf{n}, \tag{5.51}$$

where $\kappa$ is the modulus of $\mathbf{t}'$; it is a positive scalar, called the **curvature** of a curve $\Gamma$ at any point $\mathbf{r}$. The line through a point of $\Gamma$ in the direction of $\mathbf{n}$ is called the **normal** to $\Gamma$ at that point. Since $\mathbf{t}$ is a function of $s$, $\kappa$ and $\mathbf{n}$ are also functions of $s$, and usually vary as $s$ varies.

If $\mathbf{t}$ and $\mathbf{t}+\delta\mathbf{t}$ are the unit tangent vectors corresponding to lengths $s$ and $s+\delta s$, they can be represented, as in Fig. 5.6, by displacements

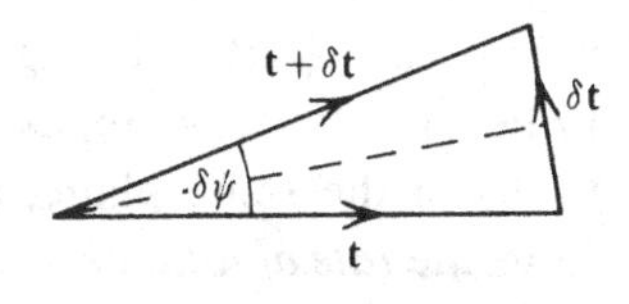


Fig. 5.6

along two sides of an isosceles triangle; if $\delta\psi$ is the angle between the two unit vectors, measured in radians,

$$\begin{aligned}|\delta\mathbf{t}|&=2\sin\tfrac{1}{2}\delta\psi\\&=\delta\psi+O(\delta\psi^3),\end{aligned}$$

using a standard approximation for the sine of a small angle [Reference 5.7]. Letting $\delta\psi\to 0$, and using (5.51),

$$\begin{aligned}\kappa=|\mathbf{t}'|&=\lim_{\delta s\to 0}\frac{|\delta\mathbf{t}|}{\delta s}\\&=\lim_{\delta s\to 0}\frac{\delta\psi}{\delta s},\end{aligned}$$

since the existence of this limit ensures that terms of order $(\delta\psi)^3/\delta s$

vanish in the limit $\delta s \to 0$. So we can write

$$\kappa = \frac{d\psi}{ds} \tag{5.52}$$

and $\kappa$ tells us the *rate of rotation of the tangent vector per unit length* along the curve. The reciprocal of the curvature $\kappa$ is called the **radius of curvature** $r_c$; thus

$$r_c = \kappa^{-1} = \frac{ds}{d\psi}; \tag{5.53}$$

this inverse exists if $\mathbf{t}' \neq \mathbf{0}$, so that $\kappa \neq 0$ by (5.51).

In (5.52), $\kappa$ is a function of $s$ which can be calculated from (5.51) by differentiating $\mathbf{t}$. We can therefore regard (5.52) as a first order differential equation for a function $\psi(s)$; provided that $\kappa(s)$ is integrable over the range $s_0 \leqslant s \leqslant s_1$, we can integrate the equation to give

$$\psi(s_1) - \psi(s_0) = \int_{s_0}^{s_1} \kappa(s)\, ds. \tag{5.54}$$

This quantity does not have any particular significance except for **plane curves**, for which $\mathbf{r}$ lies in a given plane, usually taken to be the plane $z = 0$. Then $\mathbf{t} = \mathbf{r}'$ lies in the same plane, at all points on the curve. So $\kappa$ in (5.52) tells us the *rate of rotation of* $\mathbf{t}$ *about the fixed axis normal to the plane*, and the integral (5.54) tells us the angle through which $\mathbf{t}$ rotates between points with $s = s_0$ and $s = s_1$. If we regard $s_0$ as fixed, (5.54) defines $\psi(s_1)$ as a function of $s_1$; the derivative of this function is given by (5.52). The variables $s$ and $\psi$ are known as the **intrinsic coordinates** of a plane curve, and the relation (5.54) is called the **intrinsic equation** of the curve.

**Example 5.14**

Find the curvature of the Archimedean spiral $\rho = a\phi$ of Example 5.12. Find the intrinsic coordinate $\psi$ as a function of $\phi$.

The tangent vector of the spiral $\rho = a\phi$ is given in Example 5.12. So

$$\begin{aligned}\frac{d\mathbf{t}}{d\phi} = &-\phi(1+\phi^2)^{-\frac{3}{2}}[(\cos\phi - \phi\sin\phi)\mathbf{i} + (\sin\phi + \phi\cos\phi)\mathbf{j}] \\ &+ (1+\phi^2)^{-\frac{1}{2}}[-(2\sin\phi + \phi\cos\phi)\mathbf{i} + (2\cos\phi - \phi\sin\phi)\mathbf{j}],\end{aligned}$$

giving

$$\left|\frac{\mathrm{d}\mathbf{t}}{\mathrm{d}\phi}\right|^2 = (1+\phi^2)^{-2}(2+\phi^2)^2.$$

But $\mathrm{d}s/\mathrm{d}\phi = a(1+\phi^2)^{\frac{1}{2}}$, so that (5.51) gives

$$\kappa = \left|\frac{\mathrm{d}\mathbf{t}}{\mathrm{d}s}\right| = \frac{|\mathrm{d}\mathbf{t}/\mathrm{d}\phi|}{\mathrm{d}s/\mathrm{d}\phi} = a^{-1}(1+\phi^2)^{-\frac{3}{2}}(2+\phi^2).$$

To find the intrinsic equation, we do not use (5.54), but note that in Fig. 5.7, $\psi = \phi + \alpha$. Now if $\mathbf{u}_r$ is the unit vector in the direction of $\mathbf{r}$,

$$\cos\alpha = \mathbf{t}\cdot\mathbf{u}_r.$$

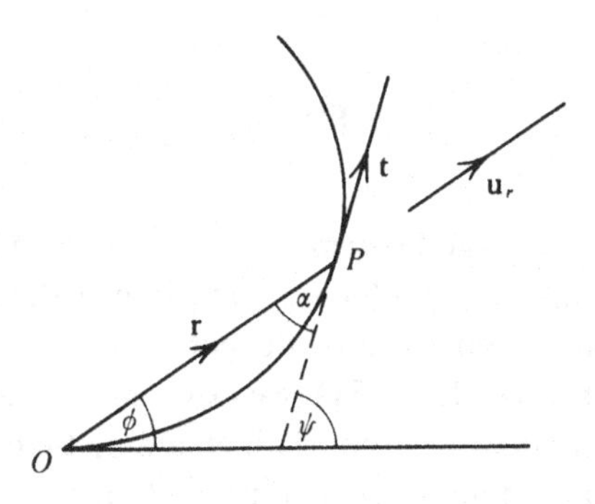


Fig. 5.7

The results of Example 5.12 give

$$\mathbf{t} = (1+\phi^2)^{-\frac{1}{2}}[(\cos\phi - \phi\sin\phi)\mathbf{i} + (\sin\phi + \phi\cos\phi)\mathbf{j}]$$

and

$$\mathbf{u}_r = \cos\phi\ \mathbf{i} + \sin\phi\ \mathbf{j}.$$

So

$$\cos\alpha = (1+\phi^2)^{-\frac{1}{2}},$$

giving

$$\begin{aligned}\psi &= \phi + \cos^{-1}(1+\phi^2)^{-\frac{1}{2}} \\ &= \phi + \tan^{-1}\phi.\end{aligned}$$

Since $s$ has already been found as a function of $\phi$, we have an implicit relation between $s$ and $\psi$, the intrinsic coordinates.

For any point $P$ on a curve $\Gamma$ where $\mathbf{r}''$ exists and is non-zero, unit tangent and normal vectors $\mathbf{t}$ and $\mathbf{n}$ have been defined as two mutually perpendicular unit vectors. It is natural to define a third unit vector $\mathbf{b}$, known as the **unit binormal vector**, to complete a right-handed triad $(\mathbf{t}, \mathbf{n}, \mathbf{b})$, as shown in Fig. 5.8. The line through a point $P$ on $\Gamma$ in the

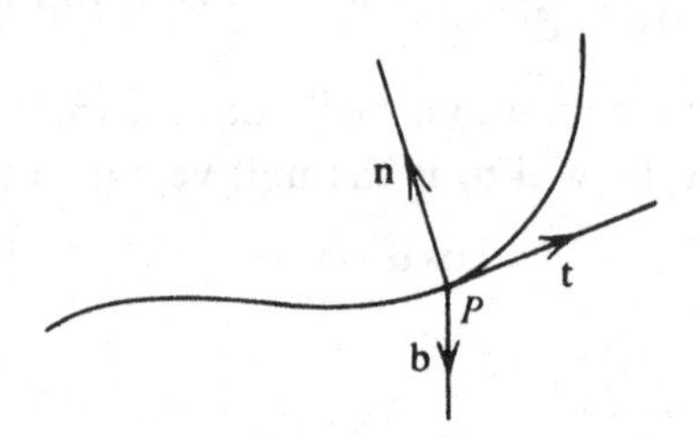


Fig. 5.8

direction of $\mathbf{b}$ is called the **binormal**. The plane through $P$, parallel to $\mathbf{b}$ and $\mathbf{n}$, is called the **normal plane** to the curve at $P$; $\mathbf{t}$ is normal to this plane. As the point $P$ moves along the curve, the rate of change of the tangent vector $\mathbf{t}$ is given by (5.51); we now find the rates of change of $\mathbf{n}$ and $\mathbf{b}$, the other vectors in the triad, assuming that their derivatives exist. Differentiating the orthogonality relation

$$\mathbf{t} \cdot \mathbf{b} = 0$$

gives, using (5.51),

$$\kappa \mathbf{n} \cdot \mathbf{b} + \mathbf{t} \cdot \mathbf{b}' = 0.$$

But $\mathbf{n} \cdot \mathbf{b} = 0$, by definition, so that $\mathbf{b}' \cdot \mathbf{t} = 0$. But differentiating $\mathbf{b} \cdot \mathbf{b} = 1$ gives $\mathbf{b}' \cdot \mathbf{b} = 0$ so that $\mathbf{b}'$ is orthogonal to both $\mathbf{t}$ and $\mathbf{b}$; hence it must be of the form

$$\mathbf{b}' = -\tau \mathbf{n}; \tag{5.55}$$

here $\tau$ is a scalar, which measures the rate at which $\mathbf{b}$, and hence $\mathbf{n}$, rotate in the normal plane; it therefore measures the 'rate of twisting' or the **torsion** of the curve at any point. The sign in (5.55) is chosen so that positive values of $\tau$ correspond to a rotation which gives a right-handed screw motion when combined with motion in direction $\mathbf{t}$.

The rates of rotation of the triad $(\mathbf{t}, \mathbf{n}, \mathbf{b})$ must be of the form (5.36), with $(\mathbf{i}, \mathbf{j}, \mathbf{k})$ replaced by $(\mathbf{t}, \mathbf{n}, \mathbf{b})$ and $u$ replaced by $s$. Equations (5.51) and (5.55) can then be identified with the first and third equations of (5.36), giving

$$\omega_3 = \kappa, \qquad \omega_2 = 0, \qquad \omega_1 = \tau,$$

which can be substituted in the second equation. The complete set of equations, known as the **Serret–Frenet formulae**, is therefore

$$\mathbf{t}' = \kappa \mathbf{n}, \tag{5.56a}$$

$$\mathbf{n}' = \tau \mathbf{b} - \kappa \mathbf{t}, \tag{5.56b}$$

$$\mathbf{b}' = -\tau \mathbf{n}. \tag{5.56c}$$

The identification of $\omega_r$ $(r = 1, 2, 3)$ with $(\tau, 0, \kappa)$ tells us that $\kappa$ and $\tau$ are, respectively, the rates of rotation per unit distance of the triad $(\mathbf{t}, \mathbf{n}, \mathbf{b})$ about $\mathbf{b}$ and $\mathbf{t}$ respectively, and that there is no rotation about $\mathbf{n}$. Since we have assumed that $\mathbf{t}'$ and $\mathbf{b}'$ exist, $\mathbf{n}'$ also exists, because $\mathbf{n} = \mathbf{b} \wedge \mathbf{t}$ and so $\mathbf{n}' = \mathbf{b}' \wedge \mathbf{t} + \mathbf{b} \wedge \mathbf{t}'$.

We have already seen that the curvature $\kappa$ is given by (5.51) to be

$$\kappa = |\mathbf{t}'| = |\mathbf{r}''|. \tag{5.57}$$

Differentiating (5.51) with respect to $s$ and using (5.56b) gives

$$\mathbf{r}''' = \kappa' \mathbf{n} + \kappa(\tau \mathbf{b} - \kappa \mathbf{t});$$

so, using (5.51) again,

$$\begin{aligned}[\mathbf{r}', \mathbf{r}'', \mathbf{r}'''] &= [\mathbf{t}, \kappa \mathbf{n}, \kappa' \mathbf{n} + \kappa(\tau \mathbf{b} - \kappa \mathbf{t})] \\ &= \kappa^2 \tau [\mathbf{t}, \mathbf{n}, \mathbf{b}] \\ &= \kappa^2 \tau.\end{aligned}$$

So the torsion can be calculated from the formula

$$\tau = \kappa^{-2} [\mathbf{r}', \mathbf{r}'', \mathbf{r}'''], \tag{5.58}$$

with $\kappa$ given by (5.57).

The curvature and torsion can also be found in terms of the derivatives $\dot{\mathbf{r}} = \partial \mathbf{r}/\partial u$, $\ddot{\mathbf{r}} = \partial^2 \mathbf{r}/\partial u^2$ and $\dddot{\mathbf{r}} = \partial^3 \mathbf{r}/\partial u^3$, with respect to a general parameter $u$. From (5.47) and (5.48),

$$\dot{\mathbf{r}} = \dot{s} \mathbf{t}.$$

Differentiating with respect to $u$ gives, using (5.51),

$$\begin{aligned}\ddot{\mathbf{r}} &= \ddot{s} \mathbf{t} + \dot{s} \dot{\mathbf{t}} \\ &= \ddot{s} \mathbf{t} + \dot{s}^2 \kappa \mathbf{n}.\end{aligned}$$

Forming the vector product with $\dot{\mathbf{r}} = \dot{s} \mathbf{t}$ now gives

$$\dot{\mathbf{r}} \wedge \ddot{\mathbf{r}} = \dot{s}^3 \kappa \mathbf{b}, \tag{5.59}$$

so that, taking the modulus,

$$\kappa = \dot{s}^{-3}|\dot{\mathbf{r}} \wedge \ddot{\mathbf{r}}| = |\dot{\mathbf{r}}|^{-3}|\dot{\mathbf{r}} \wedge \ddot{\mathbf{r}}|. \tag{5.60}$$

Also, differentiating (5.59) again, using (5.20) and (5.55),

$$\dot{\mathbf{r}} \wedge \dddot{\mathbf{r}} = -\dot{s}^4\kappa\tau\mathbf{n} + \mathbf{b}\mathrm{d}(\dot{s}^3\kappa)/\mathrm{d}u;$$

taking the scalar product with $\ddot{\mathbf{r}}$ now gives

$$[\dot{\mathbf{r}}, \ddot{\mathbf{r}}, \dddot{\mathbf{r}}] = \dot{s}^6\kappa^2\tau = \tau|\dot{\mathbf{r}} \wedge \ddot{\mathbf{r}}|^2,$$

using (5.60). So the torsion is given by

$$\tau = \frac{[\dot{\mathbf{r}}, \ddot{\mathbf{r}}, \dddot{\mathbf{r}}]}{|\dot{\mathbf{r}} \wedge \ddot{\mathbf{r}}|^2}. \tag{5.61}$$

**Example 5.15**

Find the curvature of the curve $y = f(x)$ in the plane $z = 0$. The position vector is, with $x$ as parameter,

$$\mathbf{r} = x\mathbf{i} + f(x)\mathbf{j}.$$

So

$$\frac{\mathrm{d}\mathbf{r}}{\mathrm{d}x} = \mathbf{i} + \dot{f}(x)\mathbf{j},$$

where $\dot{f}(x) = \mathrm{d}f/\mathrm{d}x$. Using (5.43) with $u = x$,

$$\frac{\mathrm{d}s}{\mathrm{d}x} = (1 + \dot{f}^2)^{\frac{1}{2}},$$

so that

$$\mathbf{t} = \frac{\mathrm{d}\mathbf{r}}{\mathrm{d}s} = (1 + \dot{f}^2)^{-\frac{1}{2}}[\mathbf{i} + \dot{f}(x)\mathbf{j}].$$

Differentiating again,

$$\kappa\mathbf{n} = \frac{\mathrm{d}\mathbf{t}}{\mathrm{d}s} = \frac{\mathrm{d}\mathbf{t}/\mathrm{d}x}{\mathrm{d}s/\mathrm{d}x}$$

$$= (1 + \dot{f}^2)^{-\frac{1}{2}}\{(1 + \dot{f}^2)^{-\frac{1}{2}}\ddot{f}\mathbf{j} - \dot{f}\ddot{f}(1 + \dot{f}^2)^{-\frac{3}{2}}[\mathbf{i} + \dot{f}\mathbf{j}]\}.$$

We could evaluate $\kappa$ by taking the modulus of this vector, but it is simpler to use (3.5) with $\mathbf{a} = \mathbf{i} + \dot{f}\mathbf{j}$ to give

$$\kappa\mathbf{t} \wedge \mathbf{n} = (1 + \dot{f}^2)^{-1}(\mathbf{i} + \dot{f}\mathbf{j}) \wedge (1 + \dot{f}^2)^{-\frac{1}{2}}\ddot{f}\mathbf{j}$$

$$= (1 + \dot{f}^2)^{-\frac{3}{2}}\ddot{f}\mathbf{k}.$$

The binormal $\mathbf{b} = \mathbf{t} \wedge \mathbf{n}$ is thus $\pm\mathbf{k}$, orthogonal to the plane of the curve, and the curvature, which is positive, is

$$\kappa = |\ddot{f}|(1 + \dot{f}^2)^{-\frac{3}{2}}.$$

An alternative derivation of this formula is suggested in Problems 5.3, Question 7.

**Example 5.16**

Find the curvature and torsion of the circular helix of Examples 5.2 and 5.13.

We already have

$$\mathbf{r}' = -\frac{a}{c}\sin\theta\,\mathbf{i} + \frac{a}{c}\cos\theta\,\mathbf{j} + \frac{b}{c}\mathbf{k}$$

and

$$\frac{\mathrm{d}s}{\mathrm{d}\theta} = c \equiv (a^2+b^2)^{\frac{1}{2}}.$$

So

$$\mathbf{r}'' = \frac{\mathrm{d}\mathbf{r}'/\mathrm{d}\theta}{\mathrm{d}s/\mathrm{d}\theta}$$

$$= -\frac{a}{c^2}\cos\theta\,\mathbf{i} - \frac{a}{c^2}\sin\theta\,\mathbf{j};$$

from (5.51) and (5.57),

$$\mathbf{n} = -\cos\theta\,\mathbf{i} - \sin\theta\,\mathbf{j}$$

and

$$\kappa = \frac{a}{c^2},$$

a positive constant.

Differentiating $\mathbf{r}''$,

$$\mathbf{r}''' = \frac{\mathrm{d}\mathbf{r}''/\mathrm{d}\theta}{\mathrm{d}s/\mathrm{d}\theta} = \frac{a}{c^3}\sin\theta\,\mathbf{i} - \frac{a}{c^3}\cos\theta\,\mathbf{j}.$$

Using (5.58),

$$\tau = \kappa^{-2}\left[\frac{b}{c}\mathbf{k}, -\frac{a}{c^2}(\cos\theta\,\mathbf{i} + \sin\theta\,\mathbf{j}), \frac{a}{c^3}(\sin\theta\,\mathbf{i} - \cos\theta\,\mathbf{j})\right]$$

$$= \left(\frac{c^4}{a^2}\right)\left(\frac{b}{c}\right)\left(\frac{a^2}{c^5}\right)\{[\mathbf{k},\mathbf{i},\mathbf{j}]\cos^2\theta - [\mathbf{k},\mathbf{j},\mathbf{i}]\sin^2\theta\}$$

$$= \frac{b}{c^2},$$

which is also constant along the curve.

Now consider a particular point $\mathbf{r}_1$ on a curve, and suppose that we know the tangent $\mathbf{t}_1$, the normal $\mathbf{n}_1$ and the curvature $\kappa_1$ (assumed non-zero) at the given point. We shall now show that there is a unique circle passing through $\mathbf{r}_1$ with these given values of $\mathbf{t}_1$, $\mathbf{n}_1$, $\kappa_1$; this circle is known as the **circle of curvature** at $\mathbf{r}_1$. Suppose that $\mathbf{a}$ is the position vector of the centre of such a circle; then the circle must be the intersection of a sphere, centre $\mathbf{a}$, and a plane through $\mathbf{r}_1$. So we can write the equations of the circle as

$$(\mathbf{r}-\mathbf{a})\cdot(\mathbf{r}-\mathbf{a})=\lambda^2, \qquad (\mathbf{r}-\mathbf{r}_1)\cdot\mathbf{u}=0 \tag{5.62}$$

where $\mathbf{u}$ is a unit vector, and $\lambda$ is a real number. We have to show that $\mathbf{a}$, $\lambda\,(>0)$ and (apart from its sign) $\mathbf{u}$ are determined by $\mathbf{t}_1$, $\mathbf{n}_1$ and $\kappa_1$.

In (5.62), $\mathbf{r}$ is the position vector of a point on the circle. We can differentiate (5.62) twice with respect to the arc length $s$ on the circle, and use (5.48) and (5.51) to obtain

$$\mathbf{t}\cdot(\mathbf{r}-\mathbf{a})=0, \qquad \mathbf{t}\cdot\mathbf{u}=0$$

and

$$\kappa\mathbf{n}\cdot(\mathbf{r}-\mathbf{a})=-\mathbf{t}\cdot\mathbf{t}=-1, \qquad \kappa\mathbf{n}\cdot\mathbf{u}=0;$$

here, $\mathbf{t}$ and $\mathbf{n}$ are the unit tangent and normal vectors to the circle. We are told that $\mathbf{r}=\mathbf{r}_1$ lies on the circle (5.62), and that $\mathbf{t}=\mathbf{t}_1$, $\mathbf{n}=\mathbf{n}_1$ and $\kappa=\kappa_1\neq 0$ at this point. So substituting in all six equations gives

$$|\mathbf{r}_1-\mathbf{a}|^2=\lambda^2, \tag{5.63a}$$

$$\mathbf{t}_1\cdot(\mathbf{r}_1-\mathbf{a})=0, \tag{5.63b}$$

$$\mathbf{n}_1\cdot(\mathbf{r}_1-\mathbf{a})=-\kappa_1^{-1}, \tag{5.63c}$$

and

$$\mathbf{t}_1\cdot\mathbf{u}=\mathbf{n}_1\cdot\mathbf{u}=0. \tag{5.63d}$$

Equations (5.63d) tell us that, apart from an unimportant choice of sign, $\mathbf{u}=\mathbf{b}_1$, the binormal at $\mathbf{r}_1$. The circle (5.62) therefore lies in the plane through $\mathbf{r}_1$ parallel to $\mathbf{t}_1$ and $\mathbf{n}_1$, known as the **osculating plane** at $\mathbf{r}_1$. The displacement from the centre $\mathbf{a}$ to $\mathbf{r}_1$ is therefore normal to $\mathbf{b}_1$, or

$$(\mathbf{r}_1-\mathbf{a})\cdot\mathbf{b}_1=0.$$

Since by (5.63b), $(\mathbf{r}_1-\mathbf{a})$ is also normal to $\mathbf{t}_1$, and by (5.63a), has magnitude $\lambda$, we know that

$$\mathbf{r}_1-\mathbf{a}=\pm\lambda\,\mathbf{n}_1.$$

Substituting in (5.63c) and remembering that $\lambda$ and $\kappa_1$ are positive, we find

$$\lambda = \kappa_1^{-1};$$

therefore

$$\mathbf{a} = \mathbf{r}_1 + \kappa_1^{-1}\mathbf{n},$$

completing the determination of **a**, **u**, $\lambda$ in (5.62). The radius $\lambda$ of the circle of curvature is therefore the radius of curvature $r_c$ defined by (5.53), while the centre of the circle is on the normal to the curve at $\mathbf{r}_1$, at the correct radial distance $r_c = \kappa_1^{-1}$ from $\mathbf{r}_1$.

Since **t**, **n** and $\kappa$ determine $\mathbf{r}'$ and $\mathbf{r}''$ through (5.48) and (5.51), the circle of curvature at a point on a curve has the same values of **r**, $\mathbf{r}'$ and $\mathbf{r}''$ as the curve itself, and is said to have **triple contact** with the curve at the point.

## *Problems 5.3*

1 A **cycloid** is the locus in the $(x, y)$ plane of a point on the circumference of a circle, radius $a$, rolling on the line $y = 0$. If the origin $O$ is on the cycloid, and $\theta$ is the angle through which the circle has rotated in a clockwise direction from the position corresponding to $O$, show that the equation of the cycloid with $y \geqslant 0$ is

$$\mathbf{r} = a(\theta - \sin\theta)\mathbf{i} + a(1 - \cos\theta)\mathbf{j}.$$

Find the unit tangent and normal vectors **t** and **n** to the cycloid for any value of $\theta$ in the range $0 < \theta < 2\pi$, and show that the curvature is $\kappa = (4a)^{-1} \operatorname{cosec} \frac{1}{2}\theta$. Show also that the distance along the curve from $O$ to the point with $\theta = \theta_1$ is $4a(1 - \cos\frac{1}{2}\theta_1)$. Integrate the function $y = a(1 - \cos\theta)$ along the curve from $\theta = 0$ to $\theta = \theta_1$.

2 A **catenary** in the $(x, y)$ plane is defined by the equation $y = c\cosh(x/c)$. Show that the intrinsic equation of the curve can be written $s = c\tan\psi$. Show also that the radius of curvature at any point $P$ is equal to the distance along the normal from $P$ to the $x$-axis. Integrate the function $y^2$ along the curve from $x = 0$ to $x = x_1$.

3 Find the curvature, the torsion, and the equation of the osculating plane at any point of the curve

$$\mathbf{r} = a\theta\cos\theta\,\mathbf{i} + a\theta\sin\theta\,\mathbf{j} + b\theta\mathbf{k}.$$

4 Find the equation of the normal plane, the curvature $\kappa$ and the torsion $\tau$ of the **twisted cubic** curve

$$\mathbf{r}=6u\mathbf{i}+3u^2\mathbf{j}+u^3\mathbf{k}.$$

Integrate the function $e^u$ along the curve from $u=0$ to $u=u_1$.

5 Show that both the curvature and the torsion of the curve

$$\mathbf{r}=3a\ \text{sech}\ \theta\mathbf{i}+3a\tanh\theta\mathbf{j}+8a\tan^{-1}(\exp\theta)\mathbf{k}$$

are constant. Find the equation of the circle of curvature at the point $\theta=\theta_1$.

6 Show that the curvature and torsion of the cubic curve

$$\mathbf{r}=3u\mathbf{i}+3u^2\mathbf{j}+2u^3\mathbf{k}$$

are equal at every point. Find the equations of the tangent, normal and binormal at the point $u=u_1$.

7 Use Equation (5.60), with $x$ as parameter, to establish the formula for the curvature of the curve $y=f(x)$, where $f(x)$ is twice differentiable.

8 If the curvature $\kappa(s)$ and the torsion $\tau(s)$ are infinitely differentiable functions of $s$, show that $\mathbf{t}$, $\mathbf{n}$ and $\mathbf{b}$ are infinitely differentiable.

## *5.4 Surface integrals*

In §5.3 we have defined the integral of a function along a line, expressing it as the integrals (5.49) over the single parameter $u$ or $s$. In a similar way, when a surface is defined, as in §5.1, in terms of *two* parameters $u_1$, $u_2$, an integral over the surface can be defined as a double integral over the variables $u_1$, $u_2$. Before defining these double integrals, we discuss in more detail how points on a surface are specified in terms of $u_1$, $u_2$.

A surface $\sigma$, defined in terms of parameters $u_1$ and $u_2$, is shown in Fig. 5.9. We have already assumed that particular values of $u_1$, $u_2$ define a unique position vector $\mathbf{r}(u_1, u_2)$, corresponding to a unique point $P(u_1, u_2)$, say. If $u_1$ takes a constant value and $u_2$ is allowed to vary, then the position vector $\mathbf{r}(u_1=\text{constant}, u_2)$ traces out the curve $AB$ on the surface; the point $Q$ on $AB$ corresponds to the second parameter value $u_2+\delta u_2$. The various values of the *fixed* parameter $u_1$ each define a curve on $\sigma$; members of this family, including $AB$ and $CD$, are drawn in Fig. 5.9; $CD$ corresponds to the fixed parameter value $u_1+\delta u_1$. In the same way, the position vector

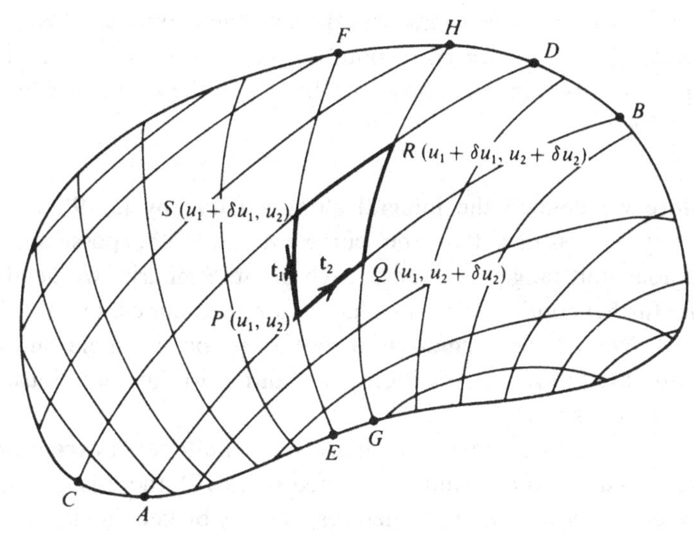


Fig. 5.9

$\mathbf{r}(u_1, u_2 = \text{constant})$, with $u_2$ fixed, traces out the curve $EF$ as $u_1$ varies, and the point $S$ on $EF$ corresponds to the value $u_1+\delta u_1$ of the first parameter. The curve $EF$ is a member of a second family of curves each defined by keeping $u_2$ fixed; $GH$ is another member of this family, corresponding to the fixed value $u_2+\delta u_2$ of the second parameter. The points of intersection of $AB$ and $CD$ with $EF$ and $GH$ are labelled $P$, $Q$, $S$, $R$, together with the corresponding values of the two parameters. The two families of curves are said to form a **mesh** over the surface $\sigma$. Apart from certain exceptional points and lines, each position vector $\mathbf{r}$ on $\sigma$ corresponds to unique values of $u_1$, $u_2$; so in general, there is only one curve of each system passing through a given point.

**Example 5.17**

As in Example 5.5, the spherical polar coordinates $\theta,\phi$ can be used to represent points on the unit sphere. In general, each point corresponds to one and only one set of values in the ranges $0 \leqslant \theta \leqslant \pi$, $0 \leqslant \phi \leqslant 2\pi$. The curves $\theta = \text{constant}$ are the 'circles of latitude', and the curves $\phi = \text{constant}$ are the 'great circles of longitude'; these two sets of curves form a mesh over the sphere. The 'poles', however, correspond to the values $\theta = 0$ and $\theta = \pi$, for any value of $\phi$; so *all* great circles with $\phi =$ constant pass through the two poles. Also, the curves $\phi=0$ and $\phi=2\pi$ correspond to the same set of points; this ambiguity can be removed, however, by taking the range of $\phi$ to be $0 \leqslant \phi < 2\pi$.

Before we defined the integral along a curve by (5.49), it was necessary to assume that the curve was smooth, possessing a continuous unit tangent vector at each point. Similarly, we need to impose further conditions on a surface before we define integrals over surfaces. We define a **smooth surface** $\sigma$ to be a simple surface satisfying the following conditions, in addition to the conditions (i), (ii) and (iii) of §5.1:

(iv) Through every point of $\sigma$, apart from a finite set of **exceptional points** belonging to the limit set defined by (5.10), there is a unique pair of curves of the mesh, defined respectively by keeping $u_1$ and $u_2$ constant. Points which are not exceptional are called **normal points** of $\sigma$.

(v) At all normal points of $\sigma$, the partial derivatives

$$\frac{\partial \mathbf{r}(u_1, u_2)}{\partial u_1}, \quad \frac{\partial \mathbf{r}(u_1, u_2)}{\partial u_2} \tag{5.64}$$

exist, and are non-zero and continuous; at points on the boundary, these derivatives may be one-sided. So at all normal points of $\sigma$, the vectors (5.64) can be normalised to define unit tangent vectors $\mathbf{t}_1$ and $\mathbf{t}_2$ to the two curves of the mesh, in the directions of $u_1$ and $u_2$ increasing, respectively; the tangent vectors at $P$ are marked in Fig. 5.9. The vectors $\mathbf{t}_1$ and $\mathbf{t}_2$ are continuous functions of $u_1$, $u_2$.

(vi) At all normal points of $\sigma$, we assume that $\mathbf{t}_1$ and $\mathbf{t}_2$ are linearly independent, so that the two curves of the mesh at a point are not tangent to each other.

Since exceptional points correspond to limiting values (5.10) of the variables $u_1$, $u_2$, we are assured that $\mathbf{t}_1$ and $\mathbf{t}_2$ exist and are continuous at interior points of the ranges, for which $a < u_1 < b$ and $\alpha(u_1) < u_2 < \beta(u_1)$. We also know that the tangent to the boundary curve can be defined at all but a finite number of exceptional points; for example a part of the boundary defined by $u_2 = \alpha(u_1)$ will have

$$\frac{\partial \mathbf{r}}{\partial u_1} + \frac{\mathrm{d}\alpha}{\mathrm{d}u_1}\frac{\partial \mathbf{r}}{\partial u_2}$$

as a tangent vector. So the boundary of a smooth surface $\sigma$ consists of a finite number of smooth curves.

**Example 5.18**

In terms of spherical polar coordinates, a hemisphere of the unit sphere $r = 1$ is defined by the ranges $0 \leqslant \theta \leqslant \frac{1}{2}\pi$, $0 \leqslant \phi \leqslant 2\pi$. The boundary is the 'equator' $\theta = \frac{1}{2}\pi$; the 'longitudes' $\phi = 0$ and $\phi = 2\pi$ coincide, but if $\theta > 0$, there are unique 'circles of latitude' $\theta =$ constant and 'longitudes' $\phi =$ constant through any point $\theta, \phi$; this is true even on the boundary $\theta = \frac{1}{2}\pi$, where one-sided non-zero derivatives (5.64) exist, and on the duplicated longitude $\phi = 0$ or $\phi = 2\pi$. The only point at which conditions (i)–(iii) above fail is the 'north pole' $N$ shown in Fig. 5.2; all the 'longitudes' $\phi =$ constant pass through this point; $\theta = 0$ is not a curve, and while each 'longitude' has a unique unit tangent vector at $N$, $\theta = 0$ does not define a unique tangential direction at $N$.

The limit set, defined by (5.10), consists of points with $\theta = 0$, $\theta = \frac{1}{2}\pi$, $\phi = 0$ or $\phi = 2\pi$. The only 'exceptional point' of the limit set is $N$, defined by $\theta = 0$. We note that $N$ is an exceptional point of the coordinate system, rather than of the sphere itself; we could, if we wished, use a different coordinate system, for which $N$ was not in the limit set, and hence not an exceptional point.

**Example 5.19**

As in Example 5.8, a part of the right circular cone $\theta = \alpha$ $(0 < \alpha < \frac{1}{2}\pi)$ is defined by the ranges $0 \leqslant r \leqslant r_1$, $0 \leqslant \phi < 2\pi$. Through each point with $r > 0$, there is one circle $r =$ constant and one line $\phi =$ constant, and the tangent vectors $\mathbf{t}_1$ and $\mathbf{t}_2$ are uniquely defined; these tangent vectors are continuous functions of $r, \phi$. The origin, the vertex of the cone, corresponds to $r = 0$, for all values of $\phi$. The equation $r = 0$ does not define a curve, and it defines no tangent vector. The origin is therefore an exceptional point. The situation is different from that in Example 5.18, however, since the vertex of the cone will be an exceptional point regardless of the choice of coordinates; in other words, it is an exceptional point of the surface, rather than of the coordinate system.

Conditions (iv)–(vi) ensure that linearly independent tangent vectors $\mathbf{t}_1$ and $\mathbf{t}_2$ exist at a point $P$. We can then define the **tangent**

**plane** at $P$, as in (3.39), to be the plane through $P$ parallel to $\mathbf{t}_1$ and $\mathbf{t}_2$. Fig. 5.10 shows the surface region $PQRS$ of Fig. 5.9, enlarged, defined by increments $\delta u_1$ and $\delta u_2$ of the parameters $u_1$, $u_2$. Corresponding to these increments, we can define displacements

$$\mathbf{PN} = \frac{\partial \mathbf{r}}{\partial u_1}\,\delta u_1, \qquad \mathbf{PL} = \frac{\partial \mathbf{r}}{\partial u_2}\,\delta u_2 \tag{5.65}$$

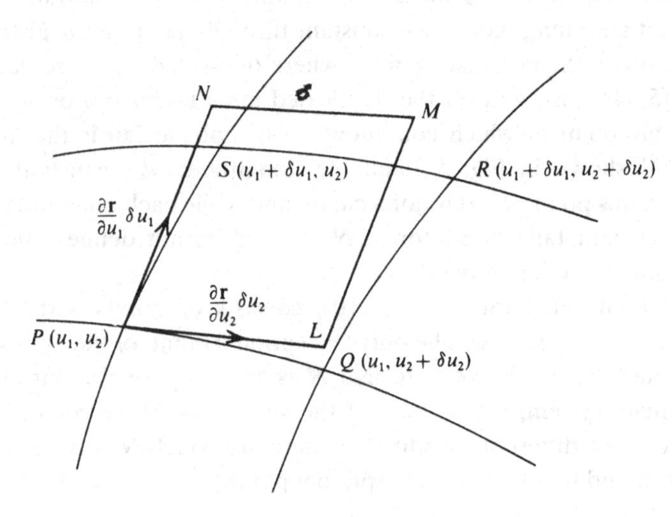


Fig. 5.10

parallel to the tangent vectors $\mathbf{t}_1$ and $\mathbf{t}_2$. If we assume that the increments $\delta u_1$ and $\delta u_2$ are positive, the area of the parallelogram $PLMN$ formed from these displacements from $P$ is

$$\delta\sigma_P = \left|\frac{\partial \mathbf{r}}{\partial u_1} \wedge \frac{\partial \mathbf{r}}{\partial u_2}\right| \delta u_1\, \delta u_2. \tag{5.66a}$$

We often use **orthogonal systems of coordinates** on $\sigma$, for which the vectors (5.65) are orthogonal at all normal points of $\sigma$; then the mesh consists of two sets of curves which are orthogonal at every normal point. Defining the moduli

$$h_r(u_1, u_2) = \left|\frac{\partial \mathbf{r}}{\partial u_r}\right| \tag{5.67}$$

for $r=1, 2$, the area (5.66a) can then be written

$$\delta\sigma_P = h_1 h_2 \delta u_1 \delta u_2 \tag{5.66b}$$

for orthogonal systems of coordinates.

When the surface $\sigma$ is a region of the plane $z=0$, so that $u_1$, $u_2$ are coordinates in the $(x, y)$ plane, then (5.1) reduces to

$$\mathbf{r} = X(u_r)\mathbf{i} + Y(u_r)\mathbf{j},$$

so that, for $r=1, 2$,

$$\frac{\partial \mathbf{r}}{\partial u_r} = \frac{\partial X}{\partial u_r}\mathbf{i} + \frac{\partial Y}{\partial u_r}\mathbf{j}.$$

Then the vector product

$$\frac{\partial \mathbf{r}}{\partial u_1} \wedge \frac{\partial \mathbf{r}}{\partial u_2}$$

is parallel to **k**, and has magnitude

$$\left| \frac{\partial X}{\partial u_1}\frac{\partial Y}{\partial u_2} - \frac{\partial X}{\partial u_2}\frac{\partial Y}{\partial u_1} \right|.$$

This is the modulus of the determinant

$$\frac{\partial(X, Y)}{\partial(u_1, u_2)} = \begin{vmatrix} \dfrac{\partial X}{\partial u_1} & \dfrac{\partial Y}{\partial u_1} \\ \dfrac{\partial X}{\partial u_2} & \dfrac{\partial Y}{\partial u_2} \end{vmatrix}, \tag{5.68}$$

which is known as the **Jacobian** of the transformation from variables $(x, y)$ to the variables $(u_1, u_2)$.

In Appendix A, we define double integrals and establish certain basic properties in Theorems A.5–A.8. Using these results, we are now in a position to define the integral over a surface $\sigma$ of any piecewise continuous function $f(u_1, u_2)$ defined at all points on $\sigma$. The parameters $u_1$, $u_2$ define a mesh over $\sigma$; by selecting a finite number of curves of each system of curves, forming a **coarse mesh,** $\sigma$ is divided into a finite number of small regions, as in Fig. 5.9. Let us represent by $P$ the point in one of these regions (including its boundary) with the minimum values of $u_1$ and $u_2$; regions which contain part of the boundary of $\sigma$ can be excluded, by Theorem A.4. Then if $\delta\sigma_P$ denotes the area (5.66) of the parallelogram formed by tangents at $P$, we construct the sum

$$\sum_P f_P \delta\sigma_P \tag{5.69}$$

over points $P$ representing all the regions not containing boundary points of $\sigma$. Now the surface $\sigma$ corresponds to ranges of $u_1$, $u_2$ of the form (5.8); a coarse mesh corresponds to division of the range $(a, b)$ of $u_1$ into intervals of finite lengths denoted by $\delta u_1$, and (for each value of $u_1$) division of the range $(\alpha(u_1), \beta(u_1))$ of $u_2$ into intervals with lengths denoted by $\delta u_2$. If $f(u_1, u_2)$ is a piecewise uniformly continuous function, then (Theorem A.3, Appendix A) as the maximum value $\text{Max}(\delta u_1, \delta u_2)$ of all the parametric increments tends to zero, the sum (5.69) tends to the double integral

$$\iint_\sigma f(u_1, u_2)\left|\frac{\partial \mathbf{r}(u_1, u_2)}{\partial u_1} \wedge \frac{\partial \mathbf{r}(u_1, u_2)}{\partial u_2}\right| du_1\, du_2 \qquad (5.70a)$$

over the surface $\sigma$. In forming this limit, we have used the expression (5.66a) for the element of area associated with $P$. If $u_1$, $u_2$ form a system of orthogonal coordinates on $\sigma$, the limit derived by using (5.66b) is

$$\iint_\sigma f(u_1, u_2) h_1(u_1, u_2) h_2(u_1, u_2)\, du_1\, du_2. \qquad (5.70b)$$

If the surface $\sigma$ is a region of the $(x, y)$ plane, we can use (5.68) to express the integral (5.70a) in the form

$$\iint_\sigma f(u_1, u_2)\left|\frac{\partial(X, Y)}{\partial(u_1, u_2)}\right| du_1\, du_2. \qquad (5.70c)$$

The double integral (5.70a) can be evaluated by performing the $u_2$ integration first, over the range (5.8b), and then performing the $u_1$ integration over the range (5.8a) [Theorem A.7, Appendix A]. The integral is then the *repeated* integral

$$\int_a^b du_1 \int_{\alpha(u_1)}^{\beta(u_1)} du_2\, f(u_1, u_2)\left|\frac{\partial \mathbf{r}(u_1, u_2)}{\partial u_1} \wedge \frac{\partial \mathbf{r}(u_1, u_2)}{\partial u_2}\right|. \qquad (5.71a)$$

For orthogonal systems of coordinates, and for integrals over plane areas, the integrand in (5.70a) simplifies to those in (5.69b) and (5.69c), giving

$$\int_a^b du_1 \int_{\alpha(u_1)}^{\beta(u_2)} du_2\, f(u_1, u_2) h_1(u_1, u_2) h_2(u_1, u_2) \qquad (5.71b)$$

and

$$\int_a^b du_1 \int_{\alpha(u_1)}^{\beta(u_1)} du_2 f(u_1, u_2)\left|\frac{\partial(X, Y)}{\partial(u_1, u_2)}\right|. \tag{5.71c}$$

The double integral (5.70a) is also equal to a repeated integral in which the integration over $u_1$ is carried out first. It may happen, however, that the ranges of integration are then not of the form (5.8). For example, if $\sigma$ is the shaded region in the $(x, y)$ plane shown in Fig. 5.11, and if $u_1 = x$, $u_2 = y$, then $\sigma$ corresponds to ranges $a \leqslant x \leqslant b$,

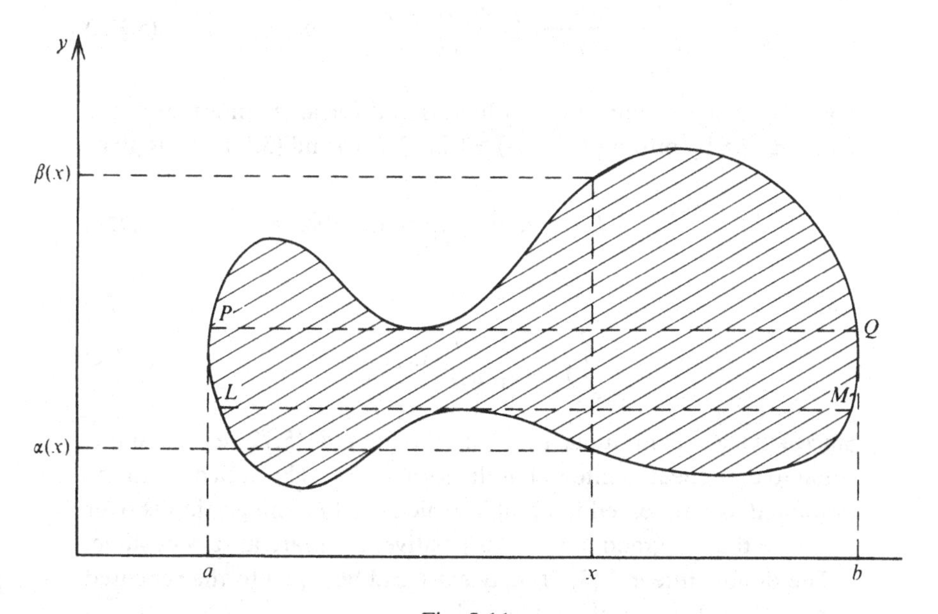


Fig. 5.11

$\alpha(x) \leqslant y \leqslant \beta(x)$. But if we integrate first over $x$, there are two ranges of $x$ corresponding to $y$-values below the line $LM$ and above the line $PQ$, so that the ranges of integration are more complicated than those in (5.8). The region $\sigma$ can, however, be cut into five finite pieces by the lines $LM$ and $PQ$, and each of these pieces corresponds to ranges of the form (5.8), with the $x$-integration performed first. The integral over $\sigma$ is the sum of the integrals over these five pieces [Theorem A.6, Appendix A]. There are, of course, other ways of dividing $\sigma$ into pieces, each with ranges of integration of the form (5.8).

Let us consider the surface integral (5.70a) of the unit function $f(u_1, u_2) \equiv 1$. From (5.69), this is the limit of the sum $\Sigma_P \, \delta\sigma_P$, where $\delta\sigma_P$ is the area of the parallelogram *PLMN* in Fig. 5.10. As the increments $\delta u_1$, $\delta u_2$ tend to zero, the region *PQRS* becomes progressively flatter, since $\mathbf{t}_1$ and $\mathbf{t}_2$ are continuous. It is natural to take the area of *PLMN* as a good approximation to the area of the surface element *PQRS* when $\delta u_1$, $\delta u_2$ are small. We therefore define the **area** of the surface $\sigma$ to be the limit of $\Sigma_P \, \delta\sigma_P$; putting $f(u_1, u_2) \equiv 1$ in (5.70a), the area is thus

$$\iint_\sigma \left| \frac{\partial \mathbf{r}(u_1, u_2)}{\partial u_1} \wedge \frac{\partial \mathbf{r}(u_1, u_2)}{\partial u_2} \right| du_1 \, du_2. \tag{5.72a}$$

For orthogonal systems of coordinates and for areas in a plane, the area is given by putting $f(u_1, u_2) \equiv 1$ in (5.70b) and (5.70c); this gives

$$\iint_\sigma h_1(u_1, u_2) h_2(u_1, u_2) \, du_1 \, du_2 \tag{5.72b}$$

and

$$\iint_\sigma \left| \frac{\partial(X, Y)}{\partial(u_1, u_2)} \right| du_1 \, du_2. \tag{5.72c}$$

Since $f \equiv 1$ is a continuous function, the integral (5.72) exists, and is equal to the repeated integral of the form (5.71). The area can also be evaluated as a repeated integral in which the first integration is over $u_1$. Since the integrand in (5.72) is positive, areas are always positive.

The double integral (5.70) may exist and be equal to the repeated integral (5.71) even if the integrand

$$f \left| \frac{\partial \mathbf{r}}{\partial u_1} \wedge \frac{\partial \mathbf{r}}{\partial u_2} \right|$$

is not bounded, but tends to infinity at some points or along some lines in $\sigma$. Unboundedness of the integrand may arise either from the unboundedness of the function $f(u_1, u_2)$, or because $\mathbf{r}(u_1, u_2)$ tends to infinity at some part of the boundary of $\sigma$. Even if the integral (5.72) is divergent, corresponding to a surface of infinite area, the integral (5.70) may exist for some class of functions $f(u_1, u_2)$, and be equal to the repeated integrals over $u_1$ and $u_2$. Conditions under which this is true need to be particularly carefully investigated, and are established in standard works on analysis [Reference 5.12]. In some classes of problems, in potential theory for example, great care needs to be taken in defining certain surface integrals.

**Example 5.20**

If $(\rho, \phi)$ are polar coordinates in the plane $z=0$, the region $\sigma$ in the plane is defined by ranges $1 \leqslant \rho \leqslant 2$, $0 \leqslant \phi \leqslant 2\pi$. If $f(x, y)$ is a bounded function of the rectangular coordinates $x$, $y$ in the region $\sigma$, write the integral of $f$ over $\sigma$ as

(i) a repeated integral over $\phi$ and then $\rho$,

(ii) a repeated integral over $x$ and then $y$.

(i) The region of integration is shown in Fig. 5.12. In terms of polar coordinates,

$$\mathbf{r}(\rho, \phi) = \rho \cos \phi \, \mathbf{i} + \rho \sin \phi \, \mathbf{j};$$

so taking $u_1 = \rho$, $u_2 = \phi$, (5.2) becomes

$$X(\rho, \phi) = \rho \cos \phi, \qquad Y(\rho, \phi) = \rho \sin \phi.$$

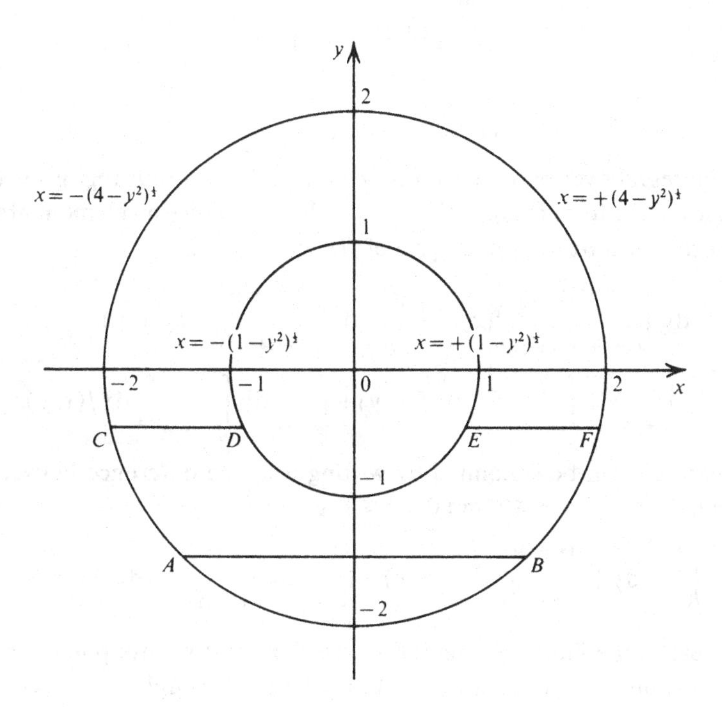


Fig. 5.12

Hence the Jacobian (5.68) is

$$\frac{\partial(X, Y)}{\partial(\rho, \phi)} = \begin{vmatrix} \cos\phi & \sin\phi \\ -\rho\sin\phi & \rho\cos\phi \end{vmatrix} = \rho.$$

In terms of $\rho$, $\phi$, the function to be integrated is $f(\rho\cos\phi, \rho\sin\phi)$. The integral (5.71c), with the correct limits inserted, is

$$\int_1^2 d\rho \int_0^{2\pi} d\phi\, \rho f(\rho\cos\phi, \rho\sin\phi).$$

Note that, since $\partial\mathbf{r}/\partial\rho$ and $\partial\mathbf{r}/\partial\phi$ are orthogonal, the factor $\rho$ could also have been obtained by using (5.71b).

(ii) Consider the range of values of $x$ for fixed values of $y$. For $-2 \leqslant y \leqslant -1$, there is a single range of values of $x$, represented by the line $AB$ in Fig. 5.12; for this range of $y$-values, the range of $x$ is

$$-(4-y^2)^{\frac{1}{2}} \leqslant x \leqslant +(4-y^2)^{\frac{1}{2}}.$$

This is also the range of $x$-values for $1 \leqslant y \leqslant 2$. For $-1 < y < 1$, the range of $x$-values is divided into two parts, represented by $CD$ and $EF$; the two ranges of $x$ are

$$-(4-y^2)^{\frac{1}{2}} \leqslant x \leqslant -(1-y^2)^{\frac{1}{2}}$$

and

$$(1-y^2)^{\frac{1}{2}} \leqslant x \leqslant (4-y^2)^{\frac{1}{2}}.$$

The integral over $\sigma$ is the sum of the integrals over four regions with ranges of the form (5.8), with $u_2 = x$ and $u_1 = y$. Since an element area is of the form $\delta x\, \delta y$, the integral of $f(x, y)$ over $\sigma$ is

$$\int_{y=-2}^{-1} dy \int_{x=-(4-y^2)^{\frac{1}{2}}}^{+(4-y^2)^{\frac{1}{2}}} dx\, f(x, y) + \int_{y=1}^{2} dy \int_{x=-(4-y^2)^{\frac{1}{2}}}^{+(4-y^2)^{\frac{1}{2}}} dx\, f(x, y)$$
$$+ \int_{y=-1}^{1} dy \int_{x=-(4-y^2)^{\frac{1}{2}}}^{-(1-y^2)^{\frac{1}{2}}} dx\, f(x, y) + \int_{y=-1}^{1} dy \int_{x=(1-y^2)^{\frac{1}{2}}}^{(4-y^2)^{\frac{1}{2}}} dx\, f(x, y).$$

This result can be simplified by writing it as the difference between integrals over $0 \leqslant \rho \leqslant 2$ and $0 \leqslant \rho \leqslant 1$, giving

$$\int_{y=-2}^{2} dy \int_{x=-(4-y^2)^{\frac{1}{2}}}^{+(4-y^2)^{\frac{1}{2}}} dx\, f(x, y) - \int_{y=-1}^{1} dy \int_{x=-(1-y^2)^{\frac{1}{2}}}^{(1-y^2)^{\frac{1}{2}}} dx\, f(x, y).$$

Clearly, the limits of integration are simpler if we use polar coordinates, but the integrand $f(x, y)$ may be more simply expressed in terms of rectangular coordinates. For such integrals, there is no definite rule for deciding whether it is best to use the coordinates with simple limits of integration, or the coordinates which keep the integrand simple.

**Example 5.21**

Using the standard notation for cylindrical and spherical polar coordinates, integrate $1/r$ over the region of the surface $\rho = a$ bounded by $z = 0$, $z = a\phi$ and $\phi = \pi$.

The boundary of the region is shown by thickened lines in Fig. 5.13, joining the points $A$, $B$ and $C$. If we integrate first over $z$, the range for a given value of $\phi$ is indicated by $DE$; the range of $z$ is then $0 \leqslant z \leqslant a\phi$. The position vector on $\rho = a$ is given by $\mathbf{r} = a\cos\phi\,\mathbf{i} + a\sin\phi\,\mathbf{j} + z\mathbf{k}$. It is easy to show that $\partial\mathbf{r}/\partial\phi$ and $\partial\mathbf{r}/\partial z$ have magnitudes $h_1 = a$, $h_2 = 1$, so that (5.66b) becomes $\delta\sigma_P = a\,\delta\phi\,\delta z$. As a function of $\rho$, $z$, the function to be integrated is $r^{-1} = (a^2 + z^2)^{-\frac{1}{2}}$. Thus the integral, in the form (5.71b), is

$$\int_{\phi=0}^{\pi} a\,\mathrm{d}\phi \int_{z=0}^{a\phi} \mathrm{d}z\,(a^2+z^2)^{-\frac{1}{2}} = a\int_0^{\pi} \mathrm{d}\phi\,\sinh^{-1}\phi$$

$$= a[\phi\sinh^{-1}\phi - (1+\phi^2)^{\frac{1}{2}}]_0^{\pi}$$

$$= a[\pi\sinh^{-1}\pi + 1 - (1+\pi^2)^{\frac{1}{2}}].$$

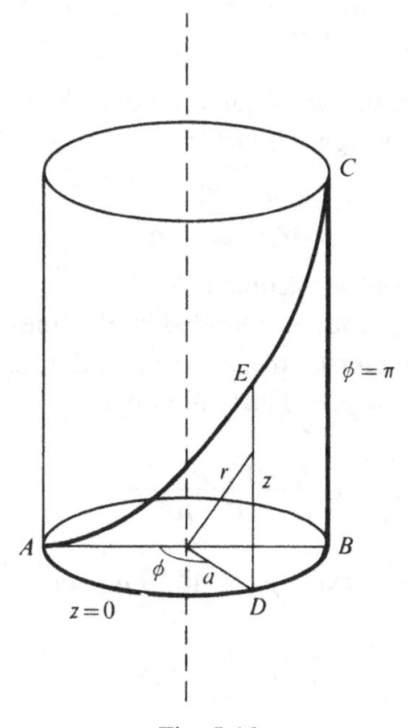


Fig. 5.13

**Example 5.22**

Integrate

$$\frac{\exp(-\rho^2/a^2)}{(\rho-a\phi)^2+\mu^2},$$

where $\rho$, $\phi$ are polar coordinates and $a$, $\mu$ are positive constants, over the region of the $(x, y)$ plane bounded by the curves $\rho=\frac{1}{4}a\pi$, $\rho=a\phi$ and $\rho=a(\phi+\frac{1}{2}\pi)$.

Use coordinates $\rho$ and $t\equiv\rho-a\phi$. The position vector in the plane

$$\begin{aligned}\mathbf{r}&=\rho\cos\phi\,\mathbf{i}+\rho\sin\phi\,\mathbf{j}\\&=\rho\cos\left(\frac{\rho-t}{a}\right)\mathbf{i}+\rho\sin\left(\frac{\rho-t}{a}\right)\mathbf{j}.\end{aligned}$$

So

$$\begin{aligned}\frac{\partial\mathbf{r}}{\partial\rho}=&\left[\cos\left(\frac{\rho-t}{a}\right)-\frac{\rho}{a}\sin\left(\frac{\rho-t}{a}\right)\right]\mathbf{i}\\&+\left[\sin\left(\frac{\rho-t}{a}\right)+\frac{\rho}{a}\cos\left(\frac{\rho-t}{a}\right)\right]\mathbf{j}\end{aligned}$$

and

$$\frac{\partial\mathbf{r}}{\partial t}=\frac{\rho}{a}\left[\sin\left(\frac{\rho-t}{a}\right)\mathbf{i}-\cos\left(\frac{\rho-t}{a}\right)\mathbf{j}\right];$$

these two vectors are not orthogonal, so that the coordinate system is not orthogonal. The weighting factor in (5.70a) or (5.71a) is

$$\left|\frac{\partial\mathbf{r}}{\partial\rho}\wedge\frac{\partial\mathbf{r}}{\partial t}\right|=\frac{\rho}{a}.$$

This is the modulus of the Jacobian (5.68).

The region of integration is bounded by the lines $\rho=\frac{1}{4}a\pi$, $t=0$ and $t=\frac{1}{2}a\pi$, the last two intersecting at $\rho=0$. So the ranges of integration are $0\leqslant\rho\leqslant\frac{1}{4}a\pi$, $0\leqslant t\leqslant\frac{1}{2}a\pi$. Thus the integral is

$$\begin{aligned}&\int_0^{\frac{1}{4}a\pi}\mathrm{d}\rho\int_0^{\frac{1}{2}a\pi}\mathrm{d}t\,\frac{\rho}{a}\,\frac{\exp(-\rho^2/a^2)}{t^2+\mu^2}\\&=-\tfrac{1}{2}a\left[\exp(-\rho^2/a^2)\right]_0^{\frac{1}{4}a\pi}\left[\mu^{-1}\tan^{-1}\frac{t}{\mu}\right]_0^{\frac{1}{2}a\pi}\\&=\frac{a}{2\mu}\left[1-\exp\left(-\frac{\pi^2}{16}\right)\right]\tan^{-1}\left(\frac{a\pi}{2\mu}\right).\end{aligned}$$

**Example 5.23**

The integral

$$I = \int_0^\infty \exp(-x^2)\,\mathrm{d}x$$

is the limit

$$I = \lim_{a\to\infty} \int_0^a \exp(-x^2)\,\mathrm{d}x.$$

It is evaluated by squaring to give the double integral

$$I^2 = \lim_{a\to 0} \int_0^a \mathrm{d}x \int_0^a \mathrm{d}y\, \exp\{-(x^2+y^2)\}.$$

In the $(x, y)$ plane, the region of integration is over the square $OABC$ shown in Fig. 5.14. Let the quarter-disc $OAC$ be denoted by $\sigma_1$, and

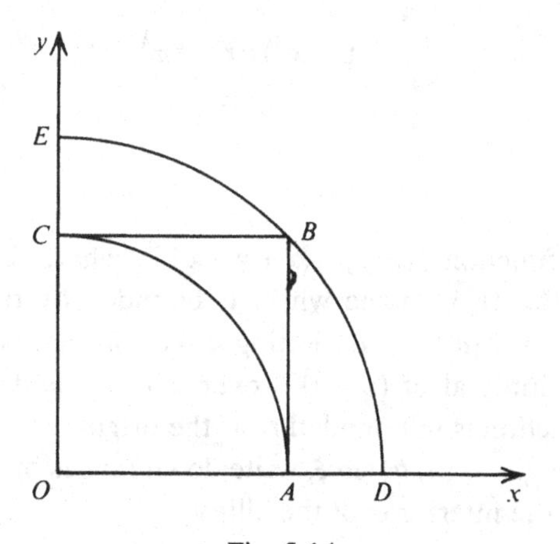


Fig. 5.14

the quarter-disc $ODE$ by $\sigma_2$. Then, since the integrand $\exp\{-(x^2+y^2)\}$ is positive everywhere,

$$\iint_{\sigma_1} \exp\{-(x^2+y^2)\}\,\mathrm{d}x\,\mathrm{d}y < \int_0^a \mathrm{d}x \int_0^a \mathrm{d}y\, \exp\{-(x^2+y^2)\}$$

$$< \iint_{\sigma_2} \exp\{-(x^2+y^2)\}\,\mathrm{d}x\,\mathrm{d}y.$$

The integral over $\sigma_1$ can be evaluated by using polar coordinates $(\rho, \phi)$ in the $(x, y)$ plane. As in Example 5.20, the Jacobian is equal to $\rho$. Thus the integral is

$$\int_0^a d\rho \int_0^{\frac{1}{2}\pi} d\phi\, \rho \exp(-\rho^2) = \tfrac{1}{4}\pi[-\exp(-\rho^2)]_0^a$$
$$= \tfrac{1}{4}\pi\{1-\exp(-\rho^2)\}.$$

Similarly the integral over $\sigma_2$ is

$$\tfrac{1}{4}\pi\{1-\exp(-2a^2)\}.$$

Thus

$$\tfrac{1}{4}\pi\{1-\exp(-a^2)\} < \int_0^a dx \int_0^a dy \exp\{-(x^2+y^2)\} < \tfrac{1}{4}\pi\{1-\exp(-2a^2)\}.$$

As $a \to \infty$, the limits of the integrals over $\sigma_1$ and $\sigma_2$ are both equal to $\frac{1}{4}\pi$. So taking this limit, we find $I^2 = \frac{1}{4}\pi$. Thus

$$\int_0^\infty \exp(-x^2)\, dx = \tfrac{1}{2}\pi^{\frac{1}{2}}.$$

## Problems 5.4

1 Integrate the function $f(x, y) = (x+y+k)^{-1}$, where $k > 0$, over the triangle $\sigma$ in the $(x, y)$ plane which is bounded by the lines $x = 0$, $y = 0$ and $x + y = a$ $(a > 0)$. By letting $k \to 0$ through positive values, show that the integral of $(x+y)^{-1}$ over $\sigma$ exists and is finite, even though the function is unbounded near the origin.

2 If $x = (\lambda/a)\cos\xi$, $y = (\lambda/b)\sin\xi$, write down ranges of $\lambda$ and $\xi$ which correspond to the interior $\sigma$ of the ellipse

$$\frac{x^2}{a^2} + \frac{y^2}{b^2} = 1.$$

Evaluate

$$\iint_\sigma x^2 y\, dx\, dy.$$

3 If $x = u^2 - v^2$, $y = 2uv$, find in terms of $u$ and $v$ the equation of the parabola

$$y^2 = 4a(x+a),$$

both when $a > 0$ and when $a < 0$. Integrate the function $(x^2+y^2)^{-\frac{1}{2}}$

over the region enclosed by the three parabolas

$$y^2 = 4a_r(x + a_r) \qquad (r = 1, 2, 3),$$

where

$$a_1 > a_2 > 0 > a_3.$$

4 Curvilinear coordinates in the $(x, y)$ plane are defined by the relations

$$x = c \cosh \xi \cos \eta,$$
$$y = c \sinh \xi \sin \eta.$$

Show that the Jacobian for the transformation from $(x, y)$ to $(u, v)$ is

$$c^2(\cosh^2 \xi - \cosh^2 \eta).$$

Using the results of Problems 5.1, Question 4, find the area of the surface whose boundary consists of the curves $\xi = \xi_1$, $\xi = \xi_2$, $\eta = \eta_1$, $\eta = \eta_2$.

5 Evaluate the integral

$$\int_0^\infty dx \int_0^\infty dy \exp(-x^2 - 2xy \cos \alpha - y^2)$$

where $-\pi < \alpha < \pi$. [*Hint*: take $(x, y)$ as oblique coordinates in a plane.]

6 Using spherical polar coordinates, show that the element of area (5.66) on the sphere $r = a$ is $a^2 \sin \theta \, \delta\theta \, \delta\phi$. Evaluate the integral of $x^2 = a^2 \sin^2 \theta \cos^2 \phi$ over the part of the surface of the sphere defined by the ranges $0 \leqslant \theta \leqslant \frac{1}{2}\pi$, $0 \leqslant \phi \leqslant \frac{1}{2}\pi$.

7 Find the area of the part of the cylindrical surface $x^2 + y^2 = a^2$ lying within the second surface $y^2 + z^2 = a^2$.

8 Using cylindrical polar coordinates $(\rho, \phi, z)$, a finite part $\sigma$ of the surface $z = a\phi$ is defined by the ranges $0 \leqslant \rho \leqslant a$ and $0 \leqslant \phi \leqslant \frac{1}{2}\pi$. Find the area of $\sigma$ and the integral of $x = \rho \cos \phi$ over $\sigma$.

9 A surface $\sigma$ is defined by the position vector

$$\mathbf{r} = u^2 \cos \phi \, \mathbf{i} + u^2 \sin \phi \, \mathbf{j} + 2u\phi\mathbf{k},$$

with parameter ranges $0 \leqslant u \leqslant u_1$ and $\pi \leqslant \phi \leqslant 2\pi$. Show that the integral of $(x^2 + y^2)^{-\frac{1}{2}}$ over $\sigma$ is equal to

$$\int_\pi^{2\pi} d\phi \, [u_1(u_1^2 + \phi^2 + 4)^{\frac{1}{2}} + (\phi^2 + 4) \sinh^{-1} \{u_1(\phi^2 + 4)^{-\frac{1}{2}}\}].$$

10 The region $\sigma$ is that part of the cone $z = +(x^2 + y^2)^{\frac{1}{2}}$ which is interior to the cylinder $(x - a)^2 + y^2 = a^2$. Find the area of $\sigma$ and calculate the integral of $x^2 z^2$ over $\sigma$.

## 5.5 Volume integrals

The definition of volume regions and of integrals over volume regions is very similar to that of surface regions and surface integrals. Points in space can be specified by a variety of coordinate systems, such as spherical polar or spheroidal coordinates. These coordinate systems consist of three coordinates which we denote by $u_1$, $u_2$, $u_3$, and the position vector $\mathbf{r}(u_1, u_2, u_3)$ is a function of $u_s$ $(s = 1, 2, 3)$. Subject to certain conditions which we shall state, the coordinates $\{u_s\}$ are called **curvilinear coordinates**. If $(\mathbf{i}, \mathbf{j}, \mathbf{k})$ is a fixed triad, the position vector $\mathbf{r}$ is expressible in the form (5.7), where $X(u_s)$, $Y(u_s)$, $Z(u_s)$ are now functions of the *three* curvilinear coordinates. The points corresponding to $\mathbf{r}(u_s)$ will define a **finite volume region** $\tau$, provided that the following conditions are satisfied:

(i) The variables $u_1$, $u_2$, $u_3$ can be chosen so that their ranges are of the form

$$a \leqslant u_1 \leqslant b, \tag{5.73a}$$

$$\alpha(u_1) \leqslant u_2 \leqslant \beta(u_1), \tag{5.73b}$$

$$\eta(u_1, u_2) \leqslant u_3 \leqslant \xi(u_1, u_2), \tag{5.73c}$$

where $a$ and $b$ $(>a)$ are constants, $\alpha(u_1)$ and $\beta(u_1)$ are continuous functions of $u_1$ in the range $a \leqslant u_1 \leqslant b$, and $\eta(u_1, u_2)$ and $\xi(u_1, u_2)$ are continuous functions of $u_1$ and $u_2$ in the region defined by (5.73a) and (5.73b). The functions $\alpha(u_1)$, $\beta(u_1)$, $\eta(u_1, u_2)$ and $\xi(u_1, u_2)$ may, of course, be constant.

(ii) The functions $X(u_s)$, $Y(u_s)$, $Z(u_s)$ are continuous functions of $u_s$ $(s = 1, 2, 3)$ throughout the ranges (5.73).

(iii) Except at a finite number of points, and over a finite number of curves and surfaces, there is only one set of parameter values $\{u_s\}$ corresponding to each position vector $\mathbf{r}$.

The **limit set** of points defined by the ranges (5.73) is the set of points of the region $\tau$ corresponding to parameter values

$$\{u_1, u_2, u_3;\ u_1 = a \quad \text{or } u_1 = b\}, \tag{5.74a}$$

$$\{u_1, u_2, u_3;\ u_2 = \alpha(u_1), \qquad u_2 = \beta(u_1)\} \tag{5.74b}$$

and

$$\{u_1, u_2, u_3;\ u_3 = \eta(u_1, u_2), \qquad u_3 = \xi(u_1, u_2)\}. \tag{5.74c}$$

Just as for surfaces, we define a **boundary point** of $\tau$ to be one whose parameters belong to the limit set (5.74), for *every* parametric system satisfy the conditions (i), (ii) and (iii); other points of $\tau$ are then

**interior points**. The **boundary** of $\tau$ is the set of all boundary points; in most practical situations, the boundary is a simple closed surface consisting of a finite number of smooth surfaces. Condition (ii) above ensures that the parameter regions (5.73) correspond to the *interior* of the closed surface $\sigma$, since the continuity condition prevents $\mathbf{r}=X\mathbf{i}+Y\mathbf{j}+Z\mathbf{k}$ from becoming indefinitely large.

The most suitable coordinate system for a given region $\tau$ may have parameter ranges which are more complicated than those allowed by (5.73). We then subdivide the region $\tau$ into a finite number of volume regions $\tau_1, \tau_2, \ldots, \tau_n$ each of which may share some boundary points with other regions, and each corresponding to parameter ranges of the form (5.73). Also, it is often necessary to consider **infinite volumes**, defined by allowing one or more of the ranges (5.73) to increase in such a way that $|\mathbf{r}|\to\infty$ in certain regions in 3-space; the boundary of an infinite volume will, of course, be an infinite surface, unless the volume is the whole of 3-space.

Suppose that $P(u_r)$ is a point in 3-space corresponding to particular values of $u_1$, $u_2$, $u_3$. If we allow $u_1$ to vary, but keep $u_2$, $u_3$ fixed, the position vector varies along a curve with $u_1$ as parameter, represented by $AB$ in Fig. 5.15; $Q$ is the point on this curve with parameters $(u_1+\delta u_1, u_2, u_3)$. Likewise, $CD$ and $EF$ are the curves obtained by allowing $u_2$ and $u_3$ respectively to vary, keeping the other parameters fixed; the points $R(u_1, u_2+\delta u_2, \delta u_3)$ and $S(u_1, u_2, u_3+\delta u_3)$ lie on these two curves. The set of curves represented by $AB$, $CD$, $EF$ form a **mesh** throughout the volume. The points $Q$, $R$, $S$ shown in Fig. 5.15 are obtained by making variations $\delta u_r$ in *two* of the three parameters, while $P'$ is obtained by making variations in all three. In order to define volume regions, we make the further assumptions, analogous to those made for surfaces:

(iv) Through every point of $\tau$, apart from a finite number of **exceptional points** and along a finite number of **exceptional curves**, all contained within the limit set (5.74), there is a unique trio of curves of the mesh, defined respectively by keeping the pairs $(u_2, u_3)$, $(u_3, u_1)$ and $(u_1, u_2)$ constant. Points which are not exceptional, and do not lie on exceptional curves, are called **normal points** of the volume $\tau$.

(v) At all normal points of $\tau$, the partial derivatives

$$\frac{\partial \mathbf{r}}{\partial u_s} \qquad (s=1,2,3)$$

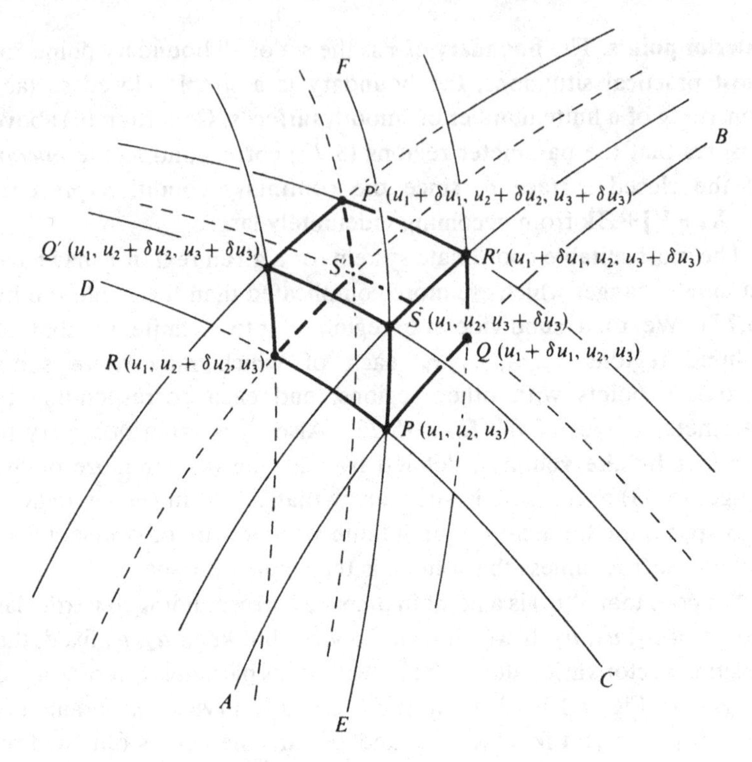


Fig. 5.15

exist and are continuous and non-zero; at points on the boundary of $\tau$, these derivatives may be one-sided. This assumption ensures the existence of *three* unit tangent vectors $\mathbf{t}_r$ $(r = 1, 2, 3)$ to the curves of the mesh at all normal points; the unit vectors $\{\mathbf{t}_r\}$ correspond to the three non-zero vectors $\{\partial\mathbf{r}/\partial u_r\}$, and are continuous functions of the coordinates.

(vi) At all normal points of $\tau$, the three tangent vectors $\{\partial\mathbf{r}/\partial u_s\}$, or alternatively $\{\mathbf{t}_s\}$, are assumed to be linearly independent. So, by (3.21),

$$\left[\frac{\partial\mathbf{r}}{\partial u_1}, \frac{\partial\mathbf{r}}{\partial u_2}, \frac{\partial\mathbf{r}}{\partial u_3}\right] \neq 0 \tag{5.75a}$$

and

$$[\mathbf{t}_1, \mathbf{t}_2, \mathbf{t}_3] \neq 0. \tag{5.75b}$$

As with the boundaries of surfaces, these assumptions ensure that a tangent plane can be defined to the boundary of the volume $\tau$, except at exceptional points and along exceptional curves. Boundaries normally consist of a finite number of smooth surfaces.

If a system of coordinates $\{u_r\}$ satisfies conditions (i)–(vi) of this section, for a finite region $\tau$ of 3-space, it satisfies all the properties usually assumed for systems of curvilinear coordinates.

If $\delta u_r$ $(r=1,2,3)$ are small positive increments, then at normal points, the vectors

$$\frac{\partial \mathbf{r}}{\partial u_r}\,\delta u_r \qquad (r=1,2,3) \tag{5.76}$$

generate, as in Fig. 5.16, a parallelepiped with one vertex at $P$, whose

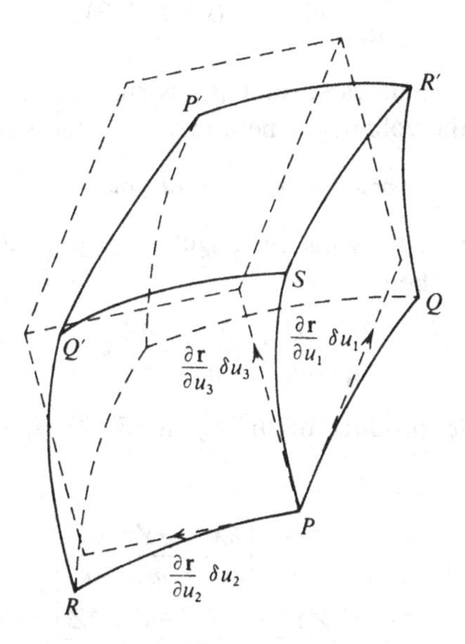


Fig. 5.16

seven other vertices lie close to the points $Q$, $R$, $S$, $Q'$, $R'$, $S'$ and $P'$; the tangent vectors $\mathbf{t}_1$, $\mathbf{t}_2$, $\mathbf{t}_3$ are parallel to the vectors (5.76), and in the same sense. The volume of the parallelepiped is, by (3.27),

$$\delta\tau_P = \left|\left[\frac{\partial \mathbf{r}}{\partial u_1}, \frac{\partial \mathbf{r}}{\partial u_2}, \frac{\partial \mathbf{r}}{\partial u_3}\right]\right|\,\delta u_1\,\delta u_2\,\delta u_3, \tag{5.77a}$$

since $\delta u_r > 0$ $(r=1,2,3)$. The condition (5.75) ensures that this volume element is non-zero except for a subset of the limit set (5.74).

Very often we use an **orthogonal system of coordinates** in 3-space, such that the vectors (5.76) are mutually orthogonal at all points, so that

$$\frac{\partial \mathbf{r}}{\partial u_r} \cdot \frac{\partial \mathbf{r}}{\partial u_s} = 0 \qquad (r \neq s) \tag{5.78}$$

everywhere. We then define, as in (5.67), the moduli

$$h_s(u_r) = \left| \frac{\partial \mathbf{r}}{\partial u_s} \right| \qquad (s = 1, 2, 3), \tag{5.79a}$$

so that

$$\frac{\partial \mathbf{r}}{\partial u_s} = h_s \mathbf{t}_s \qquad (s = 1, 2, 3). \tag{5.80}$$

Ordering the coordinates so that $\{\mathbf{t}_r\}$ is right-handed, $[\mathbf{t}_1, \mathbf{t}_2, \mathbf{t}_3] = 1$ everywhere; so the volume element (5.77a) reduces to

$$\delta\tau_P = h_1 h_2 h_3 \,\delta u_1 \,\delta u_2 \,\delta u_3. \tag{5.77b}$$

If the position vector $\mathbf{r}$ has rectangular components $X(u_r)$, $Y(u_r)$, $Z(u_r)$, as in (5.7), then

$$\frac{\partial \mathbf{r}}{\partial u_s} = \frac{\partial X}{\partial u_s}\mathbf{i} + \frac{\partial Y}{\partial u_s}\mathbf{j} + \frac{\partial Z}{\partial u_s}\mathbf{k}. \tag{5.81}$$

The scalar triple product occurring in (5.77) is, by (4.25b) and (3.28)

$$\frac{\partial(X, Y, Z)}{\partial(u_1, u_2, u_3)} \equiv \begin{vmatrix} \dfrac{\partial X}{\partial u_1} & \dfrac{\partial Y}{\partial u_1} & \dfrac{\partial Z}{\partial u_1} \\ \dfrac{\partial X}{\partial u_2} & \dfrac{\partial Y}{\partial u_2} & \dfrac{\partial Z}{\partial u_2} \\ \dfrac{\partial X}{\partial u_3} & \dfrac{\partial Y}{\partial u_3} & \dfrac{\partial Z}{\partial u_3} \end{vmatrix}; \tag{5.82}$$

this determinant is defined to be the **Jacobian** of the transformation from variables $(x, y, z)$ to variables $u_r$ $(r = 1, 2, 3)$. Its relationship to the Jacobian (5.68), for changes of variable in a plane, is clear. For orthogonal systems, the functions $h_s$ defined by (5.79a) and (5.81) are

$$h_s(u_r) = \left[\left(\frac{\partial X}{\partial u_s}\right)^2 + \left(\frac{\partial Y}{\partial u_s}\right)^2 + \left(\frac{\partial Z}{\partial u_s}\right)^2\right]^{\frac{1}{2}}. \tag{5.79b}$$

**Example 5.24**

In terms of cylindrical polar coordinates, the position vector of a general point in 3-space is

$$\mathbf{r} = \rho \cos\phi\ \mathbf{i} + \rho \sin\phi\ \mathbf{j} + z\mathbf{k}.$$

The vectors (5.81) are

$$\frac{\partial \mathbf{r}}{\partial \rho} = \cos\phi\ \mathbf{i} + \sin\phi\ \mathbf{j},$$

$$\frac{\partial \mathbf{r}}{\partial \phi} = -\rho \sin\phi\ \mathbf{i} + \rho \cos\phi\ \mathbf{j},$$

$$\frac{\partial \mathbf{r}}{\partial z} = \mathbf{k}.$$

These vectors are mutually orthogonal at all points, so the system of coordinates is orthogonal. On the $\mathbf{k}$-axis, where $\rho = 0$, the vector $\partial \mathbf{r}/\partial \phi = 0$; this is an exceptional line of points of the limit set where the volume element (5.77) and the Jacobian (5.82) are zero.

A volume $\tau$ is defined to be one quarter of a right circular cylinder, with a boundary consisting of the five surfaces $\rho = a$, $\phi = 0$, $\phi = \frac{1}{2}\pi$, $z = 0$, $z = h$, as shown in Fig. 5.17. At first sight, it appears that the

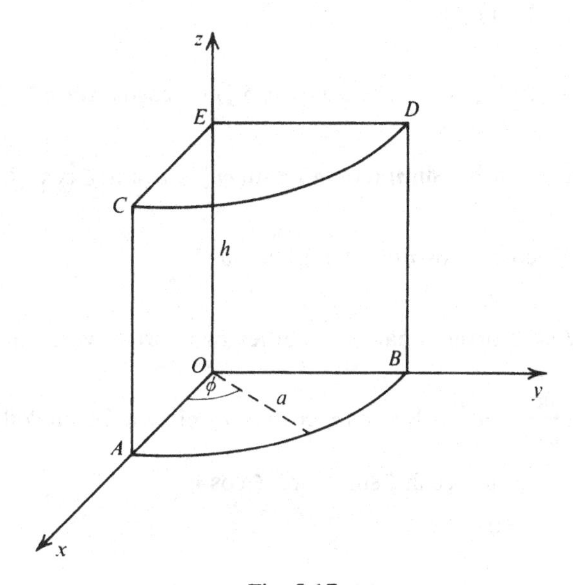


Fig. 5.17

lines $AC$, $BD$ and the arcs $AB$, $CD$ are exceptional curves. This is not so, since the derivatives $\partial\mathbf{r}/\partial\rho$, $\partial\mathbf{r}/\partial\phi$, $\partial\mathbf{r}/\partial z$ are well-defined there as one-sided derivatives; they are non-zero and mutually orthogonal on these four curves, which therefore consist of normal points only. The line $OE$ is, however, an exceptional curve of $\tau$, since $\partial\mathbf{r}/\partial\phi = 0$, and is part of the boundary of $\tau$. We might then expect $OE$ to be a line of exceptional points on the surfaces $ACEO$ and $EBDO$, violating condition (v) of the conditions we have assumed for smooth surfaces. But, on $EDBO$ for instance, $\phi$ is fixed at value $\frac{1}{2}\pi$, and the coordinates defining the surface are $\rho$, $z$. On the line $OE$, $\partial\mathbf{r}/\partial\rho$ and $\partial\mathbf{r}/\partial z$ exist (at least as one-sided derivatives), and are non-zero and orthogonal. So, while $OE$ is an exceptional line of the *volume* $\tau$, none of its points are exceptional points of the plane *surfaces* $ACEO$ and $EDBO$.

**Example 5.25**

In terms of oblate spheroidal coordinates (see Problems 5.1, Question 5) the general position vector is

$$\mathbf{r} = c\cosh\xi\cos\eta(\cos\phi\,\mathbf{i} + \sin\phi\,\mathbf{j}) + c\sinh\xi\sin\eta\,\mathbf{k}.$$

The vectors (5.81) are

$$\frac{\partial\mathbf{r}}{\partial\xi} = c\sinh\xi\cos\eta(\cos\phi\,\mathbf{i} + \sin\phi\,\mathbf{j}) + c\cosh\xi\sin\eta\,\mathbf{k},$$

$$\frac{\partial\mathbf{r}}{\partial\eta} = -c\cosh\xi\sin\eta\,(\cos\phi\,\mathbf{i} + \sin\phi\,\mathbf{j}) + c\sinh\xi\cos\eta\,\mathbf{k},$$

$$\frac{\partial\mathbf{r}}{\partial\phi} = c\cosh\xi\cos\eta(-\sin\phi\,\mathbf{i} + \cos\phi\,\mathbf{j}).$$

Clearly $\partial\mathbf{r}/\partial\phi$ is orthogonal to the other two derivatives, while

$$\begin{aligned}\frac{\partial\mathbf{r}}{\partial\xi}\cdot\frac{\partial\mathbf{r}}{\partial\eta} &= -c^2\sinh\xi\cos\eta\cosh\xi\sin\eta|\cos\phi\,\mathbf{i} + \sin\phi\,\mathbf{j}|^2 \\ &\quad + c^2\cosh\xi\sin\eta\sinh\xi\cos\eta \\ &= 0.\end{aligned}$$

Thus the coordinates form an orthogonal system.

The magnitudes (5.79a) of the three vectors are

$$h_\xi = \left|\frac{\partial \mathbf{r}}{\partial \xi}\right| = (c^2 \sinh^2 \xi \cos^2 \eta + c^2 \cosh^2 \xi \sin^2 \eta)^{\frac{1}{2}}$$

$$= c(\cosh^2 \xi - \cos^2 \eta)^{\frac{1}{2}},$$

$$h_\eta = \left|\frac{\partial \mathbf{r}}{\partial \eta}\right| = c(\cosh^2 \xi - \cos^2 \eta)^{\frac{1}{2}} = h_\xi,$$

$$h_\phi = c \cosh \xi \cos \eta.$$

So the volume element (5.77b) is

$$\delta\tau_P = c^3 \cosh \xi \cos \eta (\cosh^2 \xi - \cos^2 \eta)\, \delta\xi\, \delta\eta\, \delta\phi.$$

Given a region $\tau$ of 3-space and a system of curvilinear coordinates $\{u_r\}$ which satisfy conditions (i)–(vi), we can now define integration over $\tau$ in much the same way as we have defined surface integrals. A coarse mesh of curves is chosen, as indicated in Fig. 5.15, corresponding to a finite number of values of the three coordinates $u_1$, $u_2$, $u_3$. Apart from exceptional points and curves, each particular set of coordinates values corresponds to a point of intersection $P$ of three curves, one belonging to each family of curves. If $u_1+\delta u_1$, $u_2+\delta u_2$, $u_3+\delta u_3$ are adjacent values of the coordinates (with $\delta u_r > 0$), then the points $P, Q, R, S, Q', R', S', P'$ defined by these two sets of parameter values form a 'cell' whose volume we take to be closely approximated by (5.77) when $\delta u_1$, $\delta u_2$, $\delta u_3$ are sufficiently small. If the ranges (5.73) are divided into intervals denoted by $\delta u_r$ ($r = 1, 2, 3$), the region $\tau$ will be divided into a number of regions; if we exclude regions containing part of the boundary of $\tau$, the remainder will be cells of the kind we have discussed, in the interior of $\tau$. (The 3-variable analogue of Theorem A.4, Appendix A, justifies this exclusion.) Now let $f(u_r)$ be a piecewise continuous function in $\tau$, and let $f_P$ be the value at the point $P$ of a cell corresponding to the minimum value of each variable $u_r$. Then we form the sum

$$\sum_P f_P\, \delta\tau_P$$

over all the cells not containing boundary points of $\tau$; if we now let every increment $\delta u_r$ tend to zero, then by the analogue of Theorem

A.3 (Appendix A), this sum has the limit

$$\iiint_\tau f(u_r)\,d\tau \equiv \iiint_\tau f(u_r)\left|\left[\frac{\partial \mathbf{r}}{\partial u_1}, \frac{\partial \mathbf{r}}{\partial u_2}, \frac{\partial \mathbf{r}}{\partial u_3}\right]\right| du_1\,du_2\,du_3, \quad (5.83a)$$

using the expression (5.77) for the volume element. Since the scalar triple product is equal to the Jacobian (5.82), this volume integral can be written

$$\iiint_\tau f(u_r)\,d\tau = \iiint_\tau f(u_r)\left|\frac{\partial(X,Y,Z)}{\partial(u_1,u_2,u_3)}\right| du_1\,du_2\,du_3. \quad (5.83b)$$

For orthogonal coordinates systems, we can use the expression (5.80) for the volume element, giving the integral

$$\iiint_\tau f(u_r)h_1(u_r)h_2(u_r)h_3(u_r)\,du_1\,du_2\,du_3. \quad (5.83c)$$

By the 3-variable analogue of Theorem A.8 (Appendix A), the volume integral is equal to the repeated integral over the ranges (5.73), so that in expressions (5.83) we may make the replacement

$$\iiint_\tau du_1\,du_2\,du_3 \to \int_a^b du_1 \int_{\alpha(u_1)}^{\beta(u_2)} du_2 \int_{\eta(u_1,u_2)}^{\xi(u_1,u_2)} du_3. \quad (5.84)$$

The repeated integration may be carried out in any order, but as for surface integrals, the ranges of integration may not be of the simple form (5.73). More generally, we may wish to use a coordinate system which requires $\tau$ to be divided into a finite number of regions, each corresponding to ranges of the form (5.73).

Integrals over infinite volumes can be defined as limits of finite volume integrals, with ranges (5.73) increasing so that $|\mathbf{r}|\to\infty$ somewhere in $\tau$. It is also possible to define integrals of certain classes of functions which are unbounded in $\tau$. We again refer to textbooks of analysis for a detailed study of these classes of integral [Reference 5.12].

If $f(u_r)\equiv 1$ in (5.83), we are calculating the limit of $\Sigma_P\,\delta\tau_P$, which gives the **volume** of $\tau$. Using (5.83b) and (5.83c), the volume is

$$V_\tau = \iiint_\tau \left|\frac{\partial(X,Y,Z)}{\partial(u_1,u_2,u_3)}\right| du_1\,du_2\,du_3, \quad (5.85a)$$

or, for orthogonal coordinate systems,

$$V_\tau = \iiint_\tau h_1(u_r)h_2(u_r)h_3(u_r)\,\mathrm{d}u_1\,\mathrm{d}u_2\,\mathrm{d}u_3. \qquad (5.85b)$$

**Example 5.26**

Using standard notation for spherical and cylindrical polar coordinates, evaluate the integral of $\rho^2$ over the spherical volume $r \leqslant a$.

In terms of spherical polar coordinates, the position vector is

$$\mathbf{r} = r[\sin\theta\cos\phi\,\mathbf{i} + \sin\theta\sin\phi\,\mathbf{j} + \cos\theta\,\mathbf{k}].$$

As in (5.26),

$$\frac{\partial \mathbf{r}}{\partial r} = [\sin\theta\cos\phi\,\mathbf{i} + \sin\theta\sin\phi\,\mathbf{j} + \cos\theta\,\mathbf{k}],$$

$$\frac{\partial \mathbf{r}}{\partial \theta} = r[\cos\theta\cos\phi\,\mathbf{i} + \cos\theta\sin\phi\,\mathbf{j} - \sin\theta\,\mathbf{k}],$$

$$\frac{\partial \mathbf{r}}{\partial \phi} = r[-\sin\theta\sin\phi\,\mathbf{i} + \sin\theta\cos\phi\,\mathbf{j}].$$

These three vectors are easily seen to be mutually orthogonal, and denoting the functions (5.79) by $h_r$, $h_\theta$, $h_\phi$, we find

$$h_r = 1, \qquad h_\theta = r, \qquad h_\phi = r\sin\theta.$$

The integrand is $\rho^2 = (r\sin\theta)^2$, so the integral, given by (5.83c) and (5.84), is

$$\int_0^a \mathrm{d}r \int_0^\pi \mathrm{d}\theta \int_0^{2\pi} \mathrm{d}\phi\, r^2 \sin\theta\,(r\sin\theta)^2.$$

Now

$$\int_0^a \mathrm{d}r\, r^4 = \tfrac{1}{5}a^5$$

and

$$\int_0^\pi \sin^3\theta\,\mathrm{d}\theta = \tfrac{4}{3}.$$

Therefore the integral is

$$\tfrac{1}{5}a^5 \cdot \tfrac{4}{3} \cdot 2\pi = \tfrac{8}{15}\pi a^5.$$

**Example 5.27**

Evaluate

$$\iiint_{\tau} dx\, dy\, dz\, (1+x+2y+3z)^{-3}$$

over the **simplex** $\tau$ defined by $x \geqslant 0$, $y \geqslant 0$, $z \geqslant 0$, $x+y+z \leqslant 1$.

The ranges of integration can be taken to be

$$0 \leqslant x \leqslant 1,$$
$$0 \leqslant y \leqslant 1-x,$$
$$0 \leqslant z \leqslant 1-x-y.$$

The integral is then

$$\int_0^1 dx \int_0^{1-x} dy \int_0^{1-x-y} dz\, (1+x+2y+3z)^{-3}.$$

The $z$-integral is

$$-\tfrac{1}{6}[(1+x+2y+3z)^{-2}]_{y=0}^{1-x-y} = -\tfrac{1}{6}[(4-2x-y)^{-2}-(1+x+2y)^{-2}].$$

Integrating over $y$, we obtain

$$-\tfrac{1}{6}[(4-2x-y)^{-1}+\tfrac{1}{2}(1+x+2y)^{-1}]_{y=0}^{1-x}$$
$$=\tfrac{1}{12}[2(4-2x)^{-1}+(1+x)^{-1}-2(3-x)^{-1}-(3-x)^{-1}].$$

The final integration over $x$ gives

$$\tfrac{1}{12}[-\ln(4-2x)+\ln(1+x)+3\ln(3-x)]_{x=0}^{1} = \tfrac{1}{12}\ln\tfrac{32}{27}.$$

## *Problems 5.5*

1 A cylindrical volume region $\tau$ is defined by the conditions $0 \leqslant z \leqslant h$ and

$$\frac{x^2}{a^2}+\frac{y^2}{b^2} \leqslant 1.$$

If $x = \lambda a \cos\psi$, $y = \lambda b \sin\psi$, define ranges of the coordinates $(\lambda, \psi, z)$ corresponding to $\tau$. Define the limit set of points corresponding to these ranges, and show why certain parts of the limit set do not belong to the boundary of the region.

Integrate the integral

$$\iiint_{\tau} \left(\frac{x^2}{a^2}+\frac{y^2}{b^2}\right)^{-1} d\tau.$$

2 The boundary of a finite volume region $\tau$ is defined in terms of oblate spheroidal coordinates [Example 5.25 and Problems 5.1, Question 5] to be parts of the surfaces $\xi = a$ and $\eta = b$, where $a$ and $b$ are positive constants. Find the volume of $\tau$, and integrate $r^2 = x^2 + y^2 + z^2$ over the region $\tau$.

3 Find the volume contained within the two cylinders $x^2 + y^2 = a^2$ and $y^2 + z^2 = a^2$.

4 Evaluate

$$\iiint_{\tau} \mathrm{d}x\, \mathrm{d}y\, \mathrm{d}z\, x\, \mathrm{e}^y \cos \pi(ax + by + cz),$$

where the region $\tau$ is defined by $x \geqslant 0$, $y \geqslant 0$, $z \geqslant 0$ and $ax + by + cz \leqslant 1$.

5 In terms of spherical polar coordinates, the infinite conical region $\tau$ is defined by $\theta \leqslant \alpha$, where $\alpha < \frac{1}{2}\pi$. Show that the integral

$$\iiint_{\tau} (r^2 + a^2)^{-2} \cos^2 \theta \, \mathrm{d}\tau \ (a > 0)$$

exists as a well-defined limit, and evaluate the integral.

6 The region $\tau$ is defined by the position vector

$$\mathbf{r} = u^2 v \cos \phi \, \mathbf{i} + u^2 v \sin \phi \, \mathbf{j} + 2uv\phi \, \mathbf{k}$$

and parameter ranges $0 \leqslant u \leqslant u_1$, $0 \leqslant v \leqslant 1$, $\pi \leqslant \phi \leqslant 2\pi$. Calculate the volume of $\tau$ and evaluate the integral of $r^{-2}$ over $\tau$.

7 **Prolate spheroidal coordinates** $(\xi, \eta, \phi)$ are related to rectangular coordinates by

$$x = c \sinh \xi \sin \eta \cos \phi,$$

$$y = c \sinh \xi \sin \eta \sin \phi,$$

$$z = c \cosh \xi \cos \eta.$$

Show that the tangent vectors $\mathrm{d}\mathbf{r}/\mathrm{d}\xi$, $\mathrm{d}\mathbf{r}/\mathrm{d}\eta$, $\mathrm{d}\mathbf{r}/\mathrm{d}\phi$ are mutually orthogonal, and that the functions (5.79b) are given by

$$h_\xi = h_\eta = (\cosh^2 \xi - \cos^2 \eta)^{\frac{1}{2}},$$

$$h_\phi = c \sinh \xi \sin \eta.$$

Show that the surface $\sigma$ defined by $\xi = \alpha$, where $\alpha$ is a positive constant, is a spheroid with semi-axes of length $c \sinh \alpha$, $c \sinh \alpha$ and $c \cosh \alpha$, and that the whole surface corresponds to ranges

$$0 \leqslant \eta \leqslant \pi, \qquad 0 \leqslant \phi < 2\pi.$$

Write the integral of a function $f(\eta, \phi)$ over the surface $\sigma$ as a double integral over $\eta$, $\phi$. Also write the integral of a function $g(\xi, \eta, \phi)$, over the volume bounded by $\sigma$, as a triple integral over $\xi$, $\eta$, $\phi$.

## *5.6 Properties of Jacobians*

If curvilinear coordinates $(u_1, u_2)$ in the $(x, y)$ plane are defined by the relations $x = X(u_r)$, $y = Y(u_r)$, then the element of area (5.66a) is proportional to the Jacobian (5.68), which can also be written

$$\frac{\partial(x, y)}{\partial(u_1, u_2)}.$$

Likewise, if curvilinear coordinates $\{u_r\}$ in 3-space are defined by $x = X(u_r)$, $y = Y(u_r)$, $z = Z(u_r)$, the element of volume (5.77) is proportional to the Jacobian (5.82), which we sometimes write as

$$\frac{\partial(x, y, z)}{\partial(u_1, u_2, u_3)}.$$

More generally, let $\{t_p\}$ and $\{u_r\}$ be two sets of coordinates *either* in the plane $(p, r = 1, 2)$ *or* in 3-space $(p, r = 1, 2, 3)$, each satisfying properties (i)–(vi) of the previous sections; suppose also that in a given region, each point $\mathbf{r}$ corresponds to unique coordinate values $\{t_p\}$ and $\{u_r\}$. Then $\{t_p\}$ will be functions of the variables $u_r$,

$$t_p = T_p(u_r), \tag{5.86}$$

and these (two or three) equations can be solved uniquely to give

$$u_r = F_r(t_p) \tag{5.87}$$

as functions of $t_p$. Equations (2.53) and (2.54) are examples of the pair of functional relations (5.86) and (5.87). The Jacobian corresponding to the relationship (5.86) is then

$$\frac{\partial(t_p)}{\partial(u_r)} = \frac{\partial(T_p)}{\partial(u_r)} = \Delta\left(\frac{\partial t_p}{\partial u_r}\right), \tag{5.88a}$$

where $\Delta$ denotes either the $(2\times2)$ determinant (4.18) or the $(3\times3)$ determinant (4.25). The definition (5.88a) is identical with the definitions (5.68) and (5.82).

Let us now suppose that $\{w_s\}$ are a third set of coordinates in one-to-one correspondence with position vectors $\mathbf{r}$; then $\{u_r\}$ are functions

$$u_r = U_r(w_s) \tag{5.89}$$

of these coordinates, and we can define the Jacobian

$$\frac{\partial(u_r)}{\partial(w_s)} = \Delta\left(\frac{\partial u_r}{\partial w_s}\right) \tag{5.88b}$$

corresponding to this change of variables. Now $\{t_p\}$ are defined as functions of $\{w_s\}$ through (5.86) and (5.89), with $\{u_r\}$ as intermediate variables. The Jacobian of this transformation is

$$\frac{\partial(t_p)}{\partial(w_s)} = \Delta\left(\frac{\partial t_p}{\partial w_s}\right). \tag{5.88c}$$

However, the partial derivatives $\partial t_r/\partial w_s$, with intermediate variables $\{u_r\}$, are given by the chain rule

$$\frac{\partial t_p}{\partial w_s} = \sum_r \frac{\partial t_p}{\partial u_r}\frac{\partial u_r}{\partial w_s},$$

where the summation is over $r = 1, 2$ or $r = 1, 2, 3$. We can regard this equation as a matrix equation relating square matrices

$$\left(\frac{\partial t_p}{\partial w_s}\right), \left(\frac{\partial t_p}{\partial u_r}\right), \left(\frac{\partial u_r}{\partial w_s}\right).$$

The property (4.30) then give the determinantal relationship

$$\Delta\left(\frac{\partial t_p}{\partial w_s}\right) = \Delta\left(\frac{\partial t_p}{\partial u_r}\right)\Delta\left(\frac{\partial u_r}{\partial w_s}\right).$$

But, by (5.88), these determinants are just the Jacobians associated with the three changes of variables. So the Jacobian (5.88c) is given by

$$\frac{\partial(t_p)}{\partial(w_s)} = \frac{\partial(t_p)}{\partial(u_r)}\frac{\partial(u_r)}{\partial(w_s)}. \tag{5.90}$$

The identity (5.90) for sets of two or three variables can be compared with the change of variable rule for single variables. If a single variable $t$ is a differentiable function $t = T(u)$ of a second variable $u$, and $u$ is a differentiable function $u = U(w)$, then $t$ is a differentiable function of $w$, and

$$\frac{dt}{dw} = \frac{dt}{du}\frac{du}{dw},$$

where

$$\frac{dt}{du} = \frac{dT(u)}{du}, \frac{du}{dw} = \frac{dU(w)}{dw}.$$

The relation (5.90) between three Jacobians is a generalisation of this rule to more than one variable.

If the original and final variables in (5.90) are identical, so that $w_p = t_p$, then the definition of the Jacobian gives immediately

$$\frac{\partial(t_p)}{\partial(t_p)} = 1.$$

So (5.90) becomes

$$\frac{\partial(t_p)}{\partial(u_r)} \frac{\partial(u_r)}{\partial(t_p)} = 1; \tag{5.91}$$

therefore the Jacobian for the transformation (5.86) is the inverse of the Jacobian for the inverse transformation (5.87). For some changes of variable, it is easier to calculate the Jacobian for the inverse transformation than for the transformation itself.

Jacobians were introduced in (5.66a), (5.68) and (5.77) as factors defining elements of areas and volumes, corresponding to increments $\{\delta u_r\}$ in coordinates $\{u_r\}$. In 3-space, for example, if we take $w_1 = x$, $w_2 = y$, $w_3 = z$ in (5.89), we obtain

$$\frac{\partial(t_p)}{\partial(u_r)} = \frac{\partial(t_p)/\partial(x, y, z)}{\partial(u_r)/\partial(x, y, z)}. \tag{5.92}$$

So the Jacobian $\partial(t_p)/\partial(u_r)$ is the *ratio* of volume factors for the two sets of coordinates; we can therefore think of it intuitively as an expansion or contraction factor of elementary volumes when we change from one set of coordinates to another.

# 6

# Vector analysis

## 6.1 *Scalar and vector fields*

In almost every branch of mathematical physics, we have to deal with physical quantities which extend continuously through regions of 3-space. These regions and their boundaries do not usually have any awkward features; we therefore assume that they can be described mathematically as volume regions and smooth surfaces which satisfy the conditions set out in Chapter 5. We also assume that any curves in physical space can be validly represented as piecewise smooth curves. Volumes, areas of surfaces and lengths of curves can therefore be defined. We can also define integrals of functions along curves, over surfaces and throughout volumes, provided that the region is finite and the functions are piecewise continuous; these definitions are based on analytic theorems established in Appendix A.

In this chapter, we shall be studying the analysis of functions in 3-space which might represent the properties of, for example, fluids, gravitational or electromagnetic fields, stress and strain in solids, or wave functions in atoms, molecules and nuclei. Such a function may be unbounded when the position vector tends to certain points, curves or surfaces; for example the electrical potential of a point charge $e$ at the origin is $e/r$, which is unbounded near the origin ($r=0$). Special care must be taken in the study of functions in regions where they are unbounded; we shall concentrate our attention on the analysis of functions in regions where they are 'well-behaved'. By 'well-behaved' we not only mean that a function is continuous, but also that any derivatives we use exist and are continuous.

Let us consider a function of the position vector $\mathbf{r}$, defined throughout a volume region $\tau$; this function is called a **scalar field** if it satisfies the following properties:

(i) Let $\{u_r; r = 1, 2, 3\}$ be a set of coordinates which, with $\tau$, satisfy conditions (i)–(vi) of §5.5, so that the function is expressed as a function $\psi(u_r)$ of the coordinates. Then we assume that the derivatives

$$\frac{\partial \psi}{\partial u_r} \qquad (r = 1, 2, 3)$$

exist (at least as one-sided derivatives) and are continuous, except possibly at a finite number of points and on a finite number of curves, constituting the **exceptional set** for $\psi$; this set may overlap the exceptional set for the coordinates.

(ii) If $\{w_s\}$ is any other set of curvilinear coordinates, and the function is expressed as $g(w_s)$ in terms of these coordinates, then

$$f(u_r) = g(w_s)$$

whenever $\{u_r\}$ and $\{w_s\}$ correspond to the same point $\mathbf{r}$ in $\tau$. In other words, the value of the function at each point in Euclidean (or physical) space is independent of the coordinate system. Very often the volume $\tau$ is the whole of 3-space; a scalar field may then satisfy certain further conditions at large distances from the origin.

A **vector field** is a vector function of $\mathbf{r}$ defined in a volume region $\tau$, satisfying conditions analogous to (i) and (ii) for scalar fields: it possesses continuous derivatives in $\tau$ apart from an exceptional set of points, and at each point $\mathbf{r}$ it transforms like a vector when the coordinate system is changed. A little care is needed in understanding these conditions. At any normal point $P$, a set of curvilinear coordinates $\{u_s\}$ defines three linearly independent tangent vectors $\mathbf{t}_1(u_s)$, $\mathbf{t}_2(u_s)$, $\mathbf{t}_3(u_s)$; in general, these three vectors vary continuously from point to point, indicated by writing them as functions of the coordinates; they are parallel to the displacements **PQ**, **PR**, **PS** in Fig. 5.16. Any vector in 3-space can be expressed as a linear combination

$$\sum_{p=1}^{3} \lambda_p \mathbf{t}_p(u_s)$$

of the tangent vectors at $P$. A vector field $\mathbf{v}$ defines a particular vector at each point $\mathbf{r}$, so that the field value at $\mathbf{r}$ can be expressed

$$\mathbf{v}(\mathbf{r}) = \sum_{p=1}^{3} v_p(u_s)\mathbf{t}_p(u_s), \tag{6.1a}$$

where $\{u_s\}$ are the coordinates of $\mathbf{r}$. It is very important to note that both the components $v_p(u_s)$ and the basis vectors $\mathbf{t}_p(u_s)$ vary from point to point; the position-dependent triad $\{\mathbf{t}_p(u_s)\}$ is called a **local frame of reference**. Since $\{\partial\mathbf{t}_p/\partial u_s\}$ are already assumed to be continuous, $\mathbf{v}$ will have continuous derivatives provided:

(iii) The functions $v_p(u_s)$ in (6.1a) have continuous derivatives $\partial v_p/\partial u_s$ $(s = 1, 2, 3)$ except, as in condition (i) above, at exceptional points and on exceptional curves, the **exceptional set** for $\mathbf{v}$. Then $\{\partial\mathbf{v}/\partial u_s\}$ exist and are continuous at all normal points.

A second set of curvilinear coordinates $\{w_s\}$ will define a second set of unit tangent vectors $\mathbf{l}_q(w_s)$, say, at each normal point, and the value of the field $\mathbf{v}$ at $\mathbf{r}$ can be expressed in the form

$$\mathbf{v}(\mathbf{r}) = \sum_{q=1}^{3} v'_q(w_s)\mathbf{l}_q(w_s). \tag{6.1b}$$

At each point $\mathbf{r}$, the two sets of basis vectors $\{\mathbf{t}_p(u_s)\}$ and $\{\mathbf{l}_q(w_s)\}$ will be related by a linear transformation of the form (4.96c):

$$(\mathbf{t}_1 \quad \mathbf{t}_2 \quad \mathbf{t}_3) = (\mathbf{l}_1 \quad \mathbf{l}_2 \quad \mathbf{l}_3)M. \tag{6.2}$$

The transformation matrix $M = M(\mathbf{r})$ is a function of position, since the basis vectors are. Since both sets of basis vectors in (6.2) are linearly independent, $M(\mathbf{r})$ is non-singular at every normal point $\mathbf{r}$. Continuity of $\{\mathbf{t}_p\}$ and $\{\mathbf{l}_q\}$ implies continuity of the elements of $M(\mathbf{r})$ as functions of $\mathbf{r}$; so the determinant $\Delta[M(\mathbf{r})]$ is continuous and non-zero, and hence is of fixed sign, throughout volume regions of normal points.

The condition on a vector field $\mathbf{v}$ which is analogous to (ii) for a scalar field is:

(iv) The sets of components $\{v_p(u_s)\}$ and $\{v'_q(w_s)\}$ of a vector field $\mathbf{v}$, at a normal point $\mathbf{r}$, are related by the vector transformation law

$$v_p(u_s) = \sum m_{pq} v'_q(w_s), \tag{6.3}$$

where $M = M(\mathbf{r})$ is the transformation matrix defined by (6.2); this follows from (6.3) by a simple generalisation of the argument leading to (4.73b) or (4.99a).

In most applications, only orthogonal systems of coordinates are used. Then the transformation matrix $M(\mathbf{r})$ is an orthogonal matrix at each normal point $\mathbf{r}$; if the triads of unit vectors are chosen to be right-handed, then $\Delta[M(\mathbf{r})] = +1$ at all points.

**Example 6.1.**
Let $(r, \theta, \phi)$ be spherical polar coordinates defined relative to a triad $(\mathbf{i}, \mathbf{j}, \mathbf{k})$, and let $(\mathbf{t}_r, \mathbf{t}_\theta, \mathbf{t}_\phi)$ be the triad of tangent vectors defined at a point by the coordinate system; as in Fig. 6.1, $\mathbf{t}_r$ lies in the direction of increasing $r$, and so on. The direction cosines of $\mathbf{t}_r, \mathbf{t}_\theta, \mathbf{t}_\phi$ relative to $(\mathbf{i}, \mathbf{j}, \mathbf{k})$ are, written as column matrices,

$$\begin{pmatrix} \sin\theta\cos\phi \\ \sin\theta\sin\phi \\ \cos\theta \end{pmatrix}, \quad \begin{pmatrix} \cos\theta\cos\phi \\ \cos\theta\sin\phi \\ -\sin\theta \end{pmatrix}, \quad \begin{pmatrix} -\sin\phi \\ \cos\phi \\ 0 \end{pmatrix}.$$

The transformation matrix $R$ relating the triads, equal to $A$ in (4.96b) or $M$ in (6.3), is thus

$$R(\theta, \phi) = \begin{pmatrix} \sin\theta\cos\phi & \cos\theta\cos\phi & -\sin\phi \\ \sin\theta\sin\phi & \cos\theta\sin\phi & \cos\phi \\ \cos\theta & -\sin\theta & 0 \end{pmatrix}. \tag{6.4}$$

This is an orthogonal matrix with $\Delta(R) = +1$; so, as is evident from Fig. 6.1, the triad $(\mathbf{t}_r, \mathbf{t}_\theta, \mathbf{t}_\phi)$ is right-handed.

Suppose that the components of $\mathbf{v}$ relative to the two triads are $(v_1, v_2, v_3)$ and $(v_r, v_\theta, v_\phi)$. By (4.74), the relation between these components is

$$\begin{pmatrix} v_r \\ v_\theta \\ v_\phi \end{pmatrix} = R^{\mathrm{T}}(\theta, \phi) \begin{pmatrix} v_1 \\ v_2 \\ v_3 \end{pmatrix}.$$

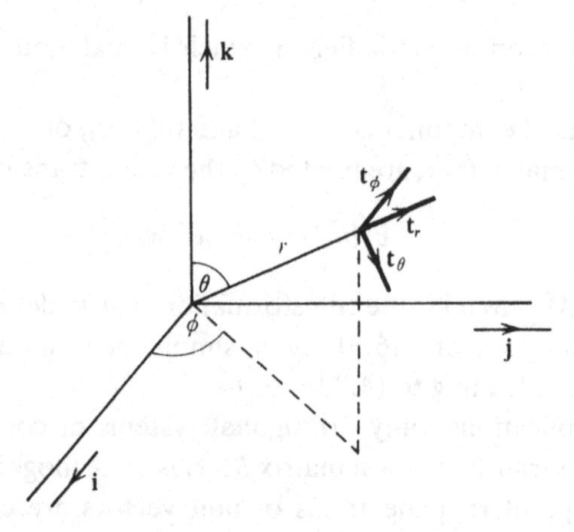


Fig. 6.1

If $\mathbf{v}$ is a vector field, then $v_1, v_2, v_3$ may be given as functions of $x, y, z$ by (2.28). The full transformation from rectangular to spherical polar coordinates therefore consists of making the substitution (2.70), and also performing the transformation (4.74) with $L$ replaced by (6.4).

Now suppose that the vector $\mathbf{v}$ is parallel to $\mathbf{r}$ at all points (a **radial vector**), so that $v_\theta = v_\phi = 0$ everywhere. Then (4.73) with $L$ replaced by (6.4) gives

$$\begin{pmatrix} v_1 \\ v_2 \\ v_3 \end{pmatrix} = R(\theta, \phi)\begin{pmatrix} v_r \\ 0 \\ 0 \end{pmatrix} = \begin{pmatrix} v_r \sin\theta\cos\phi \\ v_r \sin\theta\sin\phi \\ v_r \cos\theta \end{pmatrix}.$$

Using (2.70), this can be expressed as

$$v_1 = xv_r/r,$$

$$v_2 = yv_r/r,$$

$$v_3 = zv_r/r.$$

We frequently wish to refer to the values of a scalar or a vector field on a surface $\sigma$ or on a curve $\Gamma$. Relative to a coordinate system $\{u_r\}$ in 3-space, a surface can be defined by expressing one coordinate as a continuous function of the other two, for example

$$u_3 = f(u_1, u_2). \tag{6.5}$$

This relation restricts the general position vector $\mathbf{r}(u_1, u_2, u_3)$ to the form

$$\mathbf{r}(u_1, u_2, f(u_1, u_2)), \tag{6.6a}$$

and a scalar field $\psi(u_s)$, for instance, is restricted to the form

$$\psi(u_1, u_2, f(u_1, u_2)). \tag{6.6b}$$

In order to describe a surface in terms of certain coordinates, it is often necessary to divide the surface into a finite number of parts, as in the next example.

**Example 6.2**

Rectangular coordinates $(x, y, z)$ determine a point in space. If we wish to specify points on the sphere

$$x^2 + y^2 + z^2 = a^2$$

in terms of the two coordinates $(x, y)$, we need to express $z$ as a function of $x$ and $y$. In order to define $z$ uniquely, we need to divide the sphere into two hemispheres; one is defined by $z = +(a^2-x^2-y^2)^{\frac{1}{2}} \geqslant 0$, and the other by $z = -(a^2-x^2-y^2)^{\frac{1}{2}} < 0$. With the restriction $x^2+y^2 \leqslant a^2$, $(x, y)$ can be used as coordinates of points on the two hemispheres, for which the position vectors (6.6a) are

$$\mathbf{r}(x, y, +(a^2-x^2-y^2)^{\frac{1}{2}}), \mathbf{r}(x, y, -(a^2-x^2-y^2)^{\frac{1}{2}}).$$

More generally, we may describe a finite volume region $\tau$ by specifying parameter ranges (5.73). Suppose that the coordinates $u_1$, $u_2$ are used to parametrise the closed surface $\sigma$ bounding $\tau$; then, in general, putting $u_3 = \eta(u_1, u_2)$ will define one part of the closed surface $\sigma$ bounding $\tau$, while $u_3 = \zeta(u_1, u_2)$ defines another part. The simplest situation of this kind is illustrated in Fig. 6.2, where the whole surface is defined by $u_3 = \eta(u_1, u_2)$ and $u_3 = \zeta(u_1, u_2)$, the point

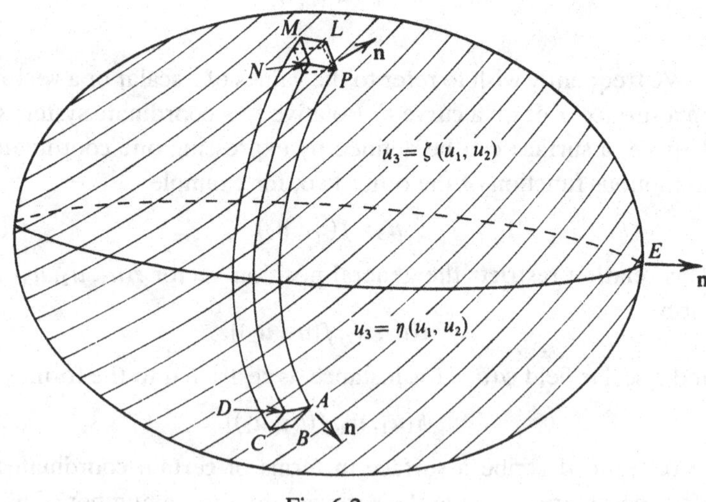


Fig. 6.2

$A$ having coordinates $(u_1, u_2, \eta(u_1, u_2))$ and the point $P$ having coordinates $(u_1, u_2, \zeta(u_1, u_2))$. In other situations, it may happen that one of the limits of $u_3$ corresponds to a point or a curve; for example, if the origin is within the volume $\tau$, and the boundary $\sigma$ is given in terms of spherical polar coordinates by $r = \zeta(\theta, \phi)$, then for given $\theta, \phi$, the limits of $r$ are 0 and $\zeta(\theta, \phi)$. But $r = 0$ is not part of the boundary, but simply an exceptional point of the coordinate system.

We have already noted that, using certain coordinate systems, the region $\tau$ does not correspond to ranges of the form (5.73); it must then be divided into a finite number of subregions each corresponding to ranges of this form. Also, it very frequently happens that the simple division of the bounding surface into two regions, as in Fig. 6.2, does not occur. In the following example, we examine a very common situation in which the bounding surface consists of six pieces.

**Example 6.3**

If the limits of each coordinate defining a volume region $\tau$ are constant, so that (5.73) is of the form

$$a \leqslant u_1 \leqslant b,$$
$$\alpha \leqslant u_2 \leqslant \beta,$$
$$\eta \leqslant u_3 \leqslant \zeta,$$

then it is best to divide the closed surface bounding $\tau$ into six parts. The definitions of these six parts, and the parameters used, are listed in Table 6.1. When $u_1 = x$, $u_2 = y$, $u_3 = z$, the region $\tau$ is a rectangular solid with six rectangular faces.

*Table 6.1*

| Defining equation (6.5) | Surface parameters | Parameter ranges |
|---|---|---|
| $u_1 = a$ | $u_2, u_3$ | $\alpha \leqslant u_2 \leqslant \beta$ |
| $u_1 = b$ | | $\eta \leqslant u_3 \leqslant \zeta$ |
| $u_2 = \alpha$ | $u_1, u_3$ | $a \leqslant u_1 \leqslant b$ |
| $u_2 = \beta$ | | $\eta \leqslant u_3 \leqslant \zeta$ |
| $u_3 = \eta$ | $u_1, u_2$ | $a \leqslant u_1 \leqslant b$ |
| $u_3 = \zeta$ | | $\alpha \leqslant u_2 \leqslant \beta$ |

As in Example 6.3, it is often desirable to choose a coordinate system so that the Equations (6.5) defining different parts of the boundary simply fix one coordinate at a constant value. This is not always possible, however; and even when coordinates can be chosen to simplify the equations defining the boundary surface, it may be preferable to use a different coordinate system, either because it is orthogonal, or because it simplifies the form of some given scalar or vector field in which we are interested.

A curve $\Gamma$ can be defined by expressing two of the coordinates $\{u_r\}$ as a function of the third, for example

$$u_2 = f(u_1), \tag{6.7a}$$

$$u_3 = g(u_1). \tag{6.7b}$$

These equations restrict the general position vector $\mathbf{r}(u_1, u_2, u_3)$ to the one-parameter form

$$\mathbf{r}(u_1, f(u_1), g(u_1)), \tag{6.8a}$$

and restrict a scalar field $\psi(u_s)$ to the range of values

$$\psi(u_1, f(u_1), g(u_1)). \tag{6.8b}$$

A vector field $\mathbf{v}(u_s)$ is restricted in the same way. In describing a particular curve, ambiguities of the type described in Example 6.2 may arise; more than one point on $\Gamma$ may correspond to a given value of $u_1$. Then $u_1$ does not satisfy the requirements of §5.3 for a simple parametrisation of the curve. When ambiguities of this kind arise, as they will later in this chapter, it is necessary to divide the curve into two or more pieces, on each of which the parameter value determines a unique position vector.

**Example 6.4**

The helix

$$\mathbf{r} = a \cos\phi \, \mathbf{i} + a \sin\phi \, \mathbf{j} + b\phi \, \mathbf{k}$$

can be described in terms of the coordinate $z = b\phi$; the conditions (6.7) are

$$x = a \cos(z/b), \; y = a \sin(z/b).$$

Since $x$ and $y$ are uniquely determined by these equations, $z$ is a parameter satisfying the conditions of §5.3.

If, however, $x$ is chosen as a parameter, $y$ and $z$ are determined by the equations

$$y = \pm(a^2 - x^2)^{\frac{1}{2}}, \qquad z = b\cos^{-1}(x/a).$$

Thus $y$ has two values, and $z$ has many values, for each value of $x$. The coordinates $y$ and $z$ are only determined as unique functions of $x$ if the helix is divided into sections corresponding to ranges of $\phi$ of the form $n\pi \leqslant \phi < (n+1)\pi$ ($n = 0, \pm 1, \pm 2, \ldots$).

As for closed surfaces, it is very often convenient to divide a closed curve into two or more sections on which the coordinate restrictions (6.7) are unambiguous. Examples of division of curves, into sections analogous to Examples 6.2 and 6.3 for surfaces, are:

(i) The use of $x$ as the parameter on the circle $x^2 + y^2 = a^2$; the two semi-circles have $y = +(a^2 - x^2)^{\frac{1}{2}}$ and $y = -(a^2 - x^2)^{\frac{1}{2}}$.

(ii) If $u_1$ and $u_2$ are coordinates in a plane, the use of $u_1$ and $u_2$ as parameters on those sections of the curves $u_1 = a$, $u_1 = b$, $u_2 = \alpha$, $u_2 = \beta$ which form a closed curve.

## 6.2 *Divergence of a vector field*

One of the physical systems which can most obviously be described by functions in a continuum is a fluid, which may be liquid or gaseous. Some of the quantities which may vary throughout a fluid are the density, which can be represented by scalar field, and the velocity of the fluid at a point, represented by a vector field $\mathbf{v}(\mathbf{r})$. Now suppose that $\delta\sigma_P$ is a plane element of area containing a point $P$, and that $\mathbf{n}$ is one of the two unit vectors normal of $\delta\sigma_P$, as in Fig. 6.3. Let $\mathbf{v}(\mathbf{r}) = \mathbf{v}_{\|}(\mathbf{r}) + \mathbf{v}_{\perp}(\mathbf{r})$, dividing the vector into two components, $\mathbf{v}_{\|}$ parallel to $\mathbf{n}$,

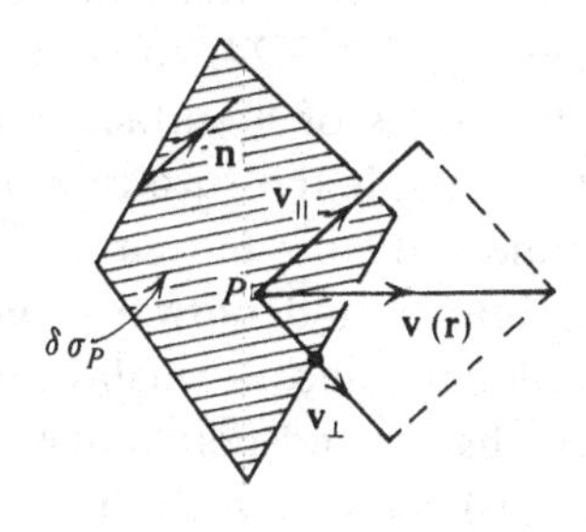


Fig. 6.3

and $\mathbf{v}_\perp$ orthogonal to $\mathbf{n}$. If $\mathbf{v}(\mathbf{r})$ is a continuous function of $\mathbf{r}$, and $\delta\sigma_P$ is small enough, the speed of flow through the element $\delta\sigma_P$ in direction $\mathbf{n}$ is approximately $\mathbf{n}\cdot\mathbf{v}_\parallel(\mathbf{r})$; the component $\mathbf{v}_\perp(\mathbf{r})$ does not contribute to the flow *through* $\delta\sigma_P$. Since $\mathbf{n}\cdot\mathbf{v}_\perp=0$, the speed of flow equals $\mathbf{n}\cdot\mathbf{v}$; hence the volume flowing through $\delta\sigma_P$ per unit time is approximately

$$\delta\sigma_P\mathbf{v}(\mathbf{r})\cdot\mathbf{n}. \tag{6.9}$$

Although we have talked about the flow of a fluid, the concept of rate of flow or **flux** is important when we consider any vector field $\mathbf{v}(\mathbf{r})$. If $\mathbf{v}_\parallel$ is in the opposite sense to $\mathbf{n}$, the flux (6.9) is negative.

Now suppose that $\sigma$ is a surface satisfying the conditions set out in §5.4, and that it is divided into elements by a coarse mesh, with $\delta\sigma_P$ representing the plane elements defined by the coarse mesh. Summing (6.9) over these elements, the total flow through these elements is approximately

$$\sum_P \delta\sigma_P\,\mathbf{v}(\mathbf{r})\cdot\mathbf{n}(\mathbf{r}), \tag{6.10}$$

where $\mathbf{n}(\mathbf{r})$ indicates that the normal vector is in general different for each surface element. If $\mathbf{v}(\mathbf{r})$ is piecewise continuous over $\sigma$, then in the limit of a fine mesh, with each element $\delta\sigma_P\to 0$, this sum becomes the surface integral

$$\iint_\sigma \mathrm{d}\sigma\,\mathbf{v}\cdot\mathbf{n}\equiv\iint_\sigma \mathrm{d}\boldsymbol{\sigma}\cdot\mathbf{v}, \tag{6.11}$$

of the form (5.70). In this limit, (6.9) accurately represents the flux of $\mathbf{v}$ through $\delta\sigma_P$, so that (6.11) defines the **flux through the surface** $\sigma$. Apart from exceptional points, we can define two unique normals $\pm\mathbf{n}(\mathbf{r})$ at each point of $\sigma$, where $\mathbf{n}(\mathbf{r})$ is a continuous function of $\mathbf{r}$; so continuity of $\mathbf{n}(\mathbf{r})$ determines two distinct 'sides' of the surface. Even when we deal with surfaces with folds or vertices where $\mathbf{n}(\mathbf{r})$ is undefined (as in Examples 5.19, 5.24, and 6.3), it will be intuitively clear which are the two sides of a surface; we will only consider surfaces for which this is so. If the surface $\sigma$ is the boundary of a continuous finite volume $\tau$ then $\sigma$ has an 'inside' and an 'outside', corresponding to the **inward** and **outward normals**. The outward normal can be distinguished, since a small displacement from $\sigma$ in that direction and sense has an end-point outside the volume $\tau$; from such a point, (continuous) curves can be drawn, containing points $\mathbf{p}$ with $|\mathbf{p}|$ arbitrary large, which do not intersect $\tau$.

We are now in a position to establish an important identity, the *divergence theorem*, for the flux (6.11) over the closed boundary $\sigma$ of a finite volume $\tau$; we assume $\tau$ contains no exceptional points, either of $\mathbf{v}$ or of the coordinate system. In the surface integral (6.11), $\mathbf{n}$ is taken to be the *outward* normal of $\sigma$. Now suppose that the volume $\tau$ is divided into cells by a coarse mesh defined by an orthogonal system of coordinates $u_1$, $u_2$, $u_3$, and that, as in Fig. 5.15, coordinate values $(u_1, u_2)$, $(u_1, u_2+\delta u_2)$, $(u_1+\delta u_1, u_2)$, $(u_1+\delta u_1, u_2+\delta u_2)$ define adjacent curves of the mesh. For the simple volume shown in Fig. 6.2, the curves $AP$, $BL$, $CM$ and $DN$ define a 'tube', with $u_3$ increasing from $\eta(u_1, u_2)$ to $\zeta(u_1, u_2)$ along $AP$; $ABCD$ and $PLMN$ are plane elements tangent to the surface $\sigma$, so that the unit normals $\mathbf{n}$ to the surface at $A$, $P$ are normal to these elements. Since $\mathbf{n}$ is chosen as the *outward* normal at every point, as shown, $n_3 \leqslant 0$ at the lower limit $u_3 = \eta(u_1, u_2)$, since $u_3$ decreases in the direction of the vector $\mathbf{n}$; likewise $n_3 \geqslant 0$ at the upper limit $u_3 = \zeta(u_1, u_2)$. The value $n_3 = 0$ is included because there are points such as $E$ in Fig. 6.2 where $\mathbf{n}$ is orthogonal to the curves with $u_1$ and $u_2$ constant; for the closed surface of Example 6.3, $n_3 = 0$ for the four parts of $\sigma$ defined by $u_1 = a$, $u_1 = b$, $u_2 = \alpha$, $u_2 = \beta$.

Let us now consider the contribution

$$\delta\sigma_P v_3(\mathbf{r}) n_3(\mathbf{r}) \tag{6.12}$$

to one of the terms in the sum (6.10). In Fig. 6.4, the upper part of the tube in Fig. 6.2 is shown, enlarged. The parallelogram $PLMN$ is

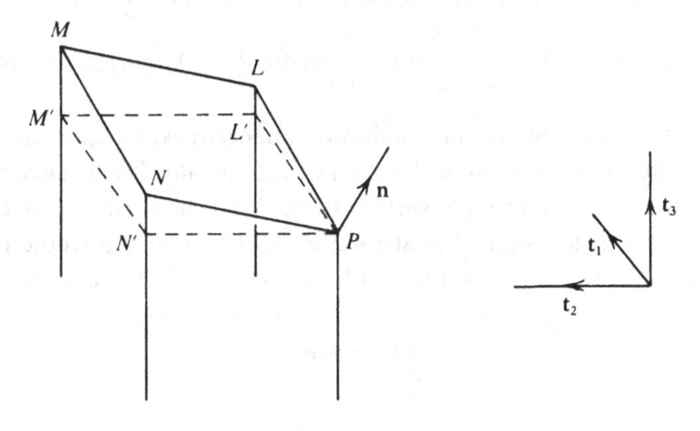


Fig. 6.4

defined by displacements **PL** and **PN** which are tangent to $\sigma$ at $P$, and which corresponds to increases $\delta u_1$, $\delta u_2$ in the parameters $u_1$, $u_2$. Their projections **PL**$'$ and **PN**$'$ orthogonal to $\mathbf{t}_3$ are therefore the displacements of magnitude $h_1\,\delta u_1$ and $h_2\,\delta u_2$ parallel to $\mathbf{t}_1$ and $\mathbf{t}_2$, where $h_1$ and $h_2$ are given by (5.67). Since the system of coordinates is assumed to be orthogonal, these displacements define the rectangle $PL'M'N'$ whose area is $h_1h_2\,\delta u_1\,\delta u_2$, by (5.66b). But since $\delta\sigma_P$ is the area of $PLMN$, $n_3\delta\sigma_P$ is just the area of $PL'M'N'$, its projection orthogonal to $\mathbf{t}_3$; so (6.12) can be expressed

$$\delta\sigma_P v_3(u_s)n_3(u_s) = h_1(u_s)h_2(u_s)v_3(u_s)\,\delta u_1\,\delta u_2, \qquad (6.13a)$$

where the coordinate values $\{u_s\}$ correspond to the point $P$; we have shown the dependence of $h_1$ and $h_2$ on $\{u_s\}$ explicitly, since they are in general functions of position. Applying the same argument to the surface element at $A$ in Fig. 6.2, but remembering that $n_3$ is negative there, the corresponding contribution from $A$ to (6.10) is

$$\delta\sigma_A v_3(u_s)n_3(u_s) = -h_1(u_s)h_2(u_s)v_3(u_s)\,\delta u_1\,\delta u_2, \qquad (6.13b)$$

where the values $\{u_s\}$ now correspond to the position vector of $A$. In (6.13a) and (6.13b), the increments $\delta u_1$, $\delta u_2$ have the same values all along the tube from $A$ to $P$ in Fig. 6.2, and the values of $u_1$ and $u_2$ are fixed along the curve from $A$ to $P$; thus *only* $u_3$ varies along the tube in expression (6.13a). Since we are assuming that the derivatives of $\{h_r(u_s)\}$ and $\{v_p(u_s)\}$ are continuous, the sum of (6.13a) and (6.13b) can therefore be expressed

$$\begin{aligned}\delta u_1\,\delta u_2\,\{[h_1(u_s)h_2(u_s)v_3(u_s)]_P - [h_1(u_s)h_2(u_s)v_3(u_s)]_A\}\\ = \delta u_1\,\delta u_2 \int_{\eta(u_1,u_2)}^{\zeta(u_1,u_2)} \mathrm{d}u_3\,\frac{\partial}{\partial u_3}[h_1(u_s)h_2(u_s)v_3(u_s)];\end{aligned} \qquad (6.14)$$

this is the 3-variable generalisation of Theorem A.2 (Appendix A), essentially the fundamental theorem of the calculus [Reference 6.1].

Considering still a simple surface of the kind shown in Fig. 6.2, we can now form the sum of contributions (6.12) over the whole of $\sigma$, and then, in the fine mesh limit with $\delta\sigma_P \to 0$, obtain the contribution

$$\iint_\sigma \mathrm{d}\sigma\, v_3 n_3 \qquad (6.15)$$

to (6.11). Since (6.14) gives the contribution from the two points $A$, $P$

corresponding to given values of $u_1$, $u_2$, summing (6.14) over all 'tubes' in Fig. 6.2 corresponds to $A$, $P$ varying over the whole surface $\sigma$, with $u_1$, $u_2$ varying over the ranges (5.8). In the fine mesh limit, the sum over $u_1$, $u_2$ becomes a double integral; the right-hand side of (6.14) then gives the surface integral (6.15); the left-hand side gives a triple integral over the ranges (5.73) defining $\tau$. Thus

$$\iint_\sigma \mathrm{d}\sigma\, v_3 n_3 = \int_a^b \mathrm{d}u_1 \int_{\alpha(u_1)}^{\beta(u_1)} \mathrm{d}u_2 \int_{\eta(u_1,u_2)}^{\zeta(u_1,u_2)} \mathrm{d}u_3 \frac{\partial}{\partial u_3} \times [h_1(u_s)h_2(u_s)v_3(u_s)]. \tag{6.16}$$

Using expression (5.77b) for the volume element $\delta\tau$, this integral can be expressed as

$$\iiint_\tau \mathrm{d}\tau \frac{1}{h_1 h_2 h_3} \frac{\partial}{\partial u_3}(h_1 h_2 v_3). \tag{6.17}$$

This result can be established for any volume $\tau$, bounding surface $\sigma$, and coordinate system $\{u_s\}$ satisfying the conditions of Chapter 5. If any region of $\sigma$ has $n_3 \equiv 0$, as in Example 6.3, it gives no contribution to (6.15), leaving the result (6.17) unchanged. As we explained in §5.5, using a particular coordinate system may necessitate dividing the volume $\tau$ into a finite number of parts, each of which corresponds to ranges of the form (5.73), and $\sigma$ may not comprise the whole boundary of these parts. Consider, for example, the division of $\tau$ into two parts, $\tau^{(1)}$ and $\tau^{(2)}$, as in Fig. 6.5, with a common surface $\sigma^{(3)}$. The

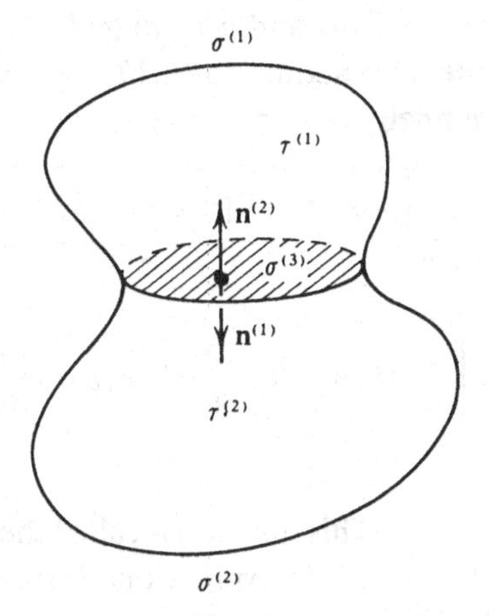


Fig. 6.5

outward unit normals $\mathbf{n}^{(1)}$ and $\mathbf{n}^{(2)}$ to the two volumes are equal and opposite over $\sigma$, as shown, so that

$$\iint_{\sigma^{(3)}} d\sigma\,(\mathbf{v}\cdot\mathbf{n}^{(1)}+\mathbf{v}\cdot\mathbf{n}^{(2)})=0; \tag{6.18}$$

this equation simply means that the flux of $\mathbf{v}$ out of $\tau^{(1)}$ over $\sigma^{(3)}$ is equal to the flux into $\tau^{(2)}$. If the original surface $\sigma$ is divided into the two sections $\sigma^{(1)}$ and $\sigma^{(2)}$, as shown, the total flux (6.11) can be written

$$\iint_{\sigma^{(1)}} d\sigma\,\mathbf{v}\cdot\mathbf{n}+\iint_{\sigma^{(2)}} d\sigma\,\mathbf{v}\cdot\mathbf{n}.$$

Adding the zero term (6.18), we obtain

$$\iint_{\sigma^{(1)}+\sigma^{(3)}} d\sigma\,\mathbf{v}\cdot\mathbf{n}^{(1)}+\iint_{\sigma^{(2)}+\sigma^{(3)}} d\sigma\,\mathbf{v}\cdot\mathbf{n}^{(2)},$$

where the unit outward normals on $\sigma^{(1)}$ and $\sigma^{(2)}$ have been written as $\mathbf{n}=\mathbf{n}^{(1)}$ and $\mathbf{n}=\mathbf{n}^{(2)}$ respectively; this is just the sum of the integrals of form (6.11) over the complete boundaries of $\tau^{(1)}$ and $\tau^{(2)}$. The two terms of the form (6.15) are, as before, equal to integrals of the form (6.17) over the volumes $\tau^{(1)}$ and $\tau^{(2)}$, so that their sum is exactly the integral (6.17) over the whole volume $\tau$ (using the 3-variable generalisation of Theorem A.6, Appendix A). This argument can clearly be extended to a volume $\tau$ divided into any finite number of non-overlapping parts, ensuring the validity of the result (6.17).

The surface integral of $v_1n_1$ and $v_2n_2$ in (6.11) can be transformed in the same way as (6.15); adding to (6.17) similar results for these surface integrals, we finally obtain

$$\iint_{\sigma} d\boldsymbol{\sigma}\cdot\mathbf{v}=\iiint_{\tau} d\tau\,\operatorname{div}\mathbf{v}, \tag{6.19}$$

where we define

$$\operatorname{div}\mathbf{v}=\frac{1}{h_1h_2h_3}\left[\frac{\partial}{\partial u_1}(h_2h_3v_1)+\frac{\partial}{\partial u_2}(h_3h_1v_2)+\frac{\partial}{\partial u_3}(h_1h_2v_3)\right], \tag{6.20}$$

at all points where it exists. This function is called the **divergence** of the vector field $\mathbf{v}$; (6.19) and (6.20) express one form of the **divergence theorem**, or **Green's theorem**.

Since $\{h_s\}$ and $\{v_s\}$ are assumed to have continuous derivatives and $h_s \neq 0$, div $\mathbf{v}$ is a continuous function of the coordinates $\{u_s\}$, except possibly at exceptional points. Now suppose that the lengths of the ranges (5.73) all tend to zero in such a way that they always contain the coordinates $\{u_s\}$ of a particular normal point $P$, and that

$$V_\tau = \iiint_\tau \mathrm{d}\tau \tag{6.21}$$

is the volume of $\tau$. Then the 3-dimensional analogue of Theorem A.8 (Appendix A) tells us that in this limit,

$$\operatorname{div} \mathbf{v}(u_s) = \lim_{\tau \to 0} \frac{1}{V_\tau} \iiint \operatorname{div} \mathbf{v} \, \mathrm{d}\tau,$$

since div $\mathbf{v}$ can be 'taken outside the integral sign'. Using the divergence theorem (6.19) then gives the expression

$$\operatorname{div} \mathbf{v} = \lim_{\tau \to 0} \frac{1}{V_\tau} \iint_\sigma \mathrm{d}\boldsymbol{\sigma} \cdot \mathbf{v} \tag{6.22}$$

for the divergence at the point $P$.

If we wish, we can regard (6.22) as the definition of the divergence, instead of (6.20). One advantage in using (6.22) is that it is independent of any coordinate system, since the integrand is a scalar product. However, the limit $\tau \to 0$ needs to be defined very carefully, since difficulties can arise if the surface $\sigma$ becomes very 'corrugated' as $\tau$ shrinks upon the point $P$; we shall not discuss this particular problem, since (6.20) is a valid and equivalent definition of divergence. The fact that $V_\tau$, the surface elements (5.66a), and the integrand $\mathbf{n} \cdot \mathbf{v}$ in (6.22) are all scalars means that (6.22) is a scalar, and is therefore independent of the choice of coordinates; it follows that div $\mathbf{v}$, evaluated by using (6.20), must be the same for all orthogonal systems of coordinates. For rectangular coordinates $(x, y, z)$ with $h_r \equiv 1$ $(r = 1, 2, 3)$, the divergence of $\mathbf{v} = v_1\mathbf{i} + v_2\mathbf{j} + v_3\mathbf{k}$ is

$$\operatorname{div} \mathbf{v} = \frac{\partial v_1}{\partial x} + \frac{\partial v_2}{\partial y} + \frac{\partial v_3}{\partial z}. \tag{6.23a}$$

The expression (6.22) allows an immediate physical interpretation of the divergence: the surface integral is the outward flux of $\mathbf{v}$ over $\sigma$; division by $V_\tau$ means that the divergence measures *outflow per unit volume* of the vector field $\mathbf{v}$ at any point. This concept can be used to give a very clear meaning to (6.19), which simply says that the outflow of $\mathbf{v}$ over the boundary $\sigma$ is the integral throughout $\tau$ of the outflow per unit volume. In fact, the divergence theorem (6.19) can be looked upon as the integral of the definition (6.22). Suppose that the volume $\tau$ is divided into small cells by a coarse mesh; if $\delta\tau_P$ is the volume of a small cell containing a point $P$, then (6.22) tells us that the outflow from the cell is approximately

$$\delta\tau_P(\operatorname{div}\mathbf{v})_P. \tag{6.24}$$

Denoting by $\delta\sigma_P$ the surface of the cell with volume $\delta\tau_P$, the sum of the outflows from all cells in $\tau$ is

$$\sum_{\text{cells}} \iint_{\delta\sigma_P} \mathbf{v}\cdot d\boldsymbol{\sigma}. \tag{6.25}$$

Now any surface element of a cell which lies within $\tau$ is shared between two cells, just as $\sigma^{(3)}$ is shared in Fig. 6.5; so in the 'sum over cells', equations of type (6.18) ensure that contributions from all boundary elements *inside* $\tau$ cancel in pairs; all that remain are the integrals over the surface elements belonging to $\sigma$. So the sum (6.25) is exactly

$$\iint_{\sigma} d\boldsymbol{\sigma}\cdot\mathbf{v}.$$

But, using (6.24), the sum of the outflows is approximately

$$\sum_{\text{cells}} \delta\tau_P(\operatorname{div}\mathbf{v})_P;$$

in the limit $\delta\tau_P \to 0$ of a fine mesh, each term accurately represents the outflow from a cell, and the sum then becomes the volume integral in (6.19). This non-rigorous derivation of the divergence theorem can be made mathematically rigorous; our purpose here is to bring out the essential physical meaning of (6.19) and (6.22) in terms of 'outflow' or 'outward flux'.

**Example 6.5**

If $(r, \theta, \phi)$ are spherical polar coordinates, the corresponding functions (5.79b) are

$$h_r = 1, \qquad h_\theta = r, \qquad h_\phi = r\sin\theta.$$

Since the system of coordinates is orthogonal, expression (6.20) with $u_1 = r$, $u_2 = \theta$, $u_3 = \phi$ and $v_1 = v_r$, $v_2 = v_\theta$, $v_3 = v_\phi$ gives

$$\operatorname{div} \mathbf{v} = \frac{1}{r^2 \sin\theta}\left[\frac{\partial}{\partial r}(r^2 \sin\theta\, v_r) + \frac{\partial}{\partial \theta}(r \sin\theta\, v_\theta) + \frac{\partial}{\partial \phi}(r v_\phi)\right]$$

$$= \frac{1}{r^2}\frac{\partial}{\partial r}(r^2 v_r) + \frac{1}{r\sin\theta}\left[\frac{\partial}{\partial\theta}(\sin\theta\, v_\theta) + \frac{\partial v_\phi}{\partial \phi}\right]. \qquad (6.23b)$$

We now use this result to evaluate div $\mathbf{v}$ when $\mathbf{v}$ is the radial vector field with components $v_r = r^3$, $v_\theta = v_\phi = 0$. We obtain

$$\operatorname{div} \mathbf{v} = \frac{1}{r^2}\frac{\partial}{\partial r}(r^5) = 5r^2.$$

We can also obtain this result by using rectangular coordinates. The final equations of Example 6.1 with $v_r = r^3$ give

$$v_1 = xr^2 = x(x^2+y^2+z^2),$$
$$v_2 = y(x^2+y^2+z^2),$$
$$v_3 = z(x^2+y^2+z^2),$$

as the rectangular components of $\mathbf{v}$. Then (6.23a) gives

$$\operatorname{div} \mathbf{v} = (3x^2+y^2+z^2) + (x^2+3y^2+z^2) + (x^2+y^2+3z^2)$$
$$= 5r^2.$$

This result exemplifies the fact that div $\mathbf{v}$ is independent of the coordinate system used.

**Example 6.6**

The components of a vector field $\mathbf{v}$ in the orthogonal directions defined by cylindrical polar coordinates $(\rho, \phi, z)$ are

$$v_\rho = \rho^2, \qquad v_\phi = \rho^2 \sin 2\phi, \qquad v_z = \lambda\rho z,$$

where $\lambda$ is constant. Check that the divergence theorem (6.19) holds for this field $\mathbf{v}$, where $\tau$ is the circular cylinder bounded by the surfaces

$$z = 0, \qquad z = h \quad \text{and} \quad \rho = a.$$

For cylindrical polar coordinates, the functions (5.79) are

$$h_\rho = 1, \qquad h_\phi = \rho, \qquad h_z = 1.$$

So (6.20) gives

$$\operatorname{div} \mathbf{v} = \frac{1}{\rho}\frac{\partial}{\partial\rho}(\rho v_\rho) + \frac{1}{\rho}\frac{\partial v_\phi}{\partial\phi} + \frac{\partial v_z}{\partial z} \qquad (6.23c)$$

so that, for the given field **v**,

$$\operatorname{div} \mathbf{v} = 3\rho + 2\rho \cos 2\phi + \lambda\rho.$$

We note that **v** is continuously differentiable everywhere, even at the end-points $\phi = 0$ and $\phi = 2\pi$ of the range of $\phi$, so that the divergence theorem is valid. Using the form (5.83c) for the volume integral over the cylinder, the right of (6.19) becomes

$$\iiint \operatorname{div} \mathbf{v}\, \mathrm{d}\tau = \int_{\rho=0}^{a} \rho\, \mathrm{d}\rho \int_{\phi=0}^{2\pi} \mathrm{d}\phi \int_{z=0}^{h} \mathrm{d}z\, \rho(3+\lambda+2\cos 2\phi)$$

since

$$\int_{\phi=0}^{2\pi} \mathrm{d}\phi\, 2\cos 2\phi = 0,$$

this integral is

$$\tfrac{1}{3}a^3 2\pi(3+\lambda)h = 2\pi a^3 h(1+\tfrac{1}{3}\lambda). \tag{6.26}$$

The integral on the left of (6.19) is the sum of surface integrals over the circular discs

$$\sigma_1 : z = 0, \qquad 0 \leqslant \rho \leqslant a, \qquad 0 \leqslant \phi \leqslant 2\pi,$$
$$\sigma_2 : z = h, \qquad 0 \leqslant \rho \leqslant a, \qquad 0 \leqslant \phi \leqslant 2\pi,$$

and the cylindrical surface

$$\sigma_3 : \rho = a, \qquad 0 \leqslant \phi \leqslant 2\pi, \quad 0 \leqslant z \leqslant h.$$

On $\sigma_1$ the outward normal component of **v** is

$$\mathbf{v} \cdot \mathbf{n} = -v_z = -\lambda\rho z = 0,$$

since $z = 0$, so that this surface gives zero contribution. On $\sigma_2$, the component is

$$\mathbf{v} \cdot \mathbf{n} = +v_z = \lambda\rho z = \lambda\rho h.$$

This surface therefore contributes

$$\int_{\rho=0}^{a} \rho\, \mathrm{d}\rho \int_{\phi=0}^{2\pi} \mathrm{d}\phi\, \lambda\rho h = \tfrac{2}{3}\pi a^3 h\lambda,$$

equal to the second term in the volume integral (6.26). On $\sigma_3$ the normal component is, since $\rho = a$,

$$\mathbf{v} \cdot \mathbf{n} = +v_\rho = \rho^2 = a^2$$

and the contribution of this surface to the left of (6.19) is thus

$$\int_{\phi=0}^{2\pi} a\, \mathrm{d}\phi \int_{z=0}^{h} \mathrm{d}z\, a^2 = 2\pi a^3 h.$$

This is equal to the first term in (6.26), so that the divergence theorem is verified.

### Problems 6.1

1 Use the results of Example 5.25 and of Problems 5.5, Question 7 to evaluate the divergence of a vector $\mathbf{v}$ in ($a$) oblate spheroidal coordinates, ($b$) prolate spheroidal coordinates.

2 Check the divergence theorem (6.19) for the vector field

$$\mathbf{v}=x^2\,e^z\mathbf{i}+y^2\,e^z\mathbf{j}+z^2\mathbf{k}$$

over the cube $\tau$ bounded by the six planes $x=0$, $y=0$, $z=0$, $x=a$, $y=a$, $z=a$.

3 A vector field $\mathbf{v}$ is defined by

$$\mathbf{v}=xz\mathbf{i}+yz\mathbf{j}+(x^2+y^2)\mathbf{k}.$$

Calculate div $\mathbf{v}$, and use the matrix (6.4) to show that the components of $\mathbf{v}$ in the orthogonal directions defined by spherical polar coordinates are

$$v_r=2r^2\sin^2\theta\cos\theta,$$

$$v_\theta=r^2\sin\theta(1-2\sin^2\theta),$$

$$v_\phi=0.$$

Check the divergence theorem (6.19) for the field $\mathbf{v}$, where the volume $\tau$ is half of the sphere $r\leqslant a$ with $z\geqslant 0$.

## 6.3 *Gradient of a scalar field; conservative fields*

The concept of 'rate of change' of a function is fundamental to the differential calculus; for functions of a single variable, the rate of change is equal to the derivative. In order to describe the rate of change of a scalar field in 3-space, we assume that it is a function $\psi(u_s)$ of a set of three curvilinear coordinates $\{u_s\}$ satisfying the conditions of §5.5; we further assume that the coordinate system is orthogonal. If $\mathbf{r}(u_s)$ is the position vector of a point $P$, then the magnitudes and directions of $\partial\mathbf{r}/\partial u_s$ are given by (5.80):

$$\frac{\partial \mathbf{r}}{\partial u_s}=h_s\mathbf{t}_s \qquad (s=1,2,3). \tag{5.80}$$

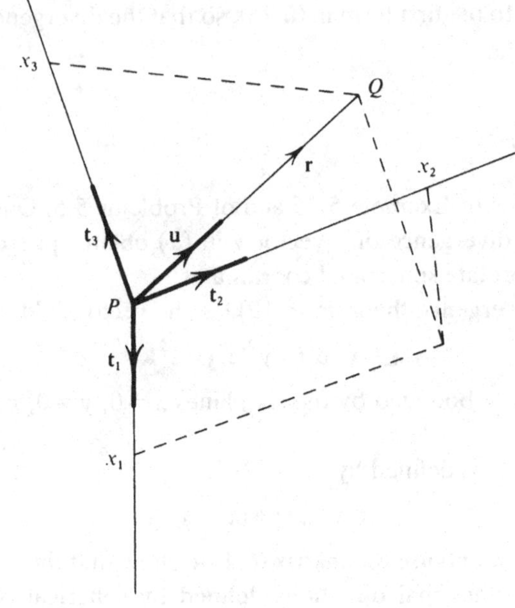


Fig. 6.6

If $\{x_s\}$ are rectangular coordinates relative to the triad $\{\mathbf{t}_s\}$, as in Fig. 6.6, then

$$\mathbf{r} = \sum_r x_r \mathbf{t}_r; \tag{6.27a}$$

these coordinates are functions

$$x_r = X_r(u_s) \qquad (r = 1, 2, 3) \tag{6.28}$$

of the curvilinear coordinates, so that

$$\mathbf{r} = \sum_r X_r(u_s)\mathbf{t}_r. \tag{6.27b}$$

Differentiating this equation with respect to $u_s$ gives

$$\frac{\partial \mathbf{r}}{\partial u_s} = \sum_r \frac{\partial X_r}{\partial u_s}\mathbf{t}_r.$$

Comparing with (5.80) above, we find that

$$\frac{\partial X_r}{\partial u_s} = h_s \delta_{rs} \qquad (r, s = 1, 2, 3), \tag{6.29}$$

where $\delta_{rs}$ is the Kronecker delta; this is simply another way of expressing the fact that $\{h_s\}$ are the rates of change of $\mathbf{r}$ with respect to the coordinates $\{u_s\}$.

A scalar field $\psi$ can be regarded as a function of the coordinates $\{x_r\}$; then the rate of change of $\psi$ in direction $\mathbf{t}_1$ is the partial derivative $\partial\psi/\partial x_1$, with $x_2$, $x_3$ fixed. Now, using the chain rule and (6.29) with $s=1$, the rate of change of $\psi$ with respect to $u_1$ is

$$\frac{\partial\psi}{\partial u_1}=\sum_r \frac{\partial\psi}{\partial x_r}\frac{\partial X_r}{\partial u_1}$$

$$=h_1\frac{\partial\psi}{\partial x_1}.$$

Similar equations hold for $\partial\psi/\partial u_2$, $\partial\psi/\partial u_3$, so the rates of change of $\psi$ in directions $\mathbf{t}_r$ $(r=1,2,3)$ are

$$\frac{\partial\psi}{\partial x_r}=\frac{1}{h_r}\frac{\partial\psi}{\partial u_r}. \tag{6.30}$$

We are now in a position to calculate the rate of change of $\psi$ in *any* direction in 3-space. The radial distance from the origin $P$ is $r=|\mathbf{r}|$; the unit vector $\mathbf{u}$ parallel to $\mathbf{r}$, shown in Fig. 6.6, is

$$\mathbf{u}=\sum_s l_s\mathbf{t}_r, \tag{6.31}$$

where $\{l_s\}$ are the direction cosines of $\mathbf{r}$. By (2.61),

$$x_s=rl_s \quad (s=1,2,3);$$

this defines $\{x_s\}$ as functions of spherical polar coordinates $(r,\theta,\phi)$, with

$$(l_1,l_2,l_3)=(\sin\theta\cos\phi,\sin\theta\sin\phi,\cos\theta),$$

and we can differentiate partially with respect to $r$ to give

$$\frac{\partial x_s}{\partial r}=l_s \qquad (s=1,2,3). \tag{6.32}$$

The chain rule and the results (6.30) and (6.32) can then be used to calculate the rate of change $\partial\psi/\partial r$ in the radial direction, known as the **directional derivative** of $\psi$ in direction $\mathbf{u}$:

$$\frac{\partial\psi}{\partial r}=\sum_s \frac{\partial\psi}{\partial x_s}\frac{\partial x_s}{\partial r}$$

$$=\sum_s l_s\frac{\partial\psi}{\partial x_s}$$

$$=\sum_s l_s\frac{1}{h_s}\frac{\partial\psi}{\partial u_s}. \tag{6.33a}$$

If we now define the **gradient vector**

$$\text{grad}\,\psi = \sum_s \frac{1}{h_s}\frac{\partial\psi}{\partial u_s}\mathbf{t}_s, \tag{6.34}$$

with components given by (6.30), then, using (6.31), (6.33a) can be written in scalar product form

$$\frac{\partial\psi}{\partial r} = \mathbf{u}\cdot\text{grad}\,\psi. \tag{6.33b}$$

Equations (6.30) and (6.33b) express the two fundamental properties of the gradient vector:

(i) the components of grad $\psi$ in the directions $\{\mathbf{t}_s\}$ defined by an orthogonal system $\{u_s\}$ are equal to the rates of change of the scalar field $\psi$ in these directions;

(ii) the directional derivative of $\psi$ in the direction of any unit vector $\mathbf{u}$ is equal to the component of grad $\psi$ in direction $\mathbf{u}$.

At each point $P$ the gradient vector (6.34) defines a particular direction in 3-space; since the component (6.33b) of grad $\psi$ is greatest when $\mathbf{u}$ is parallel to grad $\psi$, a third property of the gradient vector is;

(iii) the rate of increase of a scalar field $\psi$ is greatest in the direction of grad $\psi$, where it equals the modulus of grad $\psi$.

If $\psi_0$ is a given real number, the equation

$$\psi(u_s) = \psi_0 \tag{6.35}$$

is a single relation between the variables $u_r$. As in Example 6.2, this relation may not define one coordinate, say $u_3$, *uniquely* in terms of $u_1$, $u_2$; but if we solve (6.35) for $u_3$, the solution (or solutions) will define a surface (or surfaces) satisfying the conditions of Chapter 5, since $\psi$ is assumed to have continuous derivatives; this ensures that the gradient vector (6.34) exists, apart from exceptional points and curves. Any surface defined by (6.35), on which $\psi$ takes the constant value $\psi_0$, is called a **level surface** of the scalar field $\psi$. If $P$ and $Q$ are two points on the *same* level surface (6.35), and $\psi(P)$ and $\psi(Q)$ the corresponding values of $\psi$, then

$$\psi(P) - \psi(Q) = 0. \tag{6.36}$$

Now let $Q$ approach $P$, so that $r \to 0$ in Fig. 6.6, in such a way that $\mathbf{u}$ becomes a definite tangent vector to the level surface. In this limit, (6.36) divided by $\delta r$ becomes

$$\frac{\partial\psi}{\partial r} = 0,$$

so that (6.33b) implies

$$\mathbf{u} \cdot \operatorname{grad} \psi = 0. \tag{6.37}$$

Since **u** may be any tangent vector at $P$ to the level surface, we have established the fourth property;

(iv) at any point $P$, grad $\psi$ is orthogonal to the level surface (6.35) with $\psi_0 = \psi(P)$.

The three most frequently used coordinate systems are rectangular, cylindrical polar, and spherical polar coordinates. Since $h_1 = h_2 = h_3 = 1$ for rectangular coordinates, the components (6.30) of grad $\psi$ are simply $\partial\psi/\partial x$, $\partial\psi/\partial y$ and $\partial\psi/\partial z$, and

$$\operatorname{grad} \psi = \frac{\partial\psi}{\partial x}\mathbf{i} + \frac{\partial\psi}{\partial y}\mathbf{j} + \frac{\partial\psi}{\partial z}\mathbf{k}. \tag{6.38}$$

For cylindrical polar coordinates, $h_\rho = 1$, $h_\phi = \rho$, $h_z = 1$, so that (6.34) becomes

$$\operatorname{grad} \psi = \frac{\partial\psi}{\partial\rho}\mathbf{t}_\rho + \frac{1}{\rho}\frac{\partial\psi}{\partial\phi}\mathbf{t}_\phi + \frac{\partial\psi}{\partial z}\mathbf{t}_z. \tag{6.39}$$

For spherical polar coordinates, with $h_r = 1$, $h_\theta = r$, $h_\phi = r \sin\theta$, the gradient vector is

$$\operatorname{grad} \psi = \frac{\partial\psi}{\partial r}\mathbf{t}_r + \frac{1}{r}\frac{\partial\psi}{\partial\theta}\mathbf{t}_\theta + \frac{1}{r\sin\theta}\frac{\partial\psi}{\partial\phi}\mathbf{t}_\phi. \tag{6.40}$$

Note that the component along $\mathbf{t}_\phi$ is the same in (6.39) and (6.40), since $\rho = r \sin\theta$.

**Example 6.7**

A scalar field $\psi$ is given by

$$\psi = \frac{xyz}{x^2 + y^2}.$$

Find the components of grad $\psi$ in rectangular, cylindrical polar and spherical polar coordinates.

In rectangular coordinates, the components of (6.38) are

$$\frac{(y^2 - x^2)yz}{(x^2 + y^2)^2}, \frac{(x^2 - y^2)xz}{(x^2 + y^2)^2}, \frac{xy}{x^2 + y^2}.$$

In terms of cylindrical polar coordinates,

$$\psi = z \cos\phi \sin\phi = \tfrac{1}{2}z \sin 2\phi.$$

Then from (6.39), the components along $\mathbf{t}_\rho$, $\mathbf{t}_\phi$, $\mathbf{t}_z$ are

$$(0, \rho^{-1}z\cos 2\phi, \tfrac{1}{2}\sin 2\phi).$$

The $z$-component is equal to that given by rectangular coordinates. In terms of spherical polar coordinates,

$$\psi = r\cos\theta\cos\phi\sin\phi = \tfrac{1}{2}r\cos\theta\sin 2\phi.$$

The components of grad $\psi$ given by (6.40) are then

$$(\tfrac{1}{2}\cos\theta\sin 2\phi, -\tfrac{1}{2}\sin\theta\sin 2\phi, \cot\theta\cos 2\phi).$$

Again it is clear that the $\phi$-component is the same as that given by cylindrical polars.

For differentiable functions of one variable, integration can be regarded as an 'inverse operation' to differentiation. For a scalar field $\psi$, we can define an integral which in a rather similar way serves as the inverse of the gradient operation. Suppose that T is a curve satisfying the conditions (i)–(iii) of §5.1 and the conditions of smoothness defined early in §5.3 and that $s$ denotes the distance along T from $s = s_0$. Since $s$ measures distance in the direction of the tangent vector $\mathbf{t}$ at every point of $\Gamma$, (6.33b) gives

$$\mathbf{t}\cdot\text{grad}\,\psi = \frac{\partial\psi}{\partial s}$$

for the tangential component of grad $\psi$; since $\partial\psi/\partial s$ is assumed to be continuous, we can integrate this equation along $\Gamma$ from $s = s_0$ to $s = s_1$, giving the **path integral**

$$\int_\Gamma \mathrm{d}s\,\mathbf{t}\cdot\text{grad}\,\psi = \int_{s_0}^{s_1}\mathrm{d}s\frac{\partial\psi}{\partial s}$$
$$= \psi(s_1) - \psi(s_0).$$

Using the notation

$$\mathbf{ds} = \mathrm{d}s\,\mathbf{t}$$

for the vector increment along $\Gamma$, we have

$$\int_\Gamma \mathbf{ds}\cdot\text{grad}\,\psi = \psi(s_1) - \psi(s_0). \tag{6.41}$$

So the integral of the component of grad $\psi$ gives the difference between the values of $\psi$ at the end-points. It is very important to realise that the integral (6.41) depends only on the values of $\psi$ at the end-points, and is independent of the particular curve $\Gamma$ joining these end-points. If $\Gamma$ is infinite or semi-infinite, so that $s_1 \to +\infty$ and/or $s_0 \to -\infty$, the result (6.41) will still be valid, provided that $\psi(s_1)$ and/or $\psi(s_0)$ have unique well-defined limits for the particular curve.

If $\Gamma$ is a closed curve, so that $s = s_0$ and $s = s_1$ represent the same point, then $\psi(s_1) = \psi(s_0)$ and the integral (6.41) vanishes. The fact that $\Gamma$ is a closed curve is indicated by adding a circle to the integral sign; thus for any scalar field $\psi$,

$$\oint_\Gamma \mathbf{ds} \cdot \text{grad}\, \psi = 0 \tag{6.42}$$

around a closed curve $\Gamma$ containing no exceptional points of $\psi$.

We can write (6.41), with a sign change, in the form

$$\int_\Gamma \mathbf{v} \cdot \mathbf{ds} = \psi(s_0) - \psi(s_1), \tag{6.43}$$

where

$$\mathbf{v} = -\text{grad}\, \psi. \tag{6.44}$$

A vector field $\mathbf{v}$ of the form (6.44) is called a **conservative field**, and $\psi$ is known as the **potential** of the field; so, by definition, a field 'derivable from a potential $\psi$' is a conservative field. The converse is also true: suppose $\mathbf{v}$ is of the form (6.44), and we choose some arbitrary value $\psi_P$ for the scalar function $\psi(\mathbf{r})$ at a point $P$; then if $P$ corresponds to $s = s_0$ on any curve $\Gamma$, and an arbitrary point $Q$ corresponds to $s = s_1$, (6.43) gives the value of $\psi(\mathbf{r})$ at $Q$ to be

$$\psi_Q \equiv \psi(s_1) = \psi_P - \int_\Gamma \mathbf{v} \cdot \mathbf{ds} \tag{6.45}$$

*independent of the path* $\Gamma$ *joining* $P$ *and* $Q$. So the potential $\psi(\mathbf{r})$ is uniquely defined at all points.

If (6.43) holds for all $\Gamma$ within a simple volume $\tau$, then

$$\oint_\Gamma \mathbf{v} \cdot \mathbf{ds} = 0 \tag{6.46}$$

for any closed curve $\Gamma$. Conversely, if (6.46) holds for *all* closed curves $\Gamma$, and $\Gamma_1, \Gamma_2$ are any two curves from a point $P$ to another point $Q$, then (6.46) implies that

$$\int_{\Gamma_1} \mathbf{v} \cdot \mathbf{ds} = \int_{\Gamma_2} \mathbf{v} \cdot \mathbf{ds},$$

since $\Gamma_1$ and $\Gamma_2$ (traversed in the opposite direction) combine to form a closed curve. Again choosing an arbitrary value $\psi_P$ of $\psi(\mathbf{r})$ at $P$, (6.45) with $\Gamma=\Gamma_1$ or $\Gamma=\Gamma_2$ defines a *unique* potential $\psi(\mathbf{r})=\psi_Q$ at any point $Q$; then $\mathbf{v}$ and $\psi$ will be related by (6.44), and $\mathbf{v}$ is a conservative field. We have therefore established the equivalence of two conditions:

(i) $\mathbf{v}$ is conservative, of the form (6.44);

(ii) equation (6.46) is true for all closed curves $\Gamma$, enabling $\psi(\mathbf{r})$ to be defined by (6.45).

We note that condition (ii) need only apply to *simple* closed curves, since any closed curve can be regarded as a combination of a finite number of simple closed curves.

Path integrals (6.43) of the tangential component of a field $\mathbf{v}$ arise frequently in mathematical physics. For example, if $\mathbf{v}=\mathbf{f}$ represents a field of force, the integral in (6.43) is the **work** done by the field during motion along $\Gamma$. If $\mathbf{f}$ is of the form (6.44), then the work done in going from $P$ to $Q$ is given by (6.43), and is independent of the path $\Gamma$ traversed; it is equal to the potential difference $\psi(P)-\psi(Q)$ between the two points. For a conservative force field $\mathbf{f}=-\text{grad}\,\psi$, the work done in traversing a closed path is zero, by (6.42).

**Example 6.8**

Near to the earth's surface, the gravitational force acting on a unit mass is

$$\mathbf{f}=-g\mathbf{k} \qquad (g \text{ constant}),$$

using a triad $(\mathbf{i}, \mathbf{j}, \mathbf{k})$ with $\mathbf{k}$ vertically upwards. The gradient of the scalar field $\psi=gz$ is, by (6.38), grad $\psi=-g\mathbf{k}$. So $\mathbf{f}$ is of the form (6.44), derivable from the potential $\psi=gz$; this is the usual **gravitational potential energy**, and (6.43) tells us that the work per unit mass done by the gravitational field in moving from $P$ to $Q$ is equal to the change of gravitational potential $\psi(P)-\psi(Q)$; it is independent of the path $\Gamma$ traversed.

**Example 6.9**

A point charge $e$ at a point $O$ produces an **electrostatic potential** $\psi=e/r$ at a distance $r$ from $O$. The **electrostatic field** derived from this potential is, by (6.44),

$$\mathbf{E}=-\text{grad}\,\psi.$$

We use the expression (6.40) for grad $\psi$ in terms of spherical polar coordinates with origin at $O$; the components of grad $\psi$ are

$$(e/r^2, 0, 0),$$

so that, in terms of the position vector $\mathbf{r}$ relative to $O$,

$$\mathbf{E} = \frac{e}{r^3}\mathbf{r}.$$

$\mathbf{E}$ is the electrical force acting on a unit positive charge; it is in the radial direction, obeys the **inverse square law** or **Coulomb's law**, is proportional to $e$, and is repulsive (attractive) if $e$ is a positive (negative) charge.

The work done by the field on a unit charge in moving along a curve $\Gamma$ from a point with $r = r_0$ to a point with $r = r_1$ is, by (6.43),

$$\int_\Gamma d\mathbf{s} \cdot \mathbf{E} = \psi(r_0) - \psi(r_1)$$

$$= \frac{e}{r_0} - \frac{e}{r_1}.$$

If we let $r_0 \to 0$, so that the initial point is 'at infinity', the work done by the field when the unit charge moves from infinity to distance $r_1$ is then $-e/r_1$.

The level surfaces, on which $\psi = e/r$ is constant, are the spheres $r = r_1$, with $r_1$ fixed. The field $\mathbf{E}$ is everywhere orthogonal to these spheres, exemplifying property (iv) of the gradient. In electrostatics, the level surfaces are in general called **equipotentials**.

## ***Problems 6.2***

1 A scalar field is defined to be

$$\psi = \rho \sin 2\phi + \rho^{-1} z^2$$

in terms of cylindrical polar coordinates. Find the components of grad $\psi$ relative to rectangular, cylindrical polar and spherical polar coordinates. Check that the three sets of components define the same vector field.

2 The field of a **magnetic dipole** is derivable from a potential of the form

$$\psi = \frac{\mu \cos \theta}{r^2} \qquad (\mu \text{ constant}),$$

using spherical polar coordinates. Find the components of the magnetic field vector at any point, and show directly that it is orthogonal to the level surfaces defined by $\psi$. Draw a diagram showing the level surfaces and the magnetic field directions in terms of polar coordinates $r$, $\theta$.

3 Show that the vector field $\mathbf{v}$ with components

$$r^{-2}(\sin 2\theta \tan \phi, -2 \cos 2\theta \tan \phi, -2 \cos \theta \sec^2 \phi)$$

with respect to spherical polar coordinates is a conservative field. Find the potential from which the field is derived. Calculate the integral $\int \mathbf{v} \cdot d\mathbf{s}$ along two distinct curves $\Gamma_1$ and $\Gamma_2$ joining the points with $r = a$, $\theta = \frac{1}{2}\pi$, $\phi = 0$ and $r = a$, $\theta = \frac{1}{4}\pi$, $\phi = \frac{1}{4}\pi$. Show that these integrals are each equal to the difference between the potentials at the two points.

## 6.4 *Curl of a vector field; Stokes' theorem*

For a conservative field $\mathbf{v}$, (6.42) ensures that the integral

$$\oint_\Gamma \mathbf{v} \cdot d\mathbf{s} \tag{6.47}$$

round a closed curve $\Gamma$ is zero. For a more general vector field $\mathbf{v}$, the integral (6.47) is not usually zero. If $\mathbf{v}$ represents a non-conservative field of force (a magnetic field produced by an electric current, for example), (6.47) still represents the work done in traversing the closed curve $\Gamma$. In a different physical situation, $\mathbf{v}$ might represent the velocity at each point in a fluid; then (6.47) is the integral of the component of velocity tangential to $\Gamma$, and measures the 'swirling' of the fluid round $\Gamma$; this analogy is used in calling the integral (6.47) the **circulation** of $\mathbf{v}$ round $\Gamma$. Since we have assumed that $\Gamma$ is traversed in the direction '$s$ increasing' there is no ambiguity of sign in (6.47); if, however, the direction of integration round $\Gamma$ were reversed, the sign of (6.47) would change, since the tangent vector $\mathbf{t}$ would be reversed at all points.

In this section, we shall establish *Stokes' theorem*, which is very similar in nature to the divergence theorem; it transforms the path integral (6.47) into a surface integral over a piecewise smooth surface $\sigma$ whose boundary is the closed curve $\Gamma$. In the process, we define the 'curl' of the vector field $\mathbf{v}$ in terms of the derivatives of the components of $\mathbf{v}$; in order to establish these results, we need to assume that $\mathbf{v}$ has continuous derivatives, not only on $\Gamma$, but also on the surface $\sigma$.

To establish Stokes' theorem for an arbitrary surface $\sigma$ is quite difficult, and would obscure the essential meaning of the theorem. We shall prove the theorem in a particularly simple situation, when the coordinate system $\{u_r\}$ used has a particularly simple relation to $\sigma$, and the boundary $\Gamma$ also has a simple form. Then, in Appendix B, we establish Stokes' theorem using an arbitrary orthogonal coordinate system, but still with a simple type of boundary; we also indicate how the theorem can be extended to different and more complicated types of boundary.

The simple situation we consider is when the surface $\sigma$ is of the form $u_3 = c$, where $\{u_r; r = 1, 2, 3\}$ is an orthogonal coordinate system with no exceptional points on $\sigma$, and $c$ is a constant. Then the coordinates $u_1$, $u_2$ define an orthogonal system of curves on $\sigma$. We further assume that the ranges of $u_1$, $u_2$ are of the form

$$a \leqslant u_1 \leqslant b \tag{5.8a}$$

$$\alpha(u_1) \leqslant u_2 \leqslant \beta(u_1), \tag{5.8b}$$

as shown in Fig. 6.7; we also assume that the ranges are of this form when $u_1$ and $u_2$ are interchanged. Then a curve $\gamma$ with $u_1$ fixed $(a < u_1 < b)$ will intersect the boundary $\Gamma$ in two points $A$, $B$, corresponding to $u_2 = \alpha(u_1)$ and $u_2 = \beta(u_1)$.

The coordinate system defines an orthogonal triad $\{\mathbf{t}_r\}$ at each point of $\sigma$; $\mathbf{t}_1$ and $\mathbf{t}_2$ are tangent to $\sigma$, while $\mathbf{t}_3 = \mathbf{n}$ is the unit normal to $\sigma$ at

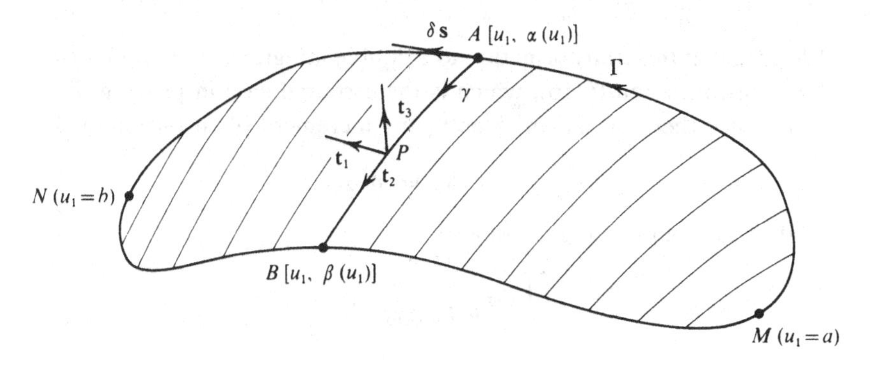


Fig. 6.7

all normal points. We choose $\mathbf{t}_1$ and $\mathbf{t}_2$ in the directions of increase of $u_1$ and $u_2$, as shown and then choose $\mathbf{t}_3 = \mathbf{n}$ to form a right-handed triad. An increment $\delta\mathbf{s}$ tangent to $\Gamma$ is, by (5.76) and (5.80), of the form

$$\delta\mathbf{s} = h_1 \delta u_1 \mathbf{t}_1 + h_2\, \delta u_2\, \mathbf{t}_2, \tag{6.48}$$

defining corresponding increments $\delta u_1$, $\delta u_2$. Writing

$$\mathbf{v} = v_1\mathbf{t}_1 + v_2\mathbf{t}_2 + v_3\mathbf{t}_3,$$

the integral (6.47) becomes

$$\oint_\Gamma (v_1 h_1\, du_1 + v_2 h_2\, du_2). \tag{6.49}$$

Consider the first term here. We have assumed that $\Gamma$ is traversed in a right-handed sense about the normal vector $\mathbf{n}$, so that in Fig. 6.7, the variable $u_1$ increases from $a$ to $b$ along $MAN$, and then decreases to the value $a$ along $NBM$. Denoting values at $A$ and $B$ by suffixes, the first term in (6.49) is therefore

$$\int_a^b du_1[(v_1h_1)_A - (v_1h_1)_B]. \tag{6.50}$$

Since we assume that $\{h_r\}$ and $\{v_r\}$ have continuous derivatives on $\sigma$, and $u_1$ is constant on $\gamma$, Theorem A.2 (Appendix A) allows us to write

$$(v_1h_1)_A - (v_1h_1)_B = -\int_{u_2=\alpha(u_1)}^{\beta(u_1)} du_2 \frac{\partial}{\partial u_2}(v_1h_1);$$

so (6.50) is equal to

$$-\int_a^b du_1 \int_{\alpha(u_1)}^{\beta(u_1)} du_2 \frac{\partial}{\partial u_2}(v_1h_1).$$

We note that this transformation to a double integral is very similar to the transformation (6.16), which is the essential step in proving the divergence theorem. As in (5.66b), the increment of surface area is

$$\delta\sigma = h_1h_2\, \delta u_1\, \delta u_2,$$

so that the double integral equals

$$-\iint_\sigma d\sigma \frac{1}{h_1h_2} \frac{\partial}{\partial u_2}(v_1h_1),$$

integrated over the whole surface $\sigma$. Treating the second term in

(6.49) in the same way, the whole integral becomes

$$\oint_\Gamma \mathbf{v}\cdot d\mathbf{s} = \iint_\sigma d\sigma \frac{1}{h_1 h_2}\left[\frac{\partial}{\partial u_1}(v_2 h_2) - \frac{\partial}{\partial u_2}(v_1 h_1)\right]; \qquad (6.51)$$

the difference in sign of the two terms arises from the right-handedness of the sense of rotation about **n**.

The result (6.51) is valid for any surface which can be divided into a finite number of sections, each with coordinate ranges of the form (5.8). For example, suppose that $\sigma$ can be divided into two such pieces $\sigma^{(1)}$ and $\sigma^{(2)}$, as in Fig. 6.8, by a curve $\Gamma^{(3)}$, so that $\sigma^{(1)}$ has

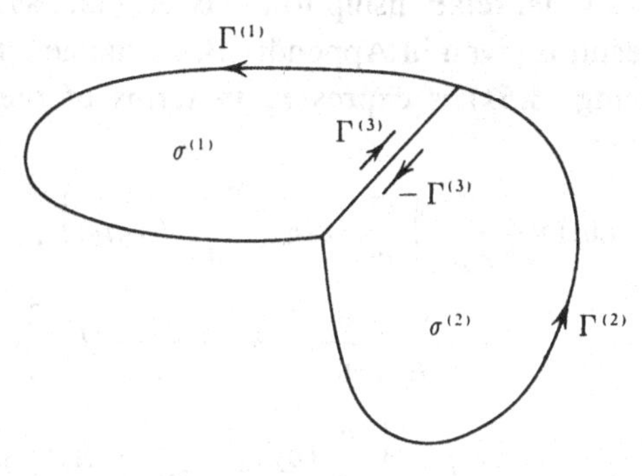


Fig. 6.8

boundary $\Gamma^{(1)}+\Gamma^{(3)}$, while $\sigma^{(2)}$ has boundary $\Gamma^{(2)}-\Gamma^{(3)}$, the minus sign indicating that $\Gamma^{(3)}$ is traversed in the opposite sense. Then, if **v** and $\{h_r\}$ have continuous derivatives on $\sigma$,

$$\oint_\Gamma \mathbf{v}\cdot d\mathbf{s} = \int_{\Gamma^{(1)}+\Gamma^{(3)}} \mathbf{v}\cdot d\mathbf{s} + \int_{\Gamma^{(2)}-\Gamma^{(3)}} \mathbf{v}\cdot d\mathbf{s}.$$

By using the result (6.49) for the surfaces $\sigma^{(1)}$ and $\sigma^{(2)}$ separately, these two integrals are equal to surface integrals (6.51) over $\sigma^{(1)}$ and $\sigma^{(2)}$; by Theorem A.6 (Appendix A), the sum of these integrals is just the integral (6.51) over the whole of $\sigma$.

It is also possible for $\sigma$ to contain an exceptional point of the coordinate system $\{u_r\}$, provided that **v** has continuous derivatives there. Consider, for example, the use of cylindrical polar coordinates when $\sigma$ is the circular disc $\rho \leqslant a$, $0 \leqslant \phi \leqslant 2\pi$ in the plane $z=0$. The boundary $\Gamma$ is the circle $\rho = a$, so that (6.46) reduces to $\delta\mathbf{s} = a\delta\phi\mathbf{t}_2$ and

(6.49) and (6.51) become

$$\int_0^{2\pi} v_\phi a\,\mathrm{d}\phi = \int_{\phi=0}^{2\pi}\mathrm{d}\phi\int_{\rho=0}^{a}\rho\,\mathrm{d}\rho\frac{1}{\rho}\frac{\partial(\rho v_\phi)}{\partial\rho};$$

only the first term in the integrand of (6.51) appears. The limits of integration correspond not just to the boundary $\Gamma$, but to the full limit set of coordinates for the disc. Provided that $\partial v_\phi/\partial\rho$ is continuous, no problem arises at the exceptional point $\rho=0$; Stokes' theorem is in fact valid for a class of fields $\mathbf{v}$ for which $\partial v_\phi/\partial\rho$ is unbounded at $\rho=0$, but we shall not discuss these details.

We now state the generalisation of the result (6.51) to surfaces $\sigma$ which have no special relationship to the coordinate system $\{u_r\}$; the proof of the result is given in Appendix B. In the general result, the integral replacing (6.51) is expressed in terms of the **curl** vector, defined by

$$\begin{aligned}\operatorname{curl}\mathbf{v} = {} & \frac{1}{h_2h_3}\left[\frac{\partial}{\partial u_2}(v_3h_3)-\frac{\partial}{\partial u_3}(v_2h_2)\right]\mathbf{t}_1 \\ & +\frac{1}{h_3h_1}\left[\frac{\partial}{\partial u_3}(v_1h_1)-\frac{\partial}{\partial u_1}(v_3h_3)\right]\mathbf{t}_2 \\ & +\frac{1}{h_1h_2}\left[\frac{\partial}{\partial u_1}(v_2h_2)-\frac{\partial}{\partial u_2}(v_1h_1)\right]\mathbf{t}_3,\end{aligned} \tag{6.52}$$

where $\{\mathbf{t}_s\}$ is the triad defined at any point by the coordinates $\{u_s\}$. The integrand in (6.51) is the third component of $\operatorname{curl}\mathbf{v}$. The integration in (6.47) prescribes a sense in which $\Gamma$ is traversed; we define $\mathbf{n}$ to be the unit normal to $\sigma$ such that motion round $\Gamma$ and in the direction of $\mathbf{n}$ is a right-handed screw motion. Then **Stokes' theorem** states that

$$\oint_\Gamma \mathbf{v}\cdot\mathrm{d}\mathbf{s} = \iint_\sigma \mathrm{d}\sigma\,\mathbf{n}\cdot\operatorname{curl}\mathbf{v}. \tag{6.53}$$

The result (6.51) was obtained by assuming that $\mathbf{n}=\mathbf{t}_3$ everywhere on $\sigma$; from (6.52), it is clear that the surface integral in (6.53) then reduces to (6.51). The theorem holds provided $\mathbf{v}$ and $\{h_r\}$ have continuous derivatives on $\sigma$; it is possible to relax these conditions to some extent, and also to allow $\sigma$ to be an infinite surface, subject to certain convergence conditions; but these extensions of Stokes' theorem are beyond the scope of this book.

Let us now consider equation (6.53) as the maximum dimension of $\sigma$ tends to zero, in such a way that $\sigma$ always contains a specific point $P$, and $\mathbf{n}$ tends continuously to a fixed (vector) value at all points of $\sigma$; then $\sigma$ becomes 'flat' as it shrinks upon the point $P$. Since $\mathbf{n} \cdot \operatorname{curl} \mathbf{v}$ is then continuous at $P$, Theorem A.8 (Appendix A) implies that its value at $P$ is

$$(\mathbf{n} \cdot \operatorname{curl} \mathbf{v})_P = \lim_{\sigma \to 0} \frac{1}{A(\sigma)} \iint_\sigma \mathrm{d}\sigma \, \mathbf{n} \cdot \operatorname{curl} \mathbf{v},$$

where $A(\sigma)$ is the area (5.72) of $\sigma$. Using (6.53), we find

$$(\mathbf{n} \cdot \operatorname{curl} \mathbf{v})_P = \lim_{\sigma \to 0} \frac{1}{A(\sigma)} \oint_\Gamma \mathbf{v} \cdot \mathrm{d}\mathbf{s}. \tag{6.54}$$

Since $\mathbf{n}$ is an arbitrary unit vector at an arbitrary point $P$, (6.54) provides an alternative definition of curl; it is analogous to the definition (6.22) of divergence. As in (6.22), the main difficulty with (6.54) is providing a precise definition of the limiting process; one must ensure that $\Gamma$ does not become very corrugated as $\sigma \to 0$. The advantages of (6.54) are essentially the same as that of (6.22): since $\mathbf{v} \cdot \mathrm{d}\mathbf{s}$ and $A(\sigma)$ are scalars, (6.54) ensures that $\mathbf{n} \cdot \operatorname{curl} \mathbf{v}$ is independent of the coordinate system, and hence that $\operatorname{curl} \mathbf{v}$ transforms as a vector under rotations. We therefore know that (6.52) is valid for all coordinate systems for which the triad $\{\mathbf{t}_r\}$ is right-handed. Equally important, (6.54) gives a direct physical meaning to a component of $\operatorname{curl} \mathbf{v}$. The integral round $\Gamma$ is the circulation of $\mathbf{v}$; as $\sigma \to 0$, this is the circulation around the axis $\mathbf{n}$, so that the limit (6.54) is directly interpretable as 'circulation per unit area around the axis $\mathbf{n}$', at any point $P$. The vector $\operatorname{curl} \mathbf{v}$ is also called the **vorticity** of $\mathbf{v}$.

In §6.2, we used the concept of 'outflow per unit volume' to give a clear meaning to the divergence theorem; in the same way, the concept of *circulation per unit area* gives a clear meaning to Stokes' theorem. Suppose we divide a surface $\sigma$ into a large number of small elements $\sigma_P$ $(n = 1, 2, 3, \ldots)$ by a coarse mesh, as in Fig. 5.9, and let the boundary of the element $\sigma_P$ be $\Gamma_P$. Provided that the closed curves $\Gamma_P$ are all traversed in the same sense, the circulation (6.47) is equal to the sum of circulations

$$\sum_P \oint_{\Gamma_P} \mathbf{v} \cdot \mathrm{d}\mathbf{s} \tag{6.55}$$

round the elements $\sigma_P$, by the argument used to extend the result (6.51): just as in the simple situation of Fig. 6.8, each element of the *mesh* contributes two equal and opposite terms to (6.55); so the sections of the mesh contribute nothing to the sum, leaving just the integral (6.47). Since the elements $\sigma_P$ are small, (6.54) ensures that (6.55) is approximately

$$\sum_P A(\sigma_P)(\mathbf{n}\cdot\operatorname{curl}\mathbf{v})_P, \tag{6.56}$$

where $P$ is a point of the element $\sigma_P$. Either through (6.55) or (6.56), the total circulation (6.47) is the sum of the circulations associated with the elements $\sigma_P$. Provided that certain analytic conditions are satisfied, it is possible to re-derive Stokes' theorem by going to the limit of a fine mesh in (6.56), giving just

$$\iint_\sigma \mathrm{d}\sigma\, \mathbf{n}\cdot\operatorname{curl}\mathbf{v}.$$

Although we have referred to curl $\mathbf{v}$ as a vector, it is in fact an axial vector if $\mathbf{v}$ is a polar vector. There are several ways to see this. If we change the sign of all basis vectors $\{\mathbf{t}_r\}$, the coordinates $\{u_r\}$ and the components $\{v_r\}$ in (6.52) are changed in sign, while $\{h_r\}$, being positive, are unchanged. Therefore the components of curl $\mathbf{v}$ retain their signs. Equally, in (6.54), $\mathbf{v}$ and $\mathrm{d}\mathbf{s}$ are changed in sign while $A(\sigma)$ is unchanged, ensuring that any component of curl $\mathbf{v}$ preserves its sign. The axial vector property stems from the rotational nature of the circulation (6.47); the sense of 'rotation from $\mathbf{i}$ to $\mathbf{j}$ about the axis $\mathbf{k}$' is unchanged if we reflect axes by the transformation (4.80).

Using rectangular coordinates $(x, y, z)$ with $h_1 = h_2 = h_3 = 1$, (6.52) becomes

$$\operatorname{curl}\mathbf{v} = \left(\frac{\partial v_z}{\partial y} - \frac{\partial v_y}{\partial z}\right)\mathbf{i} + \left(\frac{\partial v_x}{\partial z} - \frac{\partial v_z}{\partial x}\right)\mathbf{j} + \left(\frac{\partial v_y}{\partial x} - \frac{\partial v_x}{\partial y}\right)\mathbf{k}. \tag{6.57}$$

Cylindrical polar coordinates $(\rho, \phi, z)$ have $h_\rho = h_z = 1$, $h_\phi = \rho$, so that

$$\operatorname{curl}\mathbf{v} = \frac{1}{\rho}\left[\frac{\partial v_z}{\partial \phi} - \frac{\partial(\rho v_\phi)}{\partial z}\right]\mathbf{t}_\rho + \left[\frac{\partial v_\rho}{\partial z} - \frac{\partial v_z}{\partial \rho}\right]\mathbf{t}_\phi + \frac{1}{\rho}\left[\frac{\partial(\rho v_\phi)}{\partial \rho} - \frac{\partial v_\rho}{\partial \phi}\right]\mathbf{t}_z. \tag{6.58}$$

Similarly, spherical polar coordinates $(r, \theta, \phi)$ give

$$\begin{aligned}\text{curl } \mathbf{v} = &\frac{1}{r^2 \sin\theta}\left[\frac{\partial}{\partial\theta}(r\sin\theta v_\phi) - \frac{\partial}{\partial\phi}(rv_\theta)\right]\mathbf{t}_r \\ &+\frac{1}{r\sin\theta}\left[\frac{\partial v_r}{\partial\phi} - \frac{\partial}{\partial r}(r\sin\theta v_\phi)\right]\mathbf{t}_\theta \\ &+\frac{1}{r}\left[\frac{\partial}{\partial r}(rv_\theta) - \frac{\partial v_r}{\partial\theta}\right]\mathbf{t}_\phi.\end{aligned} \tag{6.59}$$

**Example 6.10**

In terms of rectangular coordinates, a vector field $\mathbf{v}(\mathbf{r})$ is defined by

$$\mathbf{v} = 2xyz^2\mathbf{i} + (x^2+y^2)z^2\mathbf{j} + 4x^2y^2\mathbf{k}.$$

Use (6.58) to find the components of curl $\mathbf{v}$ relative to cylindrical polar coordinates.

If $(\mathbf{t}_\rho, \mathbf{t}_\phi, \mathbf{t}_z)$ are the triad of unit vectors at any point, defined by cylindrical polar coordinates, then

$$\begin{aligned}\mathbf{i} &= \mathbf{t}_\rho \cos\phi - \mathbf{t}_\phi \sin\phi, \\ \mathbf{j} &= \mathbf{t}_\rho \sin\phi + \mathbf{t}_\phi \cos\phi, \\ \mathbf{k} &= \mathbf{t}_z.\end{aligned}$$

Substituting these expressions and $x = \rho\cos\phi$, $y = \rho\sin\phi$ into the formula for $\mathbf{v}$, we find

$$\begin{aligned}\mathbf{v} = &\rho^2z^2(\sin 2\phi\cos\phi + \sin\phi)\mathbf{t}_\rho \\ &-\rho^2z^2(\sin 2\phi\sin\phi - \cos\phi)\mathbf{t}_\phi \\ &+\rho^4\sin^2 2\phi\,\mathbf{t}_z\end{aligned}$$

or

$$\begin{aligned}\mathbf{v} = &2\rho^2z^2\sin\phi(\cos 2\phi + 2)\mathbf{t}_\rho \\ &+\rho^2z^2\cos\phi\cos 2\phi\mathbf{t}_\phi + \rho^4\sin^2 2\phi\mathbf{t}_z.\end{aligned}$$

The components of curl $\mathbf{v}$ are given by (6.58). They are

$$\begin{aligned}(\text{curl } \mathbf{v})_\rho &= \frac{1}{\rho}\left[\rho^4\frac{\partial}{\partial\phi}(\sin^2 2\phi) - \rho^3\cos\phi\cos 2\phi\frac{\partial}{\partial z}(z^2)\right] \\ &= 2\rho^2[\rho\sin 4\phi - z\cos\phi\cos 2\phi], \\ (\text{curl } \mathbf{v})_\phi &= 2\rho^2\sin\phi(\cos 2\phi + 2)\frac{\partial}{\partial z}(z^2) - \sin^2 2\phi\frac{\partial}{\partial\rho}(\rho^4) \\ &= 2\rho^2[z\sin\phi(\cos 2\phi + 2) - 2\rho\sin^2 2\phi],\end{aligned}$$

$$(\text{curl } \mathbf{v})_z = \frac{1}{\rho}\left[z^2 \cos\phi \cos 2\phi \frac{\partial}{\partial\rho}(\rho^3) - \rho^2 z^2 \frac{\partial}{\partial\phi}\{\sin\phi(\cos 2\phi + 2)\}\right]$$

$$= 0.$$

An alternative method of finding these components is to calculate curl $\mathbf{v}$ in rectangular coordinates, and then to transform to cylindrical polars. [See Problems 6.3, Question 1.]

**Example 6.11**

A vector field $\mathbf{v}$ is given by

$$\mathbf{v} = (x^2 - y^2)\, e^z \mathbf{i} + 2xy\, e^z \mathbf{j} + (x^2 + y^2 + z^2)\mathbf{k},$$

and a closed curve $\Gamma$ consists of the edges of the rectangle with corners at points $\mathbf{i}$, $\mathbf{i}+\mathbf{j}$, $\mathbf{j}+\mathbf{k}$ and $\mathbf{k}$; the order of these points indicates the sense in which $\Gamma$ is traversed. Calculate the integral of $\mathbf{n} \cdot \text{curl } \mathbf{v}$ over the rectangle $\sigma$ bounded by $\Gamma$, and show explicitly that it is equal to the circulation (6.47).

The rectangle is shown in Fig. 6.9, and is labelled $ABCD$. The normal vector $\mathbf{n}$, associated with $\Gamma$ by the right-hand rule, is constant

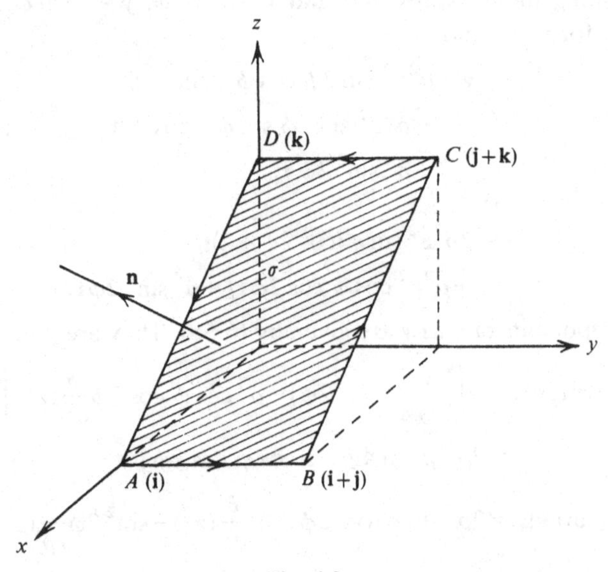


Fig. 6.9

over $\sigma$, and is $\mathbf{n}=(\mathbf{i}+\mathbf{k})/\sqrt{2}$. So, in terms of the components of curl $\mathbf{v}$,

$$\mathbf{n}\cdot\text{curl }\mathbf{v}=[(\text{curl }\mathbf{v})_1+(\text{curl }\mathbf{v})_3]/\sqrt{2}.$$

If we use $(x, y)$ to specify points on $\sigma$, the element of area is

$$\delta\sigma=\sqrt{2}\,\delta x\,\delta y.$$

Since $\sigma$ corresponds to the range $(0, 1)$ of both $x$ and $y$, the surface integral of $\mathbf{n}\cdot$ curl $\mathbf{v}$ equals

$$\int_0^1 \mathrm{d}x \int_0^1 \mathrm{d}y\,[(\text{curl }\mathbf{v})_1+(\text{curl }\mathbf{v})_3].$$

Now, using (6.57)

$$\begin{aligned}(\text{curl }\mathbf{v})_1&=\frac{\partial(y^2)}{\partial y}-\frac{\partial(2xy\,\mathrm{e}^z)}{\partial z}\\&=2y-2xy\,\mathrm{e}^z,\\(\text{curl }\mathbf{v})_3&=\frac{\partial(2xy\,\mathrm{e}^z)}{\partial x}-\frac{\partial(-y^2\,\mathrm{e}^z)}{\partial y}\\&=4y\,\mathrm{e}^z.\end{aligned}$$

Since $z=1-x$ on the surface, the double integral becomes

$$\begin{aligned}&\int_0^1 \mathrm{d}x\int_0^1 \mathrm{d}y\,[2y-2xy\,\mathrm{e}^{1-x}+4y\,\mathrm{e}^{1-x}]\\&=\int_0^1 \mathrm{d}x\,[1-x\,\mathrm{e}^{1-x}+2\,\mathrm{e}^{1-x}].\end{aligned}$$

Carrying out the final integration, we find

$$\iint_\sigma \mathrm{d}\sigma\,\mathbf{n}\cdot\text{curl }\mathbf{v}=1+\mathrm{e}.$$

The circulation (6.47) consists of integrals along $AB$, $BC$, $CD$ and $DA$. We treat these separately.

$AB(x=1, z=0, \delta\mathbf{s}=\mathbf{j}\,\delta y;\, v_2=2xy\,\mathrm{e}^z=2y)$:

$$\int_{(AB)}\mathbf{v}\cdot\mathrm{d}\mathbf{s}=\int_0^1 v_2\,\mathrm{d}y=\int_0^1 2y\,\mathrm{d}y=1.$$

$CD(x=0, z=1, \delta\mathbf{s}=-\mathbf{j}\,\delta y;\, v_2=2xy\,\mathrm{e}^z=0)$:

$$\int_{(CD)}\mathbf{v}\cdot\mathrm{d}\mathbf{s}=-\int_0^1 v_2\,\mathrm{d}y=0.$$

$BC(z=1-x, y=1, \delta z=-\delta x, \delta \mathbf{s}=\mathbf{i}\,\delta x+\mathbf{j}\,\delta z)$:

$$\int_{(BC)} \mathbf{v}\cdot d\mathbf{s} = \int_{x=1}^{0} (v_1\,dx + v_3\,dz)$$

$$= -\int_0^1 [(x^2-1)\,e^{1-x} - \{x^2+1+(1-x)^2\}]\,dx$$

$DA(z=1-x, y=0, \delta z=-\delta x, \delta \mathbf{s}=\mathbf{i}\,\delta x+\mathbf{j}\,\delta z)$:

$$\int_{(DA)} \mathbf{v}\cdot d\mathbf{s} = \int_{x=0}^{1} (v_1\,dx + v_3\,dz)$$

$$= \int_0^1 [x^2\,e^{1-x} - \{x^2+(1-x)^2\}]\,dx.$$

The sum of the last two integrals is just

$$\int_0^1 (e^{1-x}+1)\,dx = e.$$

Adding in the integrals along $AB$, $CD$ then gives

$$\oint_\Gamma \mathbf{v}\cdot d\mathbf{s} = 1+e,$$

verifying Stokes' theorem (6.53).

Towards the end of §6.3, we showed that a field $\mathbf{v}$ is conservative in a simple volume region $\tau$ if and only if condition (6.46) holds, that the circulation (6.47) is zero for all simple closed curves $\Gamma$ lying within $\tau$. We shall now show that these conditions are equivalent to the condition

$$\operatorname{curl} \mathbf{v} = \mathbf{0} \tag{6.60}$$

at all interior points of $\tau$; we are, of course, assuming that curl $\mathbf{v}$ exists and is continuous. When condition (6.60) is satisfied throughout a region, we say that $\mathbf{v}$ is **irrotational** there.

Since $\tau$ is a simple volume region, any simple closed curve $\Gamma$ lying within $\tau$ is the boundary of some surface $\sigma$ lying within $\tau$. Applying Stokes' theorem (6.53) to such a surface $\sigma$, we see that, if (6.60) holds within $\tau$, the circulation (6.47) is zero round $\tau$. Conversely, if (6.47) is zero for all circuits $\tau$ within $\tau$, then (6.54) implies that any component of curl $\mathbf{v}$ (at an interior point of $\tau$) is the limit of zero quantities, and hence is zero. We have therefore established the equivalence of three conditions:

(i) the field $\mathbf{v}$ is conservative within a simple volume $\tau$, derivable from a potential $\psi$;
(ii) the circulation (6.47) of $\mathbf{v}$ is zero for all simple closed curves lying within $\tau$;
(iii) the field $\mathbf{v}$ is irrotational, with curl $\mathbf{v}=\mathbf{0}$ at all interior points of $\tau$.

## *Problems 6.3*

1 Use (6.57) to calculate curl $\mathbf{v}$ in rectangular coordinates, where

$$\mathbf{v}=2xyz^2\mathbf{i}+(x^2+y^2)z^2\mathbf{j}+4x^2y^2\mathbf{k}.$$

Using this result, check the values of the components of curl $\mathbf{v}$ relative to cylindrical polar coordinates, evaluated in Example 6.10.

2 A closed curve $\Gamma$ on the surface of the sphere $r=a$ consists of three arcs, defined in terms of spherical polar coordinates:

$$PQ: \phi=0, 0\leqslant\theta\leqslant\tfrac{1}{2}\pi,$$
$$QR: \theta=\tfrac{1}{2}\pi, 0\leqslant\phi\leqslant\tfrac{1}{2}\pi,$$
$$RP: \phi=\tfrac{1}{2}\pi, \tfrac{1}{2}\pi\leqslant\theta\leqslant 0.$$

Calculate the circulation (6.47) of the vector field

$$\mathbf{v}=\mathrm{e}^{-r}\sin\theta(\sin\tfrac{1}{2}\phi\ \mathbf{t}_r+\cos\tfrac{1}{2}\phi\ \mathbf{t}_\theta)+\sin\theta\ \mathbf{t}_\phi$$

round $\Gamma$ in the sense $PQRP$, and verify directly that Stokes' theorem (6.53) is satisfied for this field $\mathbf{v}$ and this closed curve $\Gamma$.

3 A vector field $\mathbf{v}$ is defined in terms of cylindrical polar coordinates to be

$$\mathbf{v}=2\rho z\cos\phi\,\mathbf{t}_\rho+(\rho^2+z^2)\mathbf{t}_\phi+\rho^2\cos^2\phi\ \mathbf{t}_z.$$

Find curl $\mathbf{v}$.

Calculate directly the integral

$$\iint_\sigma \mathrm{d}\sigma\ \mathbf{n}\cdot\text{curl}\ \mathbf{v}$$

when

(*a*) the surface $\sigma$ is the disc $z=0$, $\rho\leqslant a$, with $\mathbf{n}$ in the direction of increasing $z$;
(*b*) the surface $\sigma$ is the hemispherical cap $r=a$, $0\leqslant\theta\leqslant\tfrac{1}{2}\pi$ (using spherical polar coordinates), with $\mathbf{n}$ in the direction of increasing $r$.

Use Stokes' theorem to comment on these results.

4 By calculating curl $\mathbf{v}$, show that the following fields are conservative:
($a$) $\mathbf{v}=2\rho z^{-1}\sin 2\phi\,\mathbf{t}_\rho+2\rho z^{-1}\cos 2\phi\,\mathbf{t}_\phi-\rho^2 z^{-2}\sin 2\phi\,\mathbf{t}_z$,
($b$) $\mathbf{v}=2r\theta\sin\theta\cos^2\phi\,\mathbf{t}_r+r(\sin\theta+\theta\cos\theta)\cos^2\phi\,\mathbf{t}_\theta-r\theta\sin 2\phi\,\mathbf{t}_\phi$.
[Standard notation for polar coordinates is used.]

## 6.5 *Field operators; the Laplacian*

At the end of §6.4, we saw that a conservative field $\mathbf{v}=-\text{grad}\,\psi$ was irrotational, with curl $\mathbf{v}=\mathbf{0}$. Therefore any scalar field $\psi(\mathbf{r})$ with continuous second derivatives satisfies

$$\text{curl grad}\,\psi=\mathbf{0}. \tag{6.61a}$$

This equation can also be established directly by substituting the components

$$v_s=-\frac{1}{h_s}\frac{\partial\psi}{\partial u_s}\qquad (s=1,2,3)$$

of grad $\psi$, given by (6.34), into the formula (6.52) for curl $\mathbf{v}$. The first term, for example, gives

$$-\frac{1}{h_2h_3}\left[\frac{\partial}{\partial u_2}\left(\frac{\partial\psi}{\partial u_3}\right)-\frac{\partial}{\partial u_3}\left(\frac{\partial\psi}{\partial u_2}\right)\right]\mathbf{t}_1,$$

which is zero since the second derivatives are equal, since they are assumed continuous (Theorem A.1, Appendix A). The identity (6.61a) is simply a statement in 'differential form' that conservative fields are irrotational. The same statement can be made in 'integral form' by integrating (6.61a) over a simple surface $\sigma$ with a simple closed curve $\Gamma$ as boundary; then by Stokes' theorem,

$$\oint_\Gamma \text{grad}\,\psi\cdot d\mathbf{s}=\iint_\sigma d\sigma\,\mathbf{n}\cdot\text{curl grad}\,\psi=0. \tag{6.62}$$

This is just a re-statement of the result (6.42) that a conservative field is irrotational, with zero circulation round any closed curve.

A second identity can be derived from (6.20) if we replace $v_1, v_2, v_3$ by the components of curl $\mathbf{v}$, given by (6.52); the square bracket in (6.20) becomes

$$\left[\frac{\partial}{\partial u_1}\left\{\frac{\partial}{\partial u_2}(v_3h_3)-\frac{\partial}{\partial u_3}(v_2h_2)\right\}+\frac{\partial}{\partial u_2}\left\{\frac{\partial}{\partial u_3}(v_1h_1)-\frac{\partial}{\partial u_1}(v_3h_3)\right\}\right.$$
$$\left.+\frac{\partial}{\partial u_3}\left\{\frac{\partial}{\partial u_1}(v_2h_2)-\frac{\partial}{\partial u_2}(v_1h_1)\right\}\right],$$

which is zero if all the second partial derivatives are continuous. Thus

$$\text{div curl } \mathbf{v} = 0, \tag{6.63a}$$

which means that the outflow of curl **v** is zero at any point. Again, we can integrate (6.63a) over a volume $\tau$ bounded by a simple closed surface $\sigma$, obtaining the 'integral form' of the statement. Using the divergence theorem (6.19), we find

$$\iint_\sigma d\sigma\, \mathbf{n} \cdot \text{curl } \mathbf{v} = \iiint_\tau d\tau \text{ div curl } \mathbf{v} = 0.$$

This tells us that the flux of curl **v** out of the volume $\tau$ is zero. When the outflow of a vector field is zero, the field satisfies a **conservation law**; so this equation, or alternatively (6.63a) expresses the fact that curl **v** always satisfies a conservation law, provided that its derivatives are continuous: *the flow of vorticity out of any region is zero.*

Since (6.61a) and (6.63a) hold for a whole class of functions $\psi(\mathbf{r})$ and $\mathbf{v}(\mathbf{r})$, they are often written as

$$\text{curl grad} = \mathbf{0} \tag{6.61b}$$

and

$$\text{div curl} = \mathbf{0}. \tag{6.63b}$$

In these equations, we are regarding 'div', 'grad' and 'curl' as **field operators**, which can operate on scalar or vector fields according to the definitions (6.20), (6.34) and (6.52); since these definitions are linear in $\psi$ or in the components of **v**, they are linear operators. Equations (6.61b) and (6.63b) state that the operators 'curl grad' and 'div curl' each behave as a zero **axial vector operator**; the first converts a scalar field $\psi$ into the zero axial vector field, while the second converts a polar vector field **v** into the zero axial scalar field.

When rectangular coordinates are used, the field operators have particularly simple forms given by (6.23), (6.38) and (6.57). These equations can be written particularly neatly in terms of the vector operator **del** or **nabla**, defined to be

$$\boldsymbol{\nabla} = \frac{\partial}{\partial x}\mathbf{i} + \frac{\partial}{\partial y}\mathbf{j} + \frac{\partial}{\partial z}\mathbf{k}. \tag{6.64}$$

Since **i**, **j**, **k** are independent of the coordinates, and since (assuming continuity of all derivatives) the order of derivatives $\partial/\partial x$, $\partial/\partial y$, $\partial/\partial z$ can be changed, the del operator can be manipulated according to the ordinary rules of vectors, provided that *the order of* $\nabla$ *and any scalar or vector field is preserved*. For example, if $\psi_1$ and $\psi_2$ are two scalar fields,

$$\nabla\psi_1\psi_2=\frac{\partial}{\partial x}(\psi_1\psi_2)\mathbf{i}+\frac{\partial}{\partial y}(\psi_1\psi_2)\mathbf{j}+\frac{\partial}{\partial z}(\psi_1\psi_2)\mathbf{k},$$

which is not the same as

$$\psi_1\nabla\psi_2=\psi_1\frac{\partial\psi_2}{\partial x}\mathbf{i}+\psi_1\frac{\partial\psi_2}{\partial y}\mathbf{j}+\psi_1\frac{\partial\psi_2}{\partial z}\mathbf{k}.$$

Using the usual formula (2.36) and (3.19) for scalar and vector products, we see that (6.23), (6.38) and (6.57) can be written

$$\operatorname{div}\mathbf{v}=\nabla\cdot\mathbf{v}, \tag{6.65}$$

$$\operatorname{grad}\mathbf{v}=\nabla\psi, \tag{6.66}$$

$$\operatorname{curl}\mathbf{v}=\nabla\wedge\mathbf{v}. \tag{6.67}$$

In terms of the del operator, (6.61a) and (6.63a) become

$$\nabla\wedge\nabla\psi=\mathbf{0} \tag{6.61c}$$

and

$$\nabla\cdot\nabla\wedge\mathbf{v}=0 \tag{6.63c}$$

In this form, these equations follow at once by applying to $\nabla$ the ordinary rules of vector products and scalar triple products.

The definition (6.64) is in terms of rectangular coordinates. We shall now show that, in terms of a set $\{u_s\}$ of orthogonal curvilinear coordinates, and the corresponding local frame of reference $\{\mathbf{t}_r\}$,

$$\nabla=\sum_s\frac{\mathbf{t}_s}{h_s}\frac{\partial}{\partial u_s}. \tag{6.68}$$

Regarding $\{u_s\}$ as functions of $(x, y, z)$ and using the chain rule,

$$\begin{aligned}\frac{\partial}{\partial u_s}&=\frac{\partial x}{\partial u_s}\frac{\partial}{\partial x}+\frac{\partial y}{\partial u_s}\frac{\partial}{\partial y}+\frac{\partial z}{\partial u_s}\frac{\partial}{\partial z}\\&=\frac{\partial\mathbf{r}}{\partial u_s}\cdot\nabla=h_s\mathbf{t}_s\cdot\nabla,\end{aligned}$$

using the definition (6.64) and then (5.80). So

$$\sum_s\frac{\mathbf{t}_s}{h_s}\frac{\partial}{\partial u_s}=\sum_s\mathbf{t}_s(\mathbf{t}_s\cdot\nabla);$$

but, comparing with (2.27) and (2.29), this is just the sum of the components of $\nabla$ relative to the triad $\{\mathbf{t}_s\}$, and so is equal to $\nabla$. Note that in this proof, the differential operators have (correctly) been placed on the right of every expression.

If we allow the operator (6.68) to operate on a scalar field $\psi$, we immediately obtain the expression (6.34) for grad $\psi$, verifying (6.66) for the coordinate system $\{u_s\}$. It is not so easy to verify directly the formulae (6.65) and (6.67), as we see in Example 6.12 below. Direct verification is not, however, necessary, since (6.65)–(6.67) are true for rectangular coordinates, and (6.68) tells us how to change variables to any orthogonal system $\{u_s\}$.

**Example 6.12**

Use (6.68) to establish (6.65) directly, where div $\mathbf{v}$ is given by (6.20).

First, applying (6.68) to the functions $\{u_r\}$ gives

$$\nabla u_r = \mathbf{t}_r/h_r \qquad (r = 1, 2, 3).$$

Also, using the rule for differentiating products,

$$\nabla \cdot (\nabla u_r \wedge \nabla u_s) = (\nabla u_s) \cdot (\nabla \wedge \nabla u_r) - (\nabla u_r) \cdot (\nabla \wedge \nabla u_s) = 0.$$

In addition, we recall the relations $\mathbf{t}_3 = \mathbf{t}_1 \wedge \mathbf{t}_2$ and so on, for the orthonormal system $\{\mathbf{t}_r\}$.

Using these three formulae,

$$\begin{aligned}
\nabla \cdot \mathbf{v} &= \nabla \cdot \left(\sum_r v_r \mathbf{t}_r\right) \\
&= \nabla \cdot [v_1 h_2 h_3 (\nabla u_2) \wedge (\nabla u_3) + \text{cyclic terms}] \\
&= (\nabla u_2) \wedge (\nabla u_3) \cdot \nabla (v_1 h_2 h_3) + \text{cyclic terms} \\
&= \frac{1}{h_2 h_3} \mathbf{t}_1 \cdot \nabla (v_1 h_2 h_3) + \text{cyclic terms} \\
&= \frac{1}{h_1 h_2 h_3} \left[\frac{\partial}{\partial u_1}(v_1 h_2 h_3) + \text{cyclic terms}\right],
\end{aligned}$$

since the component of $\nabla$ is given by (6.34) or (6.68). This is the expression (6.20) for div $\mathbf{v}$.

The representation (6.65)–(6.67) can be used to establish other identities for scalar and vector fields. Three of these follow from the rules for differentiating products:

$$\operatorname{div}(\psi \mathbf{v}) = \psi \operatorname{div} \mathbf{v} + \mathbf{v} \cdot \operatorname{grad} \psi, \tag{6.69}$$

$$\operatorname{curl}(\psi \mathbf{v}) = \psi \operatorname{curl} \mathbf{v} + \operatorname{grad} \psi \wedge \mathbf{v}, \tag{6.70}$$

$$\operatorname{div}(\mathbf{v}_1 \wedge \mathbf{v}_2) = \mathbf{v}_2 \cdot \operatorname{curl} \mathbf{v}_1 - \mathbf{v}_1 \cdot \operatorname{curl} \mathbf{v}_2. \tag{6.71}$$

For example, (6.70) can be written as

$$\nabla \wedge (\psi \mathbf{v}) = \psi(\nabla \wedge \mathbf{v}) + (\nabla \psi) \wedge \mathbf{v}$$

in terms of the operator (6.68); in this form it follows at once from (5.20). The identities (6.69)–(6.71) can also be established from (6.20), (6.34) and (6.52).

If $\psi$ is a scalar field, grad $\psi$ is a vector field, and its divergence is scalar. So the operator

$$\nabla^2 \equiv \nabla \cdot \nabla = \operatorname{div} \operatorname{grad} \tag{6.72}$$

is a scalar differential operator, since it does not change the scalar character of $\psi$; this operator is called the **Laplacian**, often referred to as 'del squared'; it is probably the most important differential operator in mathematical physics. For an orthogonal coordinate system, (6.20) and (6.34) give

$$\nabla^2 \psi = \frac{1}{h_1 h_2 h_3}\left[\frac{\partial}{\partial u_1}\left(\frac{h_2 h_3}{h_1}\frac{\partial \psi}{\partial u_1}\right) + \frac{\partial}{\partial u_2}\left(\frac{h_3 h_1}{h_2}\frac{\partial \psi}{\partial u_2}\right) + \frac{\partial}{\partial u_3}\left(\frac{h_1 h_2}{h_3}\frac{\partial \psi}{\partial u_3}\right)\right]. \tag{6.73}$$

For rectangular coordinates, (6.64), (6.65) and (6.66) give

$$\nabla^2 = \nabla \cdot \nabla = \frac{\partial^2}{\partial x^2} + \frac{\partial^2}{\partial y^2} + \frac{\partial^2}{\partial z^2}; \tag{6.74}$$

this also follows by putting $h_1 = h_2 = h_3 = 1$ in (6.73). The differential equation

$$\nabla^2 \psi = 0 \tag{6.75}$$

for a scalar field is called the **Laplace equation**; it arises in many different physical theories, including those of gravitation, electricity and magnetism, fluid mechanics, and elasticity. Solutions of the Laplace equation are known as **harmonic functions**; a vast amount of work has been done studying and applying harmonic functions, but the subject is beyond the scope of this book. We shall, however, now show how Laplace's equation arises in electrostatics.

In Example 6.9, we saw that the electric field **E** produced by a charge $e$ at the origin $O$ is given by

$$\mathbf{E} = -\operatorname{grad} \psi, \tag{6.76}$$

where $\psi = e/r$, so that

$$\mathbf{E} = \frac{e}{r^2}\mathbf{t}_r.$$

Using Equation (6.23b) to calculate div $\mathbf{E}$ in spherical polar coordinates, we find

$$\operatorname{div}\mathbf{E} = \frac{1}{r^2}\frac{\partial}{\partial r}\left(r^2\frac{e}{r^2}\right) = 0,$$

except at $r = 0$. Since $\mathbf{E}$ is derivable from the potential $\psi$, it follows that $\psi = e/r$ satisfies Laplace's equation except at the origin.

Let us now consider a charge $e_1$ at the point with position vector $\mathbf{r}_1$; the potential at $\mathbf{r}$ is, by change of origin,

$$\psi_1 = \frac{e_1}{|\mathbf{r}-\mathbf{r}_1|}.$$

It is an experimental fact that the potential and electric field produced by a number of charges is obtained by summing the individual potentials or fields. So the potential produced by charges $e_1, e_2, \ldots, e_n$ at points $\mathbf{r}_1, \mathbf{r}_2, \ldots \mathbf{r}_n$ is

$$\psi(\mathbf{r}) = \sum_{l=1}^{n} \frac{e_l}{|\mathbf{r}-\mathbf{r}_l|}, \tag{6.77a}$$

and the electric field (the force acting on a unit charge) is

$$\begin{aligned}\mathbf{E}(\mathbf{r}) &= -\operatorname{grad}\psi(\mathbf{r}) \\ &= \sum_{l=1}^{n} \frac{e_l}{|\mathbf{r}-\mathbf{r}_l|^3}(\mathbf{r}-\mathbf{r}_l) \equiv \sum_{l=1}^{n} \mathbf{E}_l,\end{aligned} \tag{6.78a}$$

say. Since each term $\mathbf{E}_l$ in this sum has zero divergence except at the points $\mathbf{r}_l$,

$$\nabla^2\psi(\mathbf{r}) = -\operatorname{div}\mathbf{E}(\mathbf{r}) = 0 \qquad (\mathbf{r} \neq \mathbf{r}_l). \tag{6.79}$$

So the 'outflow' of $\mathbf{E}$ is zero except when $\mathbf{r} = \mathbf{r}_l$ ($l = 1, 2, \ldots, n$), and the potential $\psi$ then satisfies Laplace's equation (6.74). If we integrate (6.79) over any volume $\tau$ *not* containing any of the charges, the divergence theorem (6.19) gives

$$\iint_\sigma d\boldsymbol{\sigma}\cdot\mathbf{E} = \iiint_\tau d\tau \operatorname{div}\mathbf{E} = 0,$$

where $\sigma$ is the boundary of $\tau$. So the total outward flux of $\mathbf{E}$ over the boundary is zero, giving a conservation law for **E** in any region not containing charges. The conservation of flux of **E** allowed Faraday and Maxwell to think in terms of **lines of force** in the direction of **E**; a set of lines of force carry the same electric flux into and out of any region not containing charges, thereby satisfying the conservation law.

Let us consider the flux of **E** through a small sphere $|\mathbf{r}-\mathbf{r}_1|=a$ surrounding the charge $e_1$, due to the field $\mathbf{E}_1$ produced by $e_1$. The field along the outward normal has the same magnitude $e_1/a^2$ at all points (if $e_1<0$, the field is directed inwards). The integral of this flux over the surface of the sphere, of area $4\pi a^2$, is thus

$$(e_1/a^2)\times 4\pi a^2 = 4\pi e_1. \tag{6.80}$$

Now consider any simple volume region $\tau$, with boundary $\sigma$, which contains $\mathbf{r}_1$ as an interior point. Let $\sigma_a$ be a sphere of radius $a$, centre $\mathbf{r}_1$, with $a$ chosen so that $\sigma_a$ lies in the interior of $\tau$, as shown in Fig. 6.10. The volume region between $\sigma_a$ and $\sigma$ does not contain the

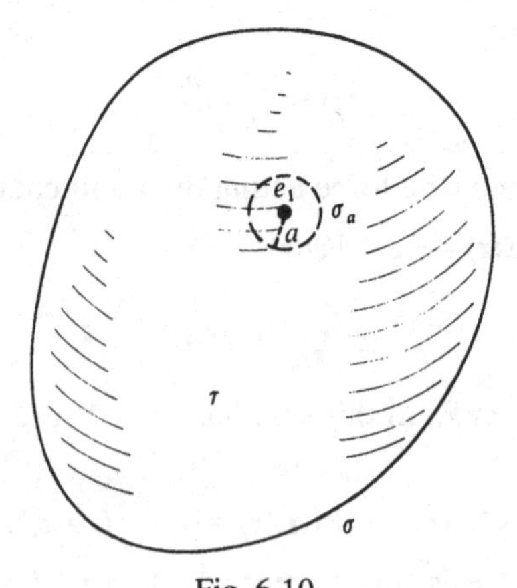


Fig. 6.10

charge $e_1$; the conservation law therefore ensures that the radially outward flux of $\mathbf{E}_1$ through $\sigma$ is equal to the outward flux (6.80) through $\sigma_a$. Now suppose that the volume $\tau$ has charges $e_1, e_2, \ldots, e_n$ at interior points $\mathbf{r}_1, \mathbf{r}_2, \ldots, \mathbf{r}_n$; each charge $e_l$ gives rise to a term $\mathbf{E}_l$ in (6.78a), and, just as for $\mathbf{E}_1$, the outward flux of $\mathbf{E}_l$ through the boundary $\sigma$ is $4\pi e_l$. Summing over $l$, the outward flux of **E** through $\sigma$

is

$$\iint_{\sigma} d\boldsymbol{\sigma} \cdot \mathbf{E} = 4\pi \sum_{l=1}^{n} e_l. \tag{6.81a}$$

This result is known as **Gauss' theorem**: *the total outward flux of* **E** *through a closed surface* $\sigma$ *is equal to* $4\pi$ *times the charge enclosed by* $\sigma$.

Instead of a set of discrete charges $\{e_l\}$ in the volume $\tau$, we can consider a volume distribution of charge of density $\rho(\mathbf{r})$ per unit volume. Then the sum of the right of (6.81a) is replaced by the volume integral over $\tau$ of the density, giving

$$\iint_{\sigma} d\boldsymbol{\sigma} \cdot \mathbf{E} = 4\pi \iiint_{\tau} \rho(\mathbf{r})\, d\tau; \tag{6.81b}$$

this equation expresses Gauss' theorem for a volume distribution of charge. If we now apply the divergence theorem (6.19) to the surface integral on the left, we see that the volume integrals of div $\mathbf{E}$ and of $4\pi\rho$ are equal; since this equality holds for every volume $\tau$, div $\mathbf{E} = 4\pi\rho$ at all points. It can be shown that, as in (6.78a), $\mathbf{E}(\mathbf{r})$ is derivable from the potential $\psi(\mathbf{r})$ at points $\mathbf{r}$ inside a volume distribution of charge; hence

$$\nabla^2\psi(\mathbf{r}) = -\text{div}\, \mathbf{E}(\mathbf{r}) = -4\pi\rho(\mathbf{r}) \tag{6.82}$$

at all points within the volume distribution; this equation is known as **Poisson's equation**. Where there is no charge, $\rho(\mathbf{r}) = 0$ and the equation reduces to Laplace's equation (6.79). Poisson's equation is the 'differential form' of Gauss' theorem (6.81b).

Although we shall not attempt a proof, we note that the potential $\psi(\mathbf{r})$ corresponding a charge density $\rho(\mathbf{r})$, and hence satisfying Poisson's equation, is found by replacing the sum in (6.77a) by an integral, giving

$$\psi(\mathbf{r}) = \iiint_{\tau'} d\tau' \frac{\rho(\mathbf{r}')}{|\mathbf{r} - \mathbf{r}'|}; \tag{6.77b}$$

the region $\tau'$ of integration is the region in which the density $\rho$ is non-zero, which is assumed to be finite. It can be shown that the integral (6.77b) is well defined, even at points $\mathbf{r}$ in the region $\tau'$; it is also valid to differentiate under the integral sign to give the formula

$$\begin{aligned}\mathbf{E}(\mathbf{r}) &= -\text{grad } \psi(\mathbf{r}) \\ &= \iiint_{\tau'} d\tau' \frac{\rho(\mathbf{r}')}{|\mathbf{r}-\mathbf{r}'|^3}(\mathbf{r}-\mathbf{r}')\end{aligned} \tag{6.78b}$$

for the field, corresponding to (6.78a) for discrete charges.

The derivation of the Laplace and Poisson equations for electrostatics applies equally to Newtonian gravitational theory, with mass density replacing charge density, and the same equations arise in other physical theories. The Laplacian is therefore a very important operator, and we shall give its explicit form in cylindrical and spherical polar coordinates. Using (6.23c) and (6.39), (6.72) becomes

$$\nabla^2 = \frac{1}{\rho}\frac{\partial}{\partial\rho}\left(\rho\frac{\partial}{\partial\rho}\right) + \frac{1}{\rho^2}\frac{\partial^2}{\partial\phi^2} + \frac{\partial^2}{\partial z^2} \tag{6.83}$$

for cylindrical polars. For spherical polars, (6.23b) and (6.40) give

$$\nabla^2 = \frac{1}{r^2}\left[\frac{\partial}{\partial r}\left(r^2\frac{\partial}{\partial r}\right) + \frac{1}{\sin\theta}\frac{\partial}{\partial\theta}\left(\sin\theta\frac{\partial}{\partial\theta}\right) + \frac{1}{\sin^2\theta}\frac{\partial^2}{\partial\phi^2}\right]. \tag{6.84}$$

We have established identities (6.61), (6.63) and (6.69)–(6.71) for the field operators div, grad and curl. A further identity is given by using the expansion (3.31) of the vector triple product,

$$\nabla \wedge (\nabla \wedge \mathbf{v}) = \nabla(\nabla \cdot \mathbf{v}) - (\nabla \cdot \nabla)\mathbf{v}, \tag{6.85a}$$

the vector $\mathbf{v}$ being kept to the right.

This formula can be expressed as

$$\text{curl curl } \mathbf{v} = \text{grad div } \mathbf{v} - \nabla^2\mathbf{v}; \tag{6.85b}$$

in the last term here, the Laplacian (6.72) operates on the *vector* $\mathbf{v}$, rather than on a scalar. Thus the derivatives in (6.68) operate on the vectors $\{\mathbf{t}_r\}$ in $\mathbf{v} = \Sigma_r v_r \mathbf{t}_r$, and $\nabla^2 = \nabla \cdot \nabla$ in (6.85a) will not have the forms (6.83) and (6.84) in polar coordinates. But since derivatives of $(\mathbf{i}, \mathbf{j}, \mathbf{k})$ are all zero, the expression (6.74) *is* valid in (6.85) when rectangular coordinates are used.

A further important identity involving two scalar fields $\psi$ and $\phi$ is obtained from the rules for differentiating products:

$$\nabla \cdot (\psi\nabla\phi) = (\nabla\psi) \cdot (\nabla\phi) + \psi\nabla^2\phi; \tag{6.86a}$$

using (6.65) and (6.66), this can be written

$$\text{div}(\psi \text{ grad } \phi) = \text{grad } \psi \cdot \text{grad } \phi + \psi\nabla^2\phi. \tag{6.86b}$$

Since $\psi$ and $\phi$ are scalars, the representations (6.74), (6.83) and (6.84) can be used in (6.86).

We have emphasised the physical importance of the vector operators, and we end this chapter by establishing conditions under which div $\mathbf{v}$ and curl $\mathbf{v}$ determine a vector field $\mathbf{v}$ uniquely; Laplace's equation occurs naturally in the proof. Let us suppose that a vector field $\mathbf{v}(\mathbf{r})$ is defined in a finite volume $\tau$ with boundary $\sigma$, and satisfies the three conditions:

(i) $\operatorname{div} \mathbf{v}(\mathbf{r}) = 4\pi\rho(\mathbf{r})$, where $\rho(\mathbf{r})$ is a scalar function defined throughout $\tau$;

(ii) $\operatorname{curl} \mathbf{v}(\mathbf{r}) = \boldsymbol{\mu}(\mathbf{r})$, where $\boldsymbol{\mu}(\mathbf{r})$ is a vector function defined throughout $\tau$;

(iii) the normal component $\mathbf{v} \cdot \mathbf{n}$ is given on the boundary $\sigma$.

Then **Helmholtz' theorem** states that *there is at most one vector field* $\mathbf{v}(\mathbf{r})$ *satisfying conditions* (i), (ii) and (iii). To establish this uniqueness theorem, let us suppose that $\mathbf{v}_1(\mathbf{r})$ and $\mathbf{v}_2(\mathbf{r})$ both satisfy these conditions; then the difference

$$\mathbf{v}_0(\mathbf{r}) = \mathbf{v}_1(\mathbf{r}) - \mathbf{v}_2(\mathbf{r}) \tag{6.87}$$

satisfies the conditions

(i)′ $\operatorname{div} \mathbf{v}_0(\mathbf{r}) = 0$ throughout $\tau$;

(ii)′ $\operatorname{curl} \mathbf{v}_0(\mathbf{r}) = \mathbf{0}$ throughout $\tau$;

(iii)′ $\mathbf{v}_0 \cdot \mathbf{n} = 0$ over the boundary $\sigma$.

Condition (ii)′ ensures that $\mathbf{v}_0(\mathbf{r})$ is derivable from a potential $\psi(r)$, with

$$\mathbf{v}_0(\mathbf{r}) = -\operatorname{grad} \psi(\mathbf{r}).$$

Now consider the integral of $\mathbf{v}_0 \cdot \mathbf{v}_0$ over the volume $\tau$; using (6.86b) with $\phi \equiv \psi$,

$$\iiint d\tau\, \mathbf{v}_0 \cdot \mathbf{v}_0 = \iiint d\tau \operatorname{grad} \psi \cdot \operatorname{grad} \psi$$

$$= \iiint d\tau[\operatorname{div}(\psi \operatorname{grad} \psi) - \psi \nabla^2 \psi].$$

Condition (i)′ gives $\nabla^2 \psi \equiv 0$, and the first term transforms by the divergence theorem to give

$$\iiint d\tau\, \mathbf{v}_0 \cdot \mathbf{v}_0 = \iint d\sigma\, \psi(\mathbf{n} \cdot \operatorname{grad} \psi)$$

$$= -\iint d\sigma\, \psi(\mathbf{n} \cdot \mathbf{v}_0) = 0,$$

using condition (iii)'. However, the integrand $\mathbf{v}_0 \cdot \mathbf{v}_0$ is non-negative; since its integral is zero, $\mathbf{v}_0 \cdot \mathbf{v}_0 = 0$ throughout $\tau$; this in turn implies that $\mathbf{v}_0(\mathbf{r}) = \mathbf{0}$ everywhere. Thus $\mathbf{v}_1(\mathbf{r}) \equiv \mathbf{v}_2(\mathbf{r})$, establishing uniqueness.

We should note that Helmholtz' theorem does not prove the existence of a vector field $\mathbf{v}$ satisfying (i), (ii) and (iii) above, but it is in fact true that a solution $\mathbf{v}$ does exist. The theorem can be extended to fields in an infinite volume $\tau$, provided that $|\mathbf{v}(\mathbf{r})| = O(r^{-2})$ as $r \to \infty$; the unique solution $\mathbf{v} = \mathbf{E}$ for an irrotational field, with curl $\mathbf{v} \equiv \mathbf{0}$, is given by (6.78b).

Helmholtz' theorem is the 3-dimensional analogue of a familiar property of one-variable functions $f(x)$ with continuous derivatives: if $df/dx$ and the value of $f(x)$ at one point are given, then $f(x)$ is determined. In 3-space, specification of div $\mathbf{v}$ and curl $\mathbf{v}$ determines the rates of change of $\mathbf{v}$ in three independent directions; the boundary condition prescribing $\mathbf{v} \cdot \mathbf{n}$ on $\sigma$ is the 3-dimensional generalisation of the condition defining $f(x)$ at one point.

## *Problems 6.4*

1 Use the operator $\boldsymbol{\nabla}$ to establish the identities (6.69) and (6.71).

2 Establish directly the equality (6.67), where $\boldsymbol{\nabla}$ and curl $\mathbf{v}$ are given by (6.68) and (6.52). [Use the methods of Example 6.12.]

3 If $\sigma$ is a simple closed surface bounding the volume $\tau$, $\psi$ a scalar field in $\tau$ and $\mathbf{a}$ any constant vector, apply the divergence theorem to $\psi\mathbf{a}$ to prove that

$$\iiint_\tau d\tau \,\text{grad}\, \psi = \iint_\sigma d\boldsymbol{\sigma}\, \psi.$$

By applying the divergence theorem to $\mathbf{v} \wedge \mathbf{a}$, where $\mathbf{v}$ is a vector field, show that

$$\iiint_\tau d\tau \,\text{curl}\, \mathbf{v} = \iint_\sigma d\boldsymbol{\sigma} \wedge \mathbf{v}.$$

APPENDIX A

# Some properties of functions of two variables

In Chapters 5 and 6, we assume certain analytic properties of derivatives and integrals of functions of two and three variables. In this appendix, we establish these properties for functions of two variables. Reference is made to these results whenever these properties, or their three-variable analogues, are assumed in the main text.

Two variables $(x, y)$ can be regarded as rectangular coordinates in a plane. A range of the coordinates, for example (5.8), can then be thought of as a region $S$ in the plane. A function $f(x, y)$ defined on $S$ is then a number uniquely determined when $(x, y)$ are given, corresponding to a point in $S$. The **partial derivative** of $f$ with respect to $x$ is defined as

$$\frac{\partial f(x, y)}{\partial x} = \lim_{\delta x \to 0} \frac{f(x+\delta x, y) - f(x, y)}{\delta x} \tag{A.1}$$

*whenever this limit exists*; it is the derivative of $f$ with respect to $x$, keeping $y$ fixed. Likewise, the partial derivative with respect to $y$ is

$$\frac{\partial f(x, y)}{\partial y} = \lim_{\delta y \to 0} \frac{f(x, y+\delta y) - f(x, y)}{\delta y} \tag{A.2}$$

whenever this limit exists. If $(x, y)$ corresponds to points on the boundary of $S$, the limits (A.1) and (A.2) may exist only as one-sided derivatives, with $\delta x$ or $\delta y$ having a fixed sign.

The derivatives (A.1) and (A.2) are themselves functions of $(x, y)$ which may be differentiable; the second partial derivatives are then denoted

$$\frac{\partial}{\partial x}\left(\frac{\partial f}{\partial x}\right) = \frac{\partial^2 f}{\partial x^2}, \qquad \frac{\partial}{\partial y}\left(\frac{\partial f}{\partial x}\right) = \frac{\partial^2 f}{\partial y\, \partial x},$$

and so on. Combining (A.2) and (A.1) we find, for example,

$$\frac{\partial^2 f(x,y)}{\partial y\,\partial x}=\lim_{\delta y\to 0}\lim_{\delta x\to 0}\frac{f(x+\delta x,y+\delta y)-f(x,y+\delta y)-f(x+\delta x,y)+f(x,y)}{\delta y\,\delta x} \tag{A.3}$$

with the limit $\delta x\to 0$ taken before the limit $\delta y\to 0$. The symmetry of the fraction on the right suggests that this derivative may be equal to $\partial^2 f/\partial x\,\partial y$, with the limit $\delta y\to 0$ taken before the limit $\delta x\to 0$, but this is not always true. We shall now establish conditions under which these two derivatives are equal.

The four points $(x, y)$, $(x+\delta x, y)$, $(x, y+\delta y)$, $(x+\delta x, y+\delta y)$ define a rectangle $R$, and we define the **second difference**

$$\delta^2 f(R)\equiv f(x+\delta x,y+\delta y)-f(x,y+\delta y)-f(x+\delta x,y)+f(x,y), \tag{A.4}$$

so that (A.3) becomes

$$\frac{\partial^2 f(x,y)}{\partial y\,\partial x}=\lim_{\delta y\to 0}\lim_{\delta x\to 0}\frac{\delta^2 f(R)}{\delta y\,\delta x}. \tag{A.5}$$

In Fig. A.1, the two rectangles denoted by $R_2$ and $R_3$ have areas $\delta x\,\delta y_2$ and $\delta x\,\delta y_3$, as shown; these rectangles combine to form a larger rectangle $R_1$, with area $\delta x\,\delta y_1$, where

$$\delta y_3=\delta y_1-\delta y_2. \tag{A.6}$$

If we apply the definition (A.4) to the rectangles $R_1$, $R_2$, $R_3$, it follows that

$$\delta^2 f(R_3)=\delta^2 f(R_1)-\delta^2 f(R_2). \tag{A.7}$$

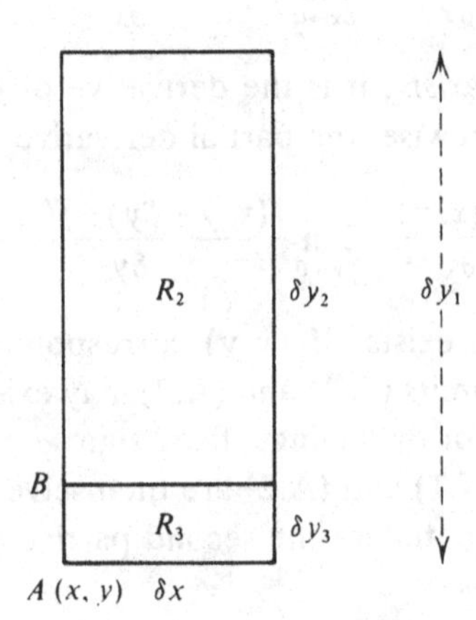


Fig. A.1

It is not hard to check the algebraic identity

$$2\frac{a_1-a_2}{b_1-b_2}-\left(\frac{a_1}{b_1}+\frac{a_2}{b_2}\right)=\frac{b_1+b_2}{b_1-b_2}\left(\frac{a_1}{b_1}-\frac{a_2}{b_2}\right);$$

putting $a_1=\delta^2 f(R_1)$, $a_2=\delta^2 f(R_2)$, $b_1=\delta y_1$, $b_2=\delta y_2$, using (A.6) and (A.7), and dividing by $\delta x$, this identity gives

$$2\frac{\delta^2 f(R_3)}{\delta x\,\delta y_3}-\left[\frac{\delta^2 f(R_1)}{\delta x\,\delta y_1}+\frac{\delta^2 f(R_2)}{\delta x\,\delta y_2}\right]$$
$$=\frac{\delta y_1+\delta y_2}{\delta y_3}\left[\frac{\delta^2 f(R_1)}{\delta x\,\delta y_1}-\frac{\delta^2 f(R_2)}{\delta x\,\delta y_2}\right]. \quad \text{(A.8)}$$

In this identity we now let $\delta x$, $\delta y_1$, $\delta y_2$, $\delta y_3$ all tend to zero, but in such a way that

$$\frac{\delta x}{\delta y_1}\to 0,\ \frac{\delta x}{\delta y_2}\to 0,\ \frac{\delta y_3}{\delta x}\to 0; \quad \text{(A.9)}$$

this is indicated in Fig. A.1 by drawing $\delta y_3$ smaller than $\delta x$, which is in turn smaller than $\delta y_1$ and $\delta y_2$. Since $\delta y_3\to 0$ faster than $\delta x$, then by (A.5)

$$\frac{\delta^2 f(R_3)}{\delta x\,\delta y_3}\to\frac{\partial^2 f_A}{\partial x\,\partial y},$$

the suffix '$A$' denoting evaluation at the point $A$; and since $\delta x\to 0$ faster than $\delta y_1$ and $\delta y_2$,

$$\frac{\delta^2 f(R_1)}{\delta x\,\delta y_1}+\frac{\delta^2 f(R_2)}{\delta x\,\delta y_2}\to\frac{\partial^2 f_A}{\partial y\,\partial x}+\frac{\partial^2 f_B}{\partial y\,\partial x}.$$

We have of course assumed that these three second derivatives exist; if we also assume that $\partial^2 f/\partial y\,\partial x$ is continuous at $A$, $\partial^2 f_A/\partial y\,\partial x=\partial^2 f_B/\partial y\,\partial x$ in the limit $B\to A$, so that the left-hand side of (A.8) has the limit

$$2\left[\frac{\partial^2 f_A}{\partial x\,\partial y}-\frac{\partial^2 f_A}{\partial y\,\partial x}\right]. \quad \text{(A.10)}$$

Now consider the right-hand side of (A.8). The square bracket is the change in $\partial^2 f(R)/\delta x\,\delta y$ when $\delta y$ changes from $\delta y_2$ to $\delta y_1$, by an amount $\delta y_3$; we therefore write the expression as

$$\frac{\delta y_1+\delta y_2}{\delta y_3}\delta\left[\frac{\delta^2 f(R_2)}{\delta x\,\delta y_2}\right].$$

In the limit, this is equivalent to

$$\frac{\delta y_1+\delta y_2}{\delta y_3}\delta\left[\frac{\partial^2 f_A}{\partial y\,\partial x}\right]. \tag{A.11}$$

Now continuity of $\partial^2 f/\partial y\,\partial x$ ensures that as $\delta y_3 \to 0$,

$$\delta\left[\frac{\partial^2 f_A}{\partial y\,\partial x}\right]\to 0. \tag{A.12}$$

Conditions (A.9) require that $(\delta y_1+\delta y_2)/\delta y_3 \to \infty$, but the continuity condition (A.12) allows us to take the limit so that (A.11) tends to zero. (For example, if the increment (A.12) is $10^{-3n}$, we can choose $\delta y_1 \simeq \delta y_2 = 10^{2n}\,\delta y_3$ and $\delta x = 10^n\,\delta y_3$; then as $n\to\infty$, (A.9) is satisfied, while (A.11) is nearly $2\cdot 10^{-n}$, which tends to zero.)

The limit of (A.8) therefore tells us that (A.10) is zero. So we have shown:

**Theorem A.1**

If the partial derivatives $\partial^2 f/\partial x\,\partial y$ and $\partial^2 f/\partial y\,\partial x$ exist at a point $(x, y)$, and if one of them is continuous, then they are equal.

Next, we establish a theorem which is a simple generalisation of the **fundamental theorem of the calculus** [Reference A.1]; this states that if $f(y)$ is a function of a real variable $y$, and if the derivative $\mathrm{d}f/\mathrm{d}y$ is continuous for $\alpha \leqslant y \leqslant \beta$, then

$$\int_\alpha^\beta \mathrm{d}y\,\frac{\mathrm{d}f}{\mathrm{d}y}=f(\beta)-f(\alpha). \tag{A.13}$$

Now suppose that a function $f(x, y)$ is defined for some range of values of $(x, y)$, perhaps of the form (5.8); the function $\partial f/\partial y$ is **uniformly continuous** in $y$ if, for any given $\varepsilon > 0$,

$$\left|\frac{\partial f(x, y+\delta y)}{\partial y}-\frac{\partial f(x, y)}{\partial y}\right|<\varepsilon \tag{A.14}$$

for all increments $\delta y$ satisfying

$$|\delta y|>\delta, \tag{A.15}$$

where $\delta$ depends on $\varepsilon$ *but does not depend on the value of x*. If $x$ is regarded as fixed, (A.14) and (A.15) express continuity of $\partial f/\partial y$ as a function of $y$; we can therefore integrate it with respect to $y$, with $x$ fixed, and use (A.13) to give:

**Theorem A.2**

If $\partial f(x, y)/\partial y$ exists and is uniformly continuous in $y$ for some range of values $\alpha \leqslant y \leqslant \beta$, where $\alpha$ and $\beta$ are finite, but may vary with $x$ over a range of values of $x$, then

$$\int_{\alpha}^{\beta} \mathrm{d}y \frac{\partial f(x, y)}{\partial y} = f(x, \beta) - f(x, \alpha) \tag{A.16}$$

throughout the range of $x$.

(We note that, for a finite closed range $a \leqslant x \leqslant b$ of $x$, continuity in $y$ is equivalent to uniform continuity.)

The definition of a *double integral* is similar to that of a single integral; integration can be defined with varying degrees of generality and sophistication, but we shall give the simplest definition, that of the **Riemann integral**, of a function $f(x, y)$ over a finite region $S$ of the $(x, y)$ plane. Since $S$ is finite, it can be contained within a rectangle $R$ of the plane corresponding to finite ranges

$$X_0 \leqslant x \leqslant X_1, \tag{A.17a}$$

$$Y_0 \leqslant y \leqslant Y_1. \tag{A.17b}$$

We define the **extension** of $f(x, y)$ in $R$ to be the function

$$\bar{f}(x, y) = \begin{Bmatrix} f(x, y) & \text{for } (x, y) \subset S \\ 0 & \text{for } (x, y) \not\subset S \end{Bmatrix}. \tag{A.18}$$

A coarse mesh of lines on $R$ is defined by $x = x_r$ $(r = 0, 1, 2, \ldots, m)$ and $y = y_s$ $(s = 0, 1, 2, \ldots, n)$, where

$$X_0 = x_0 < x_1 < x_2 < \ldots < x_m = X_1, \tag{A.19a}$$

$$Y_0 = y_0 < y_1 < y_2 < \ldots < y_n = Y_1. \tag{A.19b}$$

This mesh divides the rectangle $R$ into $mn$ smaller rectangles. If we write

$$\delta x_r = x_r - x_{r-1} \qquad (r = 1, 2, \ldots, m), \tag{A.20a}$$

$$\delta y_s = y_s - y_{s-1} \qquad (s = 1, 2, \ldots, n), \tag{A.20b}$$

then the small rectangle bounded by lines $x = x_{r-1}$, $x = x_r$, $y = y_{s-1}$, $y = y_s$ has area

$$\delta S_i = \delta x_r\, \delta y_s, \tag{A.21}$$

where $i$ denotes the pair $(r, s)$; this rectangle is labelled $R_i$. Now let $M_i$ and $m_i$ be the maximum and minimum values of $\bar{f}(x, y)$ in the rectangle $R_i$; then the **upper and lower sums** of $\tilde{f}(x, y)$ over $R$, for the mesh (A.19), are

$$M = \sum_i M_i\,\delta S_i = \sum_{r=1}^{m} \sum_{s=1}^{n} M_i\,\delta x_r\,\delta y_s, \tag{A.22a}$$

$$m = \sum_i m_i\,\delta S_i = \sum_{r=1}^{m} \sum_{s=1}^{n} m_i\,\delta x_r\,\delta y_s. \tag{A.22b}$$

Since $\tilde{f}(x, y) = 0$ outside $S$, $M_i = m_i = 0$ for rectangles wholly outside $S$, so that these do not contribute to the sums $M$ and $m$. The actual size of $R$ is therefore unimportant, so long as $S \subset R$.

The sums (A.22) depend upon the particular mesh chosen, and since $m_i \leqslant M_i$ for all $i$, $m \leqslant M$. Also, if the mesh (A.19) is subdivided into a finer mesh, $m$ increases and $M$ decreases. Now consider the limit

$$\mathrm{Max}(\delta x_r, \delta y_s) \to 0 \tag{A.23}$$

of a fine mesh: the decreasing sum $M$ and the increasing sum $m$ must each tend to a finite limit [Reference A.2]; if these limits are equal, their common value is the **double integral**

$$\iint_S f(x, y)\,\mathrm{d}S$$

of $f$ over the region $S$. Equally, this common limit is the integral of $\tilde{f}(x, y)$ over the rectangle $R$.

We now prove several theorems concerning double integrals; the first is:

**Theorem A.3**

If $f_i$ is the value of $f(x, y)$ at any point in the rectangle $R_i$, for all $i$, then the integral of $f(x, y)$ over $S$ is equal to

$$\lim \sum_i f_i\,\delta S_i. \tag{A.24}$$

The proof is simple: since $m_i \leqslant f_i \leqslant M_i$,

$$\sum_i m_i\,\delta S_i \leqslant \sum_i f_i\,\delta S_i \leqslant \sum_i M_i\,\delta S_i.$$

In the limit (A.23), the upper and lower sums tend to the same limit; hence $\Sigma_i f_i\,\delta S_i$ must tend to this limit also.

**Theorem A.4**

If $|f(x, y)| \leqslant K$ for all points $(x, y) \subset S$, and if $\Gamma$ is a finite curve lying within a finite rectangle $R$ containing $S$, then the contributions to the sums (A.22) from rectangles $R_i$ containing any points of $\Gamma$ tend to zero in the limit of a fine mesh.

**Proof.** Let the length of $\Gamma$ be $l$. Now choose the rectangle $R$ and mesh (A.19) so that all increments $\{\delta x_r\}$ and $\{\delta y_s\}$ are equal, say to a number $\delta$. Since, in general, $\Gamma$ can enter at most three other rectangles $R_i$ between complete traverses of two of the rectangles, the number $N$ of rectangles entered is limited by

$$(N-3)\delta \leqslant 4l. \tag{A.25}$$

Since $|M_i| \leqslant K$ for all $i$, the contribution to (A.22a) from rectangles intersecting $\Gamma$ has modulus satisfying

$$\begin{aligned}\left|\sum_{i(\Gamma)} M_i\, \delta S_i\right| &\leqslant \sum_{i(\Gamma)} |M_i|\, \delta^2 \\ &\leqslant NK\delta^2 \leqslant (4l+3\delta)K\delta,\end{aligned}$$

using (A.25). Since $l$ and $K$ are fixed and finite, this quantity tends to zero with the mesh size $\delta$, establishing Theorem A.4.

It is important to know that double integrals exist for certain classes of function. We shall not prove integrability of the largest possible class of functions, but content ourselves with a theorem which is applicable in many physical situations.

**Theorem A.5**

If $f(x, y)$ is bounded and piecewise uniformly continuous on $S$, then the double integral of $f$ over $S$ exists.

**Proof.** The condition **piecewise uniformly continuous** means that, except on certain curves of finite total length, the following continuity condition holds: given any positive number $\varepsilon$, we can find a positive number $\delta$ such that

$$|f(x+\delta x, y+\delta y) - f(x, y)| < \varepsilon \tag{A.26}$$

provided $|\delta x| \leqslant \delta$ and $|\delta y| \leqslant \delta$. The number $\delta$ depends on $\varepsilon$, but not on the point $(x, y)$.

Since $f$ is bounded, Theorem A.4 ensures that those rectangles $R_i$ which intersect the curves of discontinuity of $f$ do not contribute to the sums (A.22) in the limit. In all other rectangles $R_i$, $f$ is continuous, and (A.26) ensures that

$$|M_i - m_i| < \varepsilon$$

for a mesh with increments no larger than $\delta$.

So from (A.22), the difference between the upper and lower sums satisfies

$$M - m \leqslant \sum_i \varepsilon\, \delta S_i \leqslant \varepsilon (X_1 - X_0)(Y_1 - Y_0).$$

Since $\varepsilon$ can be made as small as we wish by choosing $\delta$ (the maximum mesh increment) small enough, $M - m \to 0$ in the limit of a fine mesh. So $M$ and $m$ have the same limit, which is the integral of $f$ over $S$.

One important property of integrals is their additivity.

**Theorem A.6**

If the region $S$ is divided into a finite number of non-overlapping subregions by curves of finite total length, the integral over $S$ of a bounded and uniformly continuous function $f(x, y)$ is equal to the sum of its integrals over the subregions.

**Proof.** By Theorem A.4, rectangles $R_i$ which intersect the curves do not contribute to (A.22) in the limit. The other terms in these sums can be divided into sets, each corresponding to one subregion of $S$. By Theorem A.5, the upper and lower sums corresponding to a given subregion tend to the same limit, which is the integral over the subregion. So in the limit, the integral over $S$ is the sum of the integrals of its subregions.

As exemplified in the main text, double integrals are usually evaluated as repeated integrals. It therefore is important to establish the equality of double and repeated integrals.

**Theorem A.7**

If the region $S$ is defined by

$$a \leqslant x \leqslant b, \qquad \alpha(x) \leqslant y \leqslant \beta(x), \tag{A.27}$$

and if, for each value of $x$ in $a \leqslant x \leqslant b$, the integral

$$\int_{\alpha(x)}^{\beta(x)} \mathrm{d}y\, f(x, y)$$

exists, then, provided the double integral of $f$ over $S$ also exists, it is given by

$$\iint_S \mathrm{d}S\, f(x, y) = \int_a^b \mathrm{d}x \int_{\alpha(x)}^{\beta(x)} \mathrm{d}y\, f(x, y).$$

**Proof.** Since $i$ denotes the pair $(r, s)$, the point $(x_r, y_s)$ is at one corner of the rectangle $R_i$. So we can take $f_i = f(x_r, y_s)$ in (A.24); then the integral of $f$ over $S$ is

$$\lim \sum_{r=1}^{m} \sum_{s=1}^{n} f(x_r, y_s)\, \delta x_r\, \delta y_s. \tag{A.28}$$

The coefficient of $\delta x_r$ in this double sum is

$$\sum_{s=1}^{n} f(x_r, y_s)\, \delta y_s;$$

since $f(x, y) = 0$ outside $S$, this sum is over rectangles with $\alpha(x_r) \leqslant y_s \leqslant \beta(x_r)$, with $a \leqslant x_r \leqslant b$; also, as $\max(\delta y_s) \to 0$, this sum tends to the integral

$$\int_{\alpha(x_r)}^{\beta(x_r)} \mathrm{d}y\, f(x_r, y)$$

since this integral exists. So (A.28) is the limit of

$$\sum_{r=1}^{m} \delta x_r \int_{\alpha(x_r)}^{\beta(x_r)} f(x_r, y)\, \mathrm{d}y$$

as $\max(\delta x_r) \to 0$. This limit exists, because it is the double integral; but since only points with $a \leqslant x_r \leqslant b$ contribute to the sum $\Sigma_{r=1}^{m}$, the limit is just the repeated integral

$$\int_a^b \mathrm{d}x \int_{\alpha(x)}^{\beta(x)} f(x, y)\, \mathrm{d}y. \tag{A.29}$$

If the region $S$ can be divided into a finite number of subregions, each corresponding to a range of the form (A.27), Theorem A.6 ensures that the double integral over $S$ is equal to the sum over the subregions of repeated integrals of the form (A.29). This is the method normally used in practice to evaluate double integrals.

The last result proved in this appendix is:

**Theorem A.8**

Let $S_1, S_2, S_3, \ldots$ be a sequence of regions in the $(x, y)$ plane such that

($a$) $S_n$ contains the point $(x_0, y_0)$, for all $n$;

($b$) for all points $(x, y)$ in $S_n$,

$$|x - x_0| < \delta_n, \qquad |y - y_0| < \delta_n,$$

where $\{\delta_n\}$ is a positive sequence with $\delta_n \to 0$ as $n \to \infty$;

($c$) the area of $S_n$ is

$$A_n = \iint\limits_{S_n} dS.$$

Then if $f(x, y)$ is continuous at $(x_0, y_0)$,

$$\lim_{n\to\infty} \frac{1}{A_n} \iint\limits_{S_n} f(x, y)\, dS = f(x_0, y_0).$$

**Proof.** In $S_n$, continuity of $f(x, y)$ ensures that

$$|f(x, y) - f(x_0, y_0)| < \varepsilon_n$$

for all $(x, y) \in S_n$, where $\varepsilon_n \to 0$ as $\delta_n \to 0$. Hence

$$\left|\frac{1}{A_n} \iint\limits_{S_n} f(x, y)\, dS - f(x_0, y_0)\right|$$

$$\leqslant \frac{1}{A_n} \iint\limits_{S_n} |f(x, y) - f(x_0, y_0)|\, dS$$

$$\leqslant \frac{1}{A_n} \varepsilon_n \iint\limits_{S_n} dS = \varepsilon_n.$$

We can choose $\varepsilon_n$ as small as we wish by letting $n \to \infty$ and $\delta_n \to 0$; this establishes the theorem.

APPENDIX B

# Proof of Stokes' theorem

We wish to establish Stokes' theorem for surfaces $\sigma$ which have no special relationship to the coordinate system $\{u_r\}$. As in the main text, it is convenient to make some simplifying assumptions about $\sigma$, and then to indicate how the theorem can be extended to more general surfaces. Before embarking on the main theorem we need to introduce some notation, and then to derive a formula used in establishing the theorem. The boundary to the surface $\sigma$ is the closed curve $\Gamma$. We assume that $(u_1, u_2, u_3)$ are orthogonal curvilinear coordinates, and that $\gamma$ is the curve of intersection of $\sigma$ and the surface $u_3 = c$ ($c$ constant). In Fig. B.1, we show the relationship of $\Gamma$, $\sigma$ and $\gamma$ in a particularly simple situation, when $\gamma$ is a continuous curve joining points $A$ and $B$ on the boundary $\Gamma$; we assume that $A$ and $B$ correspond to values $(u_1', u_2')$ and $(u_1'', u_2'')$ respectively of the two

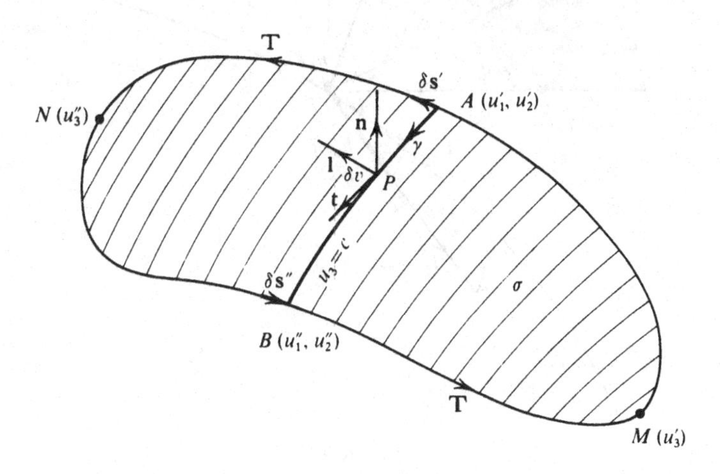


Fig. B.1

coordinates $(u_1, u_2)$. As the value $c$ of $u_3$ increases from $u_3'$ at $M$ to $u_3''$ $(>u_3')$ at $N$, the curve $\gamma$ sweeps out the whole surface $\sigma$. In general, the limiting values $u_1', u_2', u_1'', u_2''$, will depend upon the value of $u_3$.

We assume that $\sigma$ is a smooth surface, so that we can define a unit normal vector $\mathbf{n}$ to $\sigma$ at every point; it is chosen to be right-handed relative to the sense of integration round $\Gamma$; we note that $\mathbf{n}$ is *not* usually the normal vector to the curve $\gamma$. At any point $P$ of $\gamma$, a tangent vector $\mathbf{t}$ can be defined, normal to $\mathbf{n}$; a third unit vector $\mathbf{l}$ completes a right-handed triad $(\mathbf{t}, \mathbf{n}, \mathbf{l})$, as shown in Fig. B.1, $\mathbf{t}$ being chosen so that $\mathbf{l}$ is in the direction of increasing $u_3$; so a displacement tangential to $\sigma$, and normal to $\gamma$, is of the form $\mathbf{l}\,\delta v$. The increment $\delta v$ corresponds to a change $\delta u_3$ in the value of $u_3$, and to displacements $\delta\mathbf{s}'$ at $A$ and $-\delta\mathbf{s}''$ at $B$, shown in Fig. B.1. At $P$, the orthogonal system $\{u_r\}$ defines, as in §5.5, a triad $\{\mathbf{t}_r\}$; the two triads $(\mathbf{t}, \mathbf{n}, \mathbf{l})$ and $(\mathbf{t}_1, \mathbf{t}_2, \mathbf{t}_3)$ are shown in Fig. B.2. Since $\mathbf{t}$ is tangent to the surface $u_3 = c$, it is orthogonal to $\mathbf{t}_3$ and lies in the plane defined by $\mathbf{t}_1, \mathbf{t}_2$. So a displace-

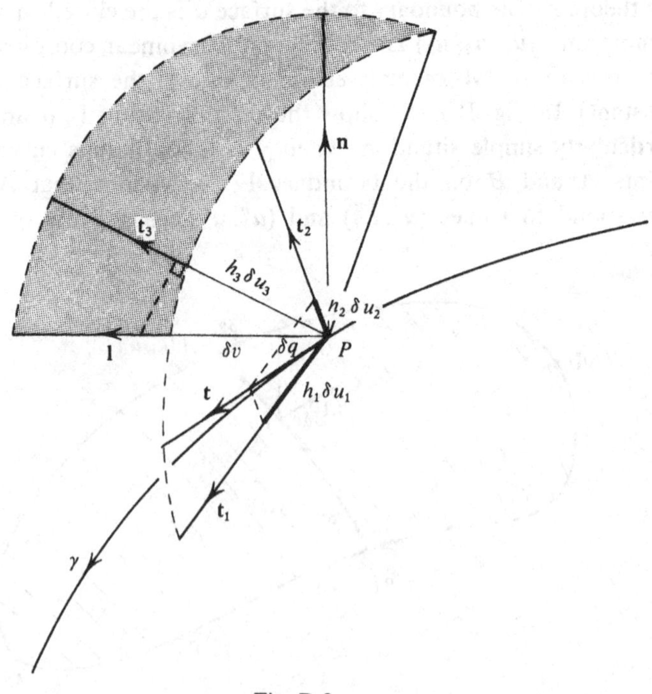


Fig. B.2

ment $\delta q$ along $\mathbf{t}$ is the sum

$$\delta q\,\mathbf{t}=h_1\,\delta u_1\,\mathbf{t}_1+h_2\,\delta u_2\,\mathbf{t}_2$$

of displacements (5.76) along $\mathbf{t}_1$ and $\mathbf{t}_2$; here $\delta u_1$ and $\delta u_2$ are the increments in $u_1$ and $u_2$ corresponding to the displacement along $\mathbf{t}$. Taking scalar products with $\mathbf{t}_1$ and $\mathbf{t}_2$ gives

$$h_1\,\delta u_1=\delta q(\mathbf{t}\cdot\mathbf{t}_1),\qquad h_2\,\delta u_2=\delta q(\mathbf{t}\cdot\mathbf{t}_2), \tag{B.1}$$

which is also evident from Fig. B.2.

We can now establish the formula needed in proving Stokes' theorem. Let us consider a function $f(u_1, u_2, u_3)$ of the three coordinates. On the surface $u_3=c$ it is the function $f(u_1, u_2, c)$ of the two variables $u_1, u_2$; on the curve $\gamma$, $u_1$ and $u_2$ can be regarded as functions

$$u_1=U_1(q),\qquad u_2=U_2(q) \tag{B.2}$$

of the distance $q$ along $\gamma$. Then, on $\gamma$, $f$ is the function

$$\tilde{f}(q)=f[U_1(q), U_2(q), c] \tag{B.3}$$

of $q$. We assume that $f$ has continuous derivatives; the chain rule then gives

$$\frac{\mathrm{d}\tilde{f}}{\mathrm{d}q}=\frac{\partial f}{\partial u_1}\frac{\partial u_1}{\partial q}+\frac{\partial f}{\partial u_2}\frac{\partial u_2}{\partial q}.$$

However, (B.1) relates $\delta q$ to the corresponding increments $\delta u_1$, $\delta u_2$, and can be written

$$\frac{\partial u_1}{\partial q}=\frac{\mathbf{t}\cdot\mathbf{t}_1}{h_1},\qquad \frac{\partial u_2}{\partial q}=\frac{\mathbf{t}\cdot\mathbf{t}_2}{h_2}.$$

Therefore the chain rule equation becomes

$$\frac{\mathrm{d}\tilde{f}}{\mathrm{d}q}=\frac{\mathbf{t}\cdot\mathbf{t}_1}{h_1}\frac{\partial f}{\partial u_1}+\frac{\mathbf{t}\cdot\mathbf{t}_2}{h_2}\frac{\partial f}{\partial u_2}. \tag{B.4}$$

In this equation, all the quantities have to be evaluated at $P$.

We now integrate (B.4) along the curve $\gamma$ from $A$ to $B$. If the values of $f(u_r)=\tilde{f}(q)$ at $A$ and $B$ are $f_A$ and $f_B$, then using Theorem A.2 of Appendix A,

$$\begin{aligned} f_B-f_A&=\int_A^B \mathrm{d}q\,\frac{\mathrm{d}\tilde{f}(q)}{\mathrm{d}q}\\ &=\int_A^B \mathrm{d}q\left[\frac{\mathbf{t}\cdot\mathbf{t}_1}{h_1}\frac{\partial f}{\partial u_1}+\frac{\mathbf{t}\cdot\mathbf{t}_2}{h_2}\frac{\partial f}{\partial u_2}\right]. \end{aligned} \tag{B.5}$$

This is the formula we shall use to establish Stokes' theorem. This key formula, as in the proof of the divergence theorem in §6.2, is essentially the fundamental theorem of the calculus.

The components of $\delta\mathbf{s}$ are $\{h_r\,\delta u_r;\, r=1,2,3\}$, so that the integral (6.47) is

$$\oint_\Gamma (v_1h_1\,du_1+v_2h_2\,du_2+v_3h_3\,du_3). \tag{B.6}$$

The integrand $v_3h_3$ in the third term

$$\oint_\Gamma v_3h_3\,du_3 \tag{B.7}$$

is evaluated on $\Gamma$ just as $\psi$ is in (6.8b), but with $u_3$ as the independent variable. In the simple situation of Fig. B.1, which we deal with first, the range of $u_3$ in passing along $\Gamma$ increases from $u_3'$ at $M$ to $u_3''$ at $N$ through $A$, and then decreases through the same range on the section of $\Gamma$ through $B$. The points $A$ and $B$ correspond to the same value of $u_3$, and the increments $\delta\mathbf{s}'$ and $-\delta\mathbf{s}''$ correspond to the same increment $\delta u_3$ of $u_3$; we have defined the increment $\mathbf{l}\delta v$ at any point $P$ to correspond also to $\delta u_3$, as in Fig. B.2. If we write

$$\mathbf{l}=l_1\mathbf{t}_1+l_2\mathbf{t}_2+l_3\mathbf{t}_3, \tag{B.8}$$

so that $(\mathbf{l}\cdot\mathbf{t}_3)=l_3$, then $\delta v$ and $\delta u_3$ are related by

$$l_3\,\delta v=h_3\,\delta u_3. \tag{B.9}$$

Using suffixes $A$ and $B$ to denote function values at these points, the integral (B.7) equals

$$\int_{u_3'}^{u_3''} du_3[(v_3h_3)_A-(v_3h_3)_B].$$

Using the formula (B.5) with $f=v_3h_3$ and using (B.9) to change variable from $u_3$ to $v$, this integral becomes

$$-\int dv\frac{l_3}{h_3}\int_A^B dq\left[\frac{\mathbf{t}\cdot\mathbf{t}_1}{h_1}\frac{\partial(v_3h_3)}{\partial u_1}+\frac{\mathbf{t}\cdot\mathbf{t}_2}{h_2}\frac{\partial(v_3h_3)}{\partial u_2}\right]. \tag{B.10}$$

Now

$$\mathbf{t}=(\mathbf{t}\cdot\mathbf{t}_1)\mathbf{t}_1+(\mathbf{t}\cdot\mathbf{t}_2)\mathbf{t}_2$$

and the normal vector $\mathbf{n}$ is given by

$$\mathbf{n}=\mathbf{l}\wedge\mathbf{t};$$

so the first two components of $\mathbf{n}$ are

$$n_1=-l_3(\mathbf{t}\cdot\mathbf{t}_2),\qquad n_2=l_3(\mathbf{t}\cdot\mathbf{t}_1).$$

Substituting these results in (B.10), we find eventually that (B.7) becomes

$$\iint \mathrm{d}v\,\mathrm{d}q\left[-\frac{n_2}{h_1h_3}\frac{\partial(v_3h_3)}{\partial u_1}+\frac{n_1}{h_2h_3}\frac{\partial(v_3h_3)}{\partial u_2}\right]. \tag{B.11}$$

In this integral, $\mathrm{d}v\,\mathrm{d}q=\mathrm{d}\sigma$ is just the increment of surface area, since $\mathbf{l}$ is normal to $\mathbf{t}$; and since the $q$-integration is along $\gamma$ in Fig. 6.7, while the $v$-integration corresponds to $\gamma$ sweeping over $\sigma$ from $M$ to $N$, (B.11) is just a surface integral over the whole of $\sigma$.

The first two terms in (B.6) can be treated in the same way, giving rise to two integrals over $\sigma$ similar to (B.11). So (6.47) is given by

$$\oint_\Gamma \mathbf{v}\cdot\mathrm{d}\mathbf{s}=\int_\sigma \mathrm{d}\sigma\left[\frac{n_1}{h_2h_3}\left\{\frac{\partial(v_3h_3)}{\partial u_2}-\frac{\partial(v_2h_2)}{\partial u_3}\right\}\right.$$
$$+\frac{n_2}{h_3h_1}\left\{\frac{\partial(v_1h_1)}{\partial u_3}-\frac{\partial(v_3h_3)}{\partial u_1}\right\}$$
$$\left.+\frac{n_3}{h_1h_2}\left\{\frac{\partial(v_2h_2)}{\partial u_1}-\frac{\partial(v_1h_1)}{\partial u_2}\right\}\right].$$

Since

$$\mathbf{n}=n_1\mathbf{t}_1+n_2\mathbf{t}_2+n_3\mathbf{t}_3,$$

the surface integral is

$$\int_\sigma \mathrm{d}\sigma\,\mathbf{n}\cdot\mathrm{curl}\,\mathbf{v},$$

where curl $\mathbf{v}$ is given by (6.52). This establishes Stokes' theorem.

In this proof, we can dispense with the assumption that the curves $\gamma$ meet the boundary $\Gamma$ in only two points. We can also allow sections of the boundary to have a constant value of $u_3$. Consider, for example, a surface $\sigma$ of the form $MNPQRS$ in Fig. B.3; the sections $NP$ and $MS$ of the boundary $\Gamma$ have $u_3$ constant, and so do not contribute to the integral (B.7). Curves on $\sigma$ with $u_3=c$, for various constants $c$, are shown as $AB$, $CD$ plus $EF$, and $GH$. The contributions to (B.7) from $TN$ and $PQ$ transform into the integral (B.11) over the surface $NPQT$; the contributions from $QR$ and $RV$ give the surface integral over $QRV$, and those from $MT$ and $VS$ give the surface integral over $MTVS$. In a similar way, Stokes' theorem can be established for any surface which consists of a finite number of pieces of the simple kind considered in the proof. We note that the surface $\sigma$ generally needs to be divided up in different ways for each of the terms in (B.6).

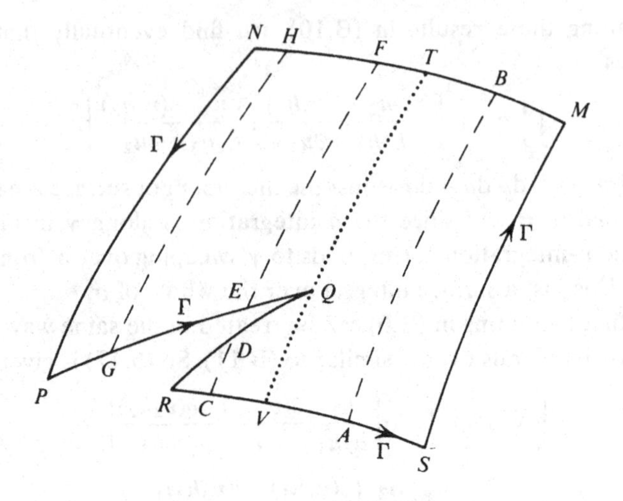


Fig. B.3

As exemplified in the main text, it is not difficult to extend the proof to a surface $\sigma$ which contains an exceptional point of the coordinate system.

# *REFERENCE LIST*


**Preface**

[P.1] The Euclidean axiomatic system has been corrected. An account of this form of axiomatic geometry is given in G. de M. Robinson, *Foundations of Geometry* (University of Toronto).

[P.2] R. Descartes, *The Geometry* (Dover).

[P.3] W. R. Hamilton, *Mathematical Papers*, eds. H. Halberstam and R. E. Ingram, vol. 3, p. 227 (C.U.P.).

[P.4] J. W. Gibbs, *Vector Analysis* (Dover).

[P.5] H. G. Grassmann, *Die Lineale Ausdehnungslehre, ein neuer Zweig der Mathematik* (Treatise, 1844).

[P.6] For example, *School Mathematics Project* (C.U.P.):
*Book 5*, Ch. 9;
*Additional Mathematics Book 1,* Ch. 1;
*Advanced Mathematics Book 1,* Ch. 1.

**Chapter 4**

[4.1] Many textbooks deal with the theory of matrices and determinants, for example:
W. Nef, *Linear Algebra* (McGraw-Hill);
P. C. Shields, *Linear Algebra* (Addison-Wesley);
R. R. Stoll and E. T. Wong, *Linear Algebra* (Academic Press);
J. S. R. Chisholm and R. Morris, *Mathematical Methods in Physics* (North-Holland), Ch. 12 and 13.

[4.2] A straightforward account of exterior forms is given by H. Flanders, *Differential Forms* (Academic Press), Ch. 2.

[4.3] J. S. R. Chisholm and R. Morris, *op. cit.*, Ch. 12, §4.5.

[4.4] Continuity is a basic concept of analysis. Here and in Chapters 5 and 6 we shall be assuming a knowledge of elementary analysis of functions of one variable and reference will be made to one of the most readable books on the subject:
J. C. Burkill, *A First Course in Mathematical Analysis* (C.U.P.).
Continuity is treated in Ch. 3 of Burkill's book.

[4.5] This approach to matrix theory is taken, for example, in E. D. Nering, *Linear Algebra and Matrix Theory* (Wiley), Ch. 2, §6 *et seq.*

## Chapter 5

[5.1] J. C. Burkill, *A First Course in Mathematical Analysis* (C.U.P.), §§3.4–3.5.
[5.2] J. C. Burkill, *ibid.*, §1.7.
[5.3] J. C. Burkill, *ibid.*, §4.5.
[5.4] J. C. Burkill, *ibid.*, §8.2.
[5.5] J. C. Burkill, *ibid.*, Ch. 5 *et seq.*
[5.6] J. C. Burkill, *ibid.*, §4.2.
[5.7] J. C. Burkill, *ibid.*, §§6.7–6.8.
[5.8] J. C. Burkill, *ibid.*, §7.4, Th. 7.41.
[5.9] J. C. Burkill, *ibid.*, §7.6.
[5.10] J. C. Burkill, *ibid.*, §7.7.
[5.11] J. C. Burkill, *ibid.*, §7.10.
[5.12] Improper double and triple integrals arise naturally in the solution of Poisson's equation (see §6.5) and similar problems. Careful definitions of such integrals and their derivatives are given in
R. Courant and D. Hilbert, *Methods of Mathematical Physics*, Vol. 2, pp. 245–61.
An alternative approach to their definition is given in
P. R. Garabedian, *Partial Differential Equations* Ch. 5. *et seq.*

## Chapter 6

[6.1] J. C. Burkill, *op. cit.*, §7.6, Th. 7.63.

## Appendix A

[A.1] See reference [6.1].
[A.2] J. C. Burkill, *op. cit.*, §2.6, Th. 2.6.

# *OUTLINE SOLUTIONS TO SELECTED PROBLEMS*

## Problems 1.1

1(*b*) $(3\mathbf{a}+3\mathbf{b})+(-1)[\mathbf{a}+3(\mathbf{b}+\mathbf{c})]$ (2E), (1.11)

$=(3\mathbf{b}+3\mathbf{a})+[(-1)\mathbf{a}+(-3)(\mathbf{b}+\mathbf{c})]$ (2B), (2E), (1C)

$=[(3\mathbf{b}+3\mathbf{a})+(-1)\mathbf{a}]+(-3)(\mathbf{b}+\mathbf{c})$ (2C)

$=(3\mathbf{b}+2\mathbf{a})+[(-3)\mathbf{b}+(-3)\mathbf{c}]$ (2B), (2D)

$=[(2\mathbf{a}+3\mathbf{b})+(-3)\mathbf{b}]+(-3)\mathbf{c}$ (2B), (2C)

$=(2\mathbf{a}+0\mathbf{b})-3\mathbf{c}$ (2C), (2D), (1.11)

$=2\mathbf{a}-3\mathbf{c}.$ (1D), (1.9)

3 $\lambda[\mathbf{a}+(-1)\mathbf{b}]$ (1.11)

$=\lambda\mathbf{a}+\lambda[(-1)\mathbf{b}]$ (2E)

$=\lambda\mathbf{a}+(-1)[\lambda\mathbf{b}]$ (1C), (1C)

$=\lambda\mathbf{a}-\lambda\mathbf{b}.$ (1.11)

## Problems 1.2

5 Position vectors of $A$, $B$, $C$, $D$ are $\mathbf{a}$, $\mathbf{b}$, $\mathbf{c}$, $\mathbf{d}$. $A'$ has position vector $\frac{1}{3}(\mathbf{b}+\mathbf{c}+\mathbf{d})$. By (1.25), point $G$ on $AA'$ with $AG:GA'=3:1$ has position vector

$$\tfrac{1}{4}\mathbf{a}+\tfrac{3}{4}[\tfrac{1}{3}(\mathbf{b}+\mathbf{c}+\mathbf{d})]$$
$$=\tfrac{1}{4}(\mathbf{a}+\mathbf{b}+\mathbf{c}+\mathbf{d}).$$

So $G$ is the centroid.

8 Position vectors of $A$, $B$, $C$ are $\mathbf{a}$, $\mathbf{b}$, $\mathbf{c}$. Those of $D$, $E$, $F$ are

$$\mathbf{d}=\frac{\nu\mathbf{b}-\mu\mathbf{c}}{\nu-\mu},\qquad \mathbf{e}=\frac{\lambda\mathbf{c}-\nu\mathbf{a}}{\lambda-\nu},\qquad \mathbf{f}=\frac{\mu\mathbf{a}-\lambda\mathbf{b}}{\mu-\lambda}.$$

$$\mathbf{DE}=\mathbf{e}-\mathbf{d}=\frac{\nu[(\mu-\nu)\mathbf{a}+(\nu-\lambda)\mathbf{b}+(\lambda-\mu)\mathbf{c}]}{(\lambda-\nu)(\nu-\mu)}.$$

Similarly, $\mathbf{DF}$ is a multiple of $(\mu-\nu)\mathbf{a}+(\nu-\lambda)\mathbf{b}+(\lambda-\mu)\mathbf{c}$, and so of $\mathbf{DE}$. So $D$, $E$, $F$ are collinear.

## Problems 2.1

2 For $\lambda>0$, $|\lambda\mathbf{a}|^2=(\lambda\mathbf{a})\cdot(\lambda\mathbf{a})=\lambda^2(\mathbf{a}\cdot\mathbf{a})=\lambda^2a^2$, by (2.11) and (2.3). Thus $|\lambda\mathbf{a}|=\lambda a>0$, and unit vector associated with $\lambda\mathbf{a}$ is $(\lambda a)^{-1}(\lambda\mathbf{a})=a^{-1}\mathbf{a}=\mathbf{u}$ in (2.4a). When $\lambda<0$, $|\lambda a|=-\lambda a>0$, and unit vector associated with $\lambda\mathbf{a}$ is $(-\lambda a)^{-1}\lambda\mathbf{a}=-a^{-1}\mathbf{a}=-\mathbf{u}$.

4 $(\mathbf{a}+\mathbf{b})\cdot\mathbf{a}-(\mathbf{a}+\mathbf{b})\cdot\mathbf{b}$ (4D)

$=\mathbf{a}\cdot\mathbf{a}+\mathbf{b}\cdot\mathbf{a}-(\mathbf{a}\cdot\mathbf{b}+\mathbf{b}\cdot\mathbf{b})$ (4B), (4D)

$=a^2-b^2.$ (2.3)

If $PQRS$ is a parallelogram and $\mathbf{PQ}=\mathbf{a}$, $\mathbf{PS}=\mathbf{b}$, and $\alpha$ is the acute angle between the diagonals,

$$PR\cdot QS\cos\alpha=|PQ^2-PS^2|.$$

## Problems 2.2

1 Position vectors of $D$, $E$, $F$ are

$$\mathbf{d}=\frac{\mu\mathbf{b}+\nu\mathbf{c}}{\mu+\nu},\qquad \mathbf{e}=\frac{\nu\mathbf{c}+\lambda\mathbf{a}}{\nu+\lambda},\qquad \mathbf{f}=\frac{\lambda\mathbf{a}+\mu\mathbf{b}}{\lambda+\mu}.$$

Points $P$, $Q$ with $AP:PD=\alpha:1-\alpha$, $BQ:QE=\beta:1-\beta$ have position vectors

$$(1-\alpha)\mathbf{a}+\alpha\frac{\mu\mathbf{b}+\nu\mathbf{c}}{\mu+\nu},\qquad (1-\beta)\mathbf{b}+\beta\frac{\nu\mathbf{c}+\lambda\mathbf{a}}{\nu+\lambda}.$$

These are same point $(\lambda\mathbf{a}+\mu\mathbf{b}+\nu\mathbf{c})/(\lambda+\mu+\nu)$ if

$$\alpha=\frac{\mu+\nu}{\lambda+\mu+\nu},\qquad \beta=\frac{\nu+\lambda}{\lambda+\mu+\nu}.$$

By symmetry, the point lies on $CF$ also.

4 Take scalar products with **a**, **b**, **c**.

5 See §2.3.

**Problems 2.3**

2
$$a=(1^2+2^2+4^2)^{\frac{1}{2}}=\sqrt{21},$$
$$b=\sqrt{5},\ c=\sqrt{11}.$$
$$\mathbf{a}\cdot\mathbf{b}=0,\text{ so angle}=\tfrac{1}{2}\pi.$$

For angle $\theta$ between **a**, **c**,

$$ac\cos\theta=\mathbf{a}\cdot\mathbf{c}=1\cdot(-1)-2.1+4\cdot(-3)=-15,$$
$$\cos\theta=-15/\sqrt{231}.$$

4
$$a=(2^2+2^2+1^2)^{\frac{1}{2}}=3.$$
$$\mathbf{i}=a^{-1}\mathbf{a}=\tfrac{1}{3}(2\mathbf{i}'+2\mathbf{j}'+\mathbf{k}').$$
$$\mathbf{b}\cdot\mathbf{i}=\tfrac{1}{3}(8+2-1)=3.$$
$$\mathbf{b}-(\mathbf{b}\cdot\mathbf{i})\mathbf{i}=4\mathbf{i}'+\mathbf{j}'-\mathbf{k}'-(2\mathbf{i}'+2\mathbf{j}'+\mathbf{k}')$$
$$=2\mathbf{i}'-\mathbf{j}'-2\mathbf{k}'.$$

Corresponding unit vector is $\frac{1}{3}(2\mathbf{i}'-\mathbf{j}'-2\mathbf{k}')$.
If $\mathbf{k}=k_1\mathbf{i}'+k_2\mathbf{j}'+k_3\mathbf{k}'$,

$$2k_1+2k_2+k_3=0,\qquad 4k_1+k_2-k_3=0.$$

Unit **k** is therefore $\pm\frac{1}{3}(\mathbf{i}'-2\mathbf{j}'+2\mathbf{k}')$.

5 Vertices with position vectors **0**, **a**, **b**, **a**+**b**, with $a=b$. Diagonal vectors are $\mathbf{a}+\mathbf{b}$, $\mathbf{a}-\mathbf{b}$, and $(\mathbf{a}+\mathbf{b})\cdot(\mathbf{a}-\mathbf{b})=a^2-b^2=0$.

**Problems 2.4**

3 $\mathbf{b}-\mathbf{a}$, $\mathbf{c}-\mathbf{a}$ linearly dependent. Any point has position vector of form $\mathbf{r}=\mathbf{a}+\lambda(\mathbf{b}-\mathbf{a})+\mu(\mathbf{c}-\mathbf{a})=\kappa\mathbf{a}+\lambda\mathbf{b}+\mu\mathbf{c}$ with $\kappa+\lambda+\mu=1$.
If $\alpha:\kappa=\beta:\lambda=\gamma:\mu$, $\mathbf{r}=(\alpha\mathbf{a}+\beta\mathbf{b}+\gamma\mathbf{c})/(\alpha+\beta+\gamma)$.
Any point $D$ on $AR$ has position vector

$$\mathbf{d}=(1-\tau)\mathbf{a}+\tau\left(\frac{\alpha\mathbf{a}+\beta\mathbf{b}+\gamma\mathbf{c}}{\alpha+\beta+\gamma}\right).$$

$D$ on $BC$ if coefficient of **a** zero, with $\mathbf{d}=(\beta\mathbf{b}+\gamma\mathbf{c})/(\beta+\gamma)$. So $BD:DC=\gamma:\beta=\beta^{-1}:\gamma^{-1}$. If $AD$, $BC$ perpendicular,

$BD:DC=\cot B:\cot C$; so we can take $\beta=\tan B$, $\gamma=\tan C$. $R$ lies on all three perpendiculars if $\alpha=\tan A$ also. Then

$$\mathbf{r}=\frac{\mathbf{a}\tan A+\mathbf{b}\tan B+\mathbf{c}\tan C}{\tan A+\tan B+\tan C}.$$

Result follows by a standard trigonometric identity.

5 Position vectors $\mathbf{a}_1, \mathbf{a}_2, \ldots, \mathbf{a}_n$.

$\mathbf{A}_r\mathbf{P}=\mathbf{p}-\mathbf{a}_r=(\mathbf{p}-\mathbf{g})-(\mathbf{a}_r-\mathbf{g})$.

$\sum_1^n|\mathbf{A}_r\mathbf{P}|^2=\sum_1^n[|\mathbf{p}-\mathbf{g}|^2+|\mathbf{a}_r-\mathbf{g}|^2]-2(\mathbf{p}-\mathbf{g})\cdot\sum_1^n(\mathbf{a}_r-\mathbf{g})$.

But $\Sigma_1^n(\mathbf{a}_r-\mathbf{g})=0$ for centroid, giving result.

7 Take origin at centre of common perpendicular, and $\mathbf{i}$ parallel to it. Then mid-points of opposite edges have position vectors of form $\pm\lambda\mathbf{i}$. Choose $\mathbf{j}$ so that opposite edges make same angle with $\mathbf{j}$; since opposite edges have equal length, their end-points (the vertices) have position vectors of form

$$\lambda\mathbf{i}+\mu\mathbf{j}+\nu\mathbf{k}, \qquad \lambda\mathbf{i}-\mu\mathbf{j}-\nu\mathbf{k}$$

and

$$-\lambda\mathbf{i}+\mu\mathbf{j}-\nu\mathbf{k}, \qquad -\lambda\mathbf{i}-\mu\mathbf{j}+\nu\mathbf{k}.$$

Symmetry between $\lambda$, $\mu$, $\nu$ ensures that other two pairs of opposite edges are similarly related.

8 Origin at centre of cube, edges parallel to $\mathbf{i}$, $\mathbf{j}$, $\mathbf{k}$. Diagonals then parallel to unit vectors $(\mathbf{i}\pm\mathbf{j}\pm\mathbf{k})/\sqrt{3}$. Let $\mathbf{u}=u_1\mathbf{i}+u_2\mathbf{j}+u_3\mathbf{k}$ be a vector in the direction of the line. Then $\cos\alpha, \ldots, \cos\delta=(u_1\pm u_2\pm u_3)/\sqrt{3}$. So $\cos^2\alpha+\ldots+\cos^2\delta=4(u_1^2+u_2^2+u_3^2)/3=\frac{4}{3}$.

**Problems 2.5**

2 $\mathbf{d}=\boldsymbol{\rho}_1-\boldsymbol{\rho}_2=\rho_1(\cos\phi_1\mathbf{i}+\sin\phi_1\mathbf{j})-\rho_2(\cos\phi_2\mathbf{i}+\sin\phi_2\mathbf{j})$,

$d^2=(\rho_1\cos\phi_1-\rho_2\cos\phi_2)^2+(\rho_1\sin\phi_1-\rho_2\sin\phi_2)^2$.

6 $\mathbf{u}_p=\sin\theta_p\cos\phi_p\,\mathbf{i}+\sin\theta_p\sin\phi_p\mathbf{j}+\cos\theta_p\mathbf{k} \qquad (p=1, 2)$.

$$\begin{aligned}\cos\alpha&=\mathbf{u}_1\cdot\mathbf{u}_2\\&=\sin\theta_1\sin\theta_2(\cos\phi_1\cos\phi_2+\sin\phi_1\sin\phi_2)+\cos\theta_1\cos\theta_2.\end{aligned}$$

7 $\mathbf{r}_p = r_p(\sin\theta_p \cos\phi_p \mathbf{i} + \sin\theta_p \sin\phi_p \mathbf{j} + \cos\theta_p \mathbf{k})$
$d^2 = |\mathbf{r}_1 - \mathbf{r}_2|^2$ gives result.

## Problems 3.1

1 $\mathbf{a}\wedge\mathbf{b} = \mathbf{j} - 2\mathbf{k}, \quad \mathbf{a}\wedge\mathbf{c} = 6\mathbf{i} + \mathbf{j} - 5\mathbf{k}.$

$(\mathbf{a}\wedge\mathbf{b})\cdot(\mathbf{a}\wedge\mathbf{c}) = 0 + 1 + 10 = 11$, not orthogonal.

If $\mathbf{d} = \mathbf{b}\wedge\mathbf{c}$, then $\mathbf{d}\wedge\mathbf{b}$ and $\mathbf{d}\wedge\mathbf{c}$ lie in the plane of $\mathbf{b}$, $\mathbf{c}$, orthogonal to $\mathbf{b}$, $\mathbf{c}$ respectively. Since $\mathbf{b}$, $\mathbf{c}$ are orthogonal, so are $\mathbf{d}\wedge\mathbf{b}$, $\mathbf{d}\wedge\mathbf{c}$. So choose $\mathbf{d} = \mathbf{b}\wedge\mathbf{c} = 3(\mathbf{i} - \mathbf{k})$.

2 $\mathbf{a}$, $\mathbf{b}$, $\mathbf{c}$ can be represented by the sides of a triangle. Each vector product is normal to the plane of the triangle, magnitude twice the area, same sense.

## Problems 3.2

1 Final part: use $(\mathbf{a}\cdot\mathbf{c})(\mathbf{b}\wedge\mathbf{c}) - (\mathbf{b}\cdot\mathbf{c})(\mathbf{a}\wedge\mathbf{c})$.

2 (ii) $(\mathbf{a}\wedge\mathbf{b})\cdot\{[\mathbf{b}\cdot(\mathbf{c}\wedge\mathbf{a})]\mathbf{c} - [\mathbf{c}\cdot(\mathbf{c}\wedge\mathbf{a})]\mathbf{b}\}$.

3 Three concurrent edges $\mathbf{b}-\mathbf{a}$, $\mathbf{c}-\mathbf{a}$, $\mathbf{d}-\mathbf{a}$.
Volume is

$$\tfrac{1}{6}|[\mathbf{b}-\mathbf{a}, \mathbf{c}-\mathbf{a}, \mathbf{d}-\mathbf{a}]|.$$

Expand and use $[\mathbf{a}, \mathbf{a}, \mathbf{d}] = 0$, $[\mathbf{b}, \mathbf{a}, \mathbf{d}] = -[\mathbf{d}, \mathbf{a}, \mathbf{b}]$ and so on.

4 Origin $O$ at end of line of length $a$; choose $\mathbf{i}$ along line, $\mathbf{k}$ along common perpendicular. Then other vertices have position vectors of form $\mathbf{p} = a\mathbf{i}$, $\mathbf{q} = x\mathbf{i} + h\mathbf{k} + y(\mathbf{i}\cos\theta + \mathbf{j}\sin\theta)$, $\mathbf{r} = x\mathbf{i} + h\mathbf{k} + (y-b)(\mathbf{i}\cos\theta + \mathbf{j}\sin\theta)$. Calculate $\tfrac{1}{6}|[\mathbf{p}, \mathbf{q}, \mathbf{r}]|$.

5 $P, Q, R, S$ have position vectors $\mathbf{0}$, $\mathbf{q}$, $\mathbf{r}$, $\mathbf{s}$, satisfying $[\mathbf{q}, \mathbf{r}, \mathbf{s}] = 0$. $A, B, C, D$ have position vectors of form $\alpha\mathbf{k}$, $\mathbf{b} = \mathbf{q} + \beta\mathbf{k}$, $\mathbf{c} = \mathbf{r} + \gamma\mathbf{k}$, $\mathbf{d} = \mathbf{s} + \delta\mathbf{k}$, and $[\alpha\mathbf{k}-\mathbf{b}, \alpha\mathbf{k}-\mathbf{c}, \alpha\mathbf{k}-\mathbf{d}] = 0$. Expanding, $[\mathbf{b}, \mathbf{c}, \mathbf{d}] = [\alpha\mathbf{k}, \mathbf{c}, \mathbf{d}] +$ cyclic terms. But $[\alpha\mathbf{k}-\mathbf{q}, \alpha\mathbf{k}-\mathbf{r}, \alpha\mathbf{k}-\mathbf{s}] = [\alpha\mathbf{k}, \mathbf{r}, \mathbf{s}] + \text{cyclic} = [\alpha\mathbf{k}, \mathbf{c}, \mathbf{d}] + \text{cyclic}$. So volumes $|[\mathbf{b}, \mathbf{c}, \mathbf{d}]|$ and $|[\alpha\mathbf{k}-\mathbf{q}, \alpha\mathbf{k}-\mathbf{r}, \alpha\mathbf{k}-\mathbf{s}]|$ are equal.

6 Let $PA=\mathbf{a}$, $OA=\alpha\mathbf{a}$, $OP=(\alpha-1)\mathbf{a}$; likewise for $OB$, $RC$, $SD$. Take $O$ as origin. For $(\alpha-1)\mathbf{a}$ to lie in plane of $\beta\mathbf{b}$, $\gamma\mathbf{c}$, $\delta\mathbf{d}$, $(\alpha-1)\beta\gamma[\mathbf{a},\ \mathbf{b},\ \mathbf{c}]-\beta\gamma\delta[\mathbf{b},\ \mathbf{c},\ \mathbf{d}]+(\alpha-1)\gamma\delta[\mathbf{c},\ \mathbf{d},\ \mathbf{a}]-(\alpha-1)\delta\beta$ $[\mathbf{d},\ \mathbf{a},\ \mathbf{b}]=0$. For non-zero triple products, this and three similar equations must be consistent, with zero $(4\times 4)$ determinant; this happens when $\alpha+\beta+\gamma+\delta=3$. Each row of the determinant should be divided by $\alpha\beta\gamma\delta$ and expressed in terms of $\mathrm{A}=\alpha^{-1}$, $\mathrm{B}=\beta^{-1}$, $\mathrm{C}=\gamma^{-1}$, $\mathrm{D}=\delta^{-1}$.

### Problems 3.3

1 Unit normal $\mathbf{u}=\frac{1}{3}(2\mathbf{i}-\mathbf{j}-2\mathbf{k})$. Equation of plane is

$$\mathbf{r}\cdot\mathbf{u}=\tfrac{1}{3}(3\mathbf{i}+2\mathbf{j}-\mathbf{k})\cdot(2\mathbf{i}-\mathbf{j}-2\mathbf{k})=2.$$

By (3.43), $p=2$.
Foot of perpendicular is at $2\mathbf{u}$.

3 Plane is parallel to both $(4\mathbf{i}-\mathbf{j}-\mathbf{k})$ and $(2\mathbf{i}+\mathbf{j}-3\mathbf{k})$ and so is normal to their vector product $\mathbf{v}=2(2\mathbf{i}+5\mathbf{j}+3\mathbf{k})$. Equation is

$$\mathbf{r}\cdot(2\mathbf{i}+5\mathbf{j}+3\mathbf{k})=(\mathbf{i}-\mathbf{j}+2\mathbf{k})\cdot(2\mathbf{i}+5\mathbf{j}+3\mathbf{k})$$

or $2x+5y+3z=3$.

5 Component of $\mathbf{r}_2-\mathbf{r}_1$ perpendicular to $\mathbf{u}$ is $\mathbf{a}=\mathbf{r}_2-\mathbf{r}_1-\{(\mathbf{r}_2-\mathbf{r}_1)\cdot\mathbf{u}\}\mathbf{u}$. This is a vector along the required line; since it passes through $\mathbf{r}_2$, equation is $\mathbf{r}=\mathbf{r}_2+\mu\mathbf{a}$.

7 Line is perpendicular to both $\mathbf{v}$ and $\mathbf{u}$, and so is in direction $\mathbf{u}\wedge\mathbf{v}$. Passing through $\mathbf{r}_2$, it is $\mathbf{r}=\mathbf{r}_2+t\mathbf{u}\wedge\mathbf{v}$.

8 Parameters $t_1$, $t_2$ must be chosen so that $\mathbf{d}=(\mathbf{r}_1+t_1\mathbf{v}_1)-(\mathbf{r}_2+t_2\mathbf{v}_2)$ is orthogonal to $\mathbf{v}_1$ and $\mathbf{v}_2$, or parallel to $\mathbf{v}_1\wedge\mathbf{v}_2$. Scalar product with $\mathbf{v}_1\wedge\mathbf{v}_2$ gives $\pm d|\mathbf{v}_1\wedge\mathbf{v}_2|=[\mathbf{r}_1-\mathbf{r}_2,\ \mathbf{v}_1,\ \mathbf{v}_2]$.

12 Plane is parallel to both $\mathbf{u}$, $\mathbf{w}$, so normal to $\mathbf{u}\wedge\mathbf{w}$. Equation is of form $\mathbf{r}\cdot(\mathbf{u}\wedge\mathbf{w})=\text{const}$. But $\mathbf{r}\cdot(\mathbf{u}\wedge\mathbf{w})=(\mathbf{r}\wedge\mathbf{u})\cdot\mathbf{w}=\mathbf{m}\cdot\mathbf{w}$. Hence result $\mathbf{r}\cdot(\mathbf{u}\wedge\mathbf{w})=\mathbf{m}\cdot\mathbf{w}$.

14 Let $A$, $B$, $C$ have position vectors $x\mathbf{i}$, $y\mathbf{j}$, $z\mathbf{k}$. Since $(x\mathbf{i}-\mathbf{a})\cdot\mathbf{a}=0$, $x=a^2/\mathbf{a}\cdot\mathbf{i}$; likewise $y=a^2/\mathbf{a}\cdot\mathbf{j}$, $z=a^2/\mathbf{a}\cdot\mathbf{k}$. By Example 3.3 result, $\text{area}=\frac{1}{2}|yz\mathbf{i}+zx\mathbf{j}+xy\mathbf{k}|=\frac{1}{2}xyz(x^{-2}+y^{-2}+z^{-2})^{\frac{1}{2}}=\frac{1}{2}xyz/a$.

**Problems 4.1**

1 $\Delta(A)\Delta(B)=(\tilde{\mathbf{a}}_1 \wedge \tilde{\mathbf{a}}_2)\cdot(\mathbf{b}_1 \wedge \mathbf{b}_2)$.

$$\Delta(AB)=(\tilde{\mathbf{a}}_1\cdot\mathbf{b}_1)(\tilde{\mathbf{a}}_2\cdot\mathbf{b}_2)-(\tilde{\mathbf{a}}_1\cdot\mathbf{b}_2)(\tilde{\mathbf{a}}_2\cdot\mathbf{b}_1).$$

These are equal by (3.33).

4
$$\begin{pmatrix} -4 & 7 & -1 \\ -2 & 4 & -1 \\ 3 & -5 & 1 \end{pmatrix}.$$

**Problems 4.2**

1
$$\tfrac{1}{2}\begin{pmatrix} \sqrt{3} & -1 \\ 1 & \sqrt{3} \end{pmatrix}, \tfrac{1}{2}\begin{pmatrix} 1 & -\sqrt{3} \\ \sqrt{3} & 1 \end{pmatrix}. \text{ Product} \begin{pmatrix} 0 & -1 \\ 1 & 0 \end{pmatrix}.$$

4 Let $\{\alpha\}=\alpha\pm 2n\pi$, $0\leqslant\alpha\pm 2n\pi<2\pi$.
(*a*) $R(\alpha)R(\beta)=R(\{\alpha+\beta\})$, $R(\alpha)L(\beta)=L(\{\alpha+\beta\})$,
$L(\alpha)R(\beta)=L(\{\alpha-\beta\})$, $L(\alpha)L(\beta)=R(\{\alpha-\beta\})$.
(*b*) $R(\alpha)$, $L(\alpha)$ have inverses $R(2\pi-\alpha)$, $L(\alpha)$.
(*c*) Multiplication is associative.

6 Let $[p/n]$ be the fractional part of $p/n$; then $\cos 2\pi[p/n]=\cos 2\pi p/n$, $\sin 2\pi[p/n]=\sin 2\pi p/n$.
(*a*) Products: $R(\alpha_p)R(\alpha_q)=R(\{\alpha_p+\alpha_q\})$,
$R(\alpha)L(\beta)=L(\{\alpha+\beta\})$,
$L(\alpha)R(\beta)=L(\{\alpha-\beta\})$, $L(\alpha)L(\beta)=R(\{\alpha-\beta\})$.
(*b*) $R(\alpha_p)$, $L(\alpha_p)$ have inverses $R(2\pi-\alpha_p)$, $L(\alpha_p)$.
(*c*) Multiplication is associative.

**Problems 4.3**

4 Rotations about axis **k** represented by

$$\begin{pmatrix} 0 & -1 & 0 \\ 1 & 0 & 0 \\ 0 & 0 & 1 \end{pmatrix}, \begin{pmatrix} -1 & 0 & 0 \\ 0 & -1 & 0 \\ 0 & 0 & 1 \end{pmatrix}, \begin{pmatrix} 0 & 1 & 0 \\ -1 & 0 & 0 \\ 0 & 0 & 1 \end{pmatrix}.$$

Similarly about **i**, **j**. Unit matrix plus eight other matrices of type

$$\begin{pmatrix} 0 & 1 & 0 \\ 0 & 0 & 1 \\ 1 & 0 & 0 \end{pmatrix}, \begin{pmatrix} 0 & 0 & -1 \\ 1 & 0 & 0 \\ 0 & -1 & 0 \end{pmatrix};$$

these represent notations through $2\pi/3$, $4\pi/3$ about the four axes $\mathbf{i}\pm\mathbf{j}\pm\mathbf{k}$, permuting the coordinate axis.

6 Use the hint. $(\mathbf{i}, \mathbf{j}, \mathbf{k})$ and $(\mathbf{i}', \mathbf{j}', \mathbf{k}')$ are right-handed triads. If $I-R$ is of rank 1, then $\mathbf{i}'=\mathbf{i}+\alpha\mathbf{u}$, $\mathbf{j}'=\mathbf{j}+\beta\mathbf{u}$, $\mathbf{k}'=\mathbf{k}+\gamma\mathbf{u}$ for some real $\alpha$, $\beta$, $\gamma$ and some unit vector $\mathbf{u}$. From $\mathbf{i}'\cdot\mathbf{i}'=\mathbf{j}'\cdot\mathbf{j}'=\mathbf{k}'\cdot\mathbf{k}'=1$, we find $\alpha(\mathbf{i}\cdot\mathbf{u})=-\frac{1}{2}\alpha^2$, $\beta(\mathbf{j}\cdot\mathbf{u})=-\frac{1}{2}\beta^2$, $\gamma(\mathbf{k}\cdot\mathbf{u})=-\frac{1}{2}\gamma^2$. Then from $[\mathbf{i}', \mathbf{j}', \mathbf{k}']=[\mathbf{i}, \mathbf{j}, \mathbf{k}]=1$ we find $[\mathbf{i}', \mathbf{j}', \mathbf{k}']=1-\frac{1}{2}(\alpha^2+\beta^2+\gamma^2)$ or $\alpha^2+\beta^2+\gamma^2=0$. So $\alpha=\beta=\gamma=0$, and $R=I$, forbidden.

## Problems 4.4

2(*b*) $\Sigma_q \Sigma_r \varepsilon_{pqr} a_q b_r = (\mathbf{a}\wedge\mathbf{b})_p$.
Apply $\Sigma_q \Sigma_r \Sigma_s \Sigma_t a_q b_r c_s d_t$ to the identity:

$$\sum_p (\mathbf{a}\wedge\mathbf{b})_p \cdot (\mathbf{c}\wedge\mathbf{d})_p = \sum_q a_q c_q \sum_r b_r d_r - \sum_q a_q d_q \sum_r b_r c_r.$$

3 Apply $\Sigma_r \Sigma_s \Sigma_t a_r b_s c_t$ to the identity; the left-hand-side becomes

$$\sum_r \sum_p \varepsilon_{pqr} a_r (\mathbf{b}\wedge\mathbf{c})_p = [\mathbf{a}\wedge(\mathbf{b}\wedge\mathbf{c})]_p.$$

## Problems 5.1

2 Define $v=u$ $(u_1\leqslant u\leqslant u_2)$ and $v=u-u_3+u_2$ $(u_3<u\leqslant u_4)$. Then $\mathbf{r}(v)=\mathbf{r}_1(u)$ for $u_1\leqslant u\leqslant u_2$ and $\mathbf{r}(v)=\mathbf{r}_2(u)$ for $u_3<u\leqslant u_4$ defines the curve as a function of $v$ over the continuous range $u_1\leqslant v\leqslant u_4-u_3+u_2$.

3 The boundary consists of the curves (i) $\phi=\phi_0$, $z=\lambda\phi_0$ $(0\leqslant\rho\leqslant\rho_0)$, (ii) $\phi=\phi_1$, $z=\lambda\phi_1$ $(0\leqslant\rho\leqslant\rho_0)$, (iii) $z=\lambda\phi$, $\rho=\rho_0$ $(\phi_0\leqslant\phi\leqslant\phi_1)$, (iv) the part of the $z$-axis $\rho=0$ $(\lambda\phi_0\leqslant z\leqslant\lambda\phi_1)$.

4 When $\xi=\xi_0$ (constant), $\eta$ is curve parameter. For all $\eta$, using $\cos^2\eta+\sin^2\eta=1$,

$$\frac{x^2}{c^2\cosh^2\xi_0}+\frac{y^2}{c^2\sinh^2\xi_0}=1,$$

an ellipse with semi-axes of length $c\cosh\xi_0$, $c\sinh\xi_0$. When

$\eta=\eta_0$, elimination of $\xi$ gives the hyperbola

$$\frac{x^2}{c^2\cos^2\eta_0}-\frac{y^2}{c^2\sin^2\eta_0}=1.$$

Ranges $0\leqslant\xi<\infty$, $0\leqslant\eta<2\pi$ allow $x$, $y$ to take both signs.

5 $\rho=c\cosh\xi\cos\eta$, $z=c\sinh\xi\sin\eta$. In the above, $(x, y)$ becomes $(\rho, z)$. $\xi=\xi_0$, ellipsoid of rotation about $z$-axis; $\eta=\eta_0$, hyperboloid of rotation; $\phi=\phi_0$, plane through $z$-axis.

## Problems 5.2

1 Write $d\mathbf{v}/du=\mathbf{v}'$, $d\mathbf{w}/du=\mathbf{w}'$.

(i) $[\mathbf{a}, \mathbf{v}', \mathbf{w}]+[\mathbf{a}, \mathbf{v}, \mathbf{w}']$,

(ii) $(\mathbf{v}+\mathbf{w})\cdot(\mathbf{v}'+\mathbf{w}')/|\mathbf{v}+\mathbf{w}|$,

(iii) $\dfrac{\mathbf{w}'\wedge\mathbf{v}+(\mathbf{a}+\mathbf{w})\wedge\mathbf{v}'}{|\mathbf{v}+\mathbf{w}|}-\dfrac{[(\mathbf{a}+\mathbf{w})\wedge\mathbf{v}][(\mathbf{v}+\mathbf{w})\cdot(\mathbf{v}'+\mathbf{w}')]}{|\mathbf{v}+\mathbf{w}|^3}$.

3 Since $\mathbf{v}\cdot\mathbf{v}$ is constant, $\mathbf{v}\cdot\mathbf{v}'=0$, so $\mathbf{v}$, $\mathbf{v}'$ orthogonal. Let $\mathbf{v}\wedge\mathbf{v}'=vv'\mathbf{n}$; then $\mathbf{n}\wedge\mathbf{v}=v\mathbf{v}'/v'$. So if $\boldsymbol{\omega}=v'\mathbf{n}/v$, $\boldsymbol{\omega}\wedge\mathbf{v}=\mathbf{v}'$.
If $\boldsymbol{\omega}'=\mathbf{0}$, $d(\mathbf{v}\cdot\boldsymbol{\omega})/du=\mathbf{v}'\cdot\boldsymbol{\omega}=(\boldsymbol{\omega}\wedge\mathbf{v})\cdot\boldsymbol{\omega}=0$.
Since $v$ and $\omega$ are constant, $\mathbf{v}$, $\boldsymbol{\omega}$ are at fixed angle $\alpha$. Define $\mathbf{i}$, $\mathbf{j}$, $\mathbf{k}$ with $\boldsymbol{\omega}=\omega\mathbf{k}$; then $\mathbf{v}$ has components $v(\sin\alpha\cos\phi, \sin\alpha\sin\phi, \cos\alpha)$. Substituting in $\mathbf{v}'=\boldsymbol{\omega}\wedge\mathbf{v}$ gives $\dot\phi=\omega$.

## Problems 5.3

1 $ds/d\theta=2a\sin\frac{1}{2}\theta$; $\mathbf{t}=\sin\frac{1}{2}\theta\,\mathbf{i}+\cos\frac{1}{2}\theta\,\mathbf{j}$; $\mathbf{n}=\cos\frac{1}{2}\theta\,\mathbf{i}-\sin\frac{1}{2}\theta\,\mathbf{j}$.

2 $\mathbf{r}=x\mathbf{i}+c\cosh(x/c)\mathbf{j}$; $ds/dx=\cosh(x/c)$, $s=c\sinh(x/c)$. So $c\tan\psi=c\,dy/dx=c\sinh(x/c)=s$.
$\mathbf{t}=\operatorname{sech}(x/c)\mathbf{i}+\tanh(x/c)\mathbf{j}$, $r_c=\kappa^{-1}=c\cosh^2(x/c)$,
$\mathbf{n}=\tanh(x/c)\mathbf{i}-\operatorname{sech}(x/c)\mathbf{j}$. Distance along $\mathbf{n}$ corresponding to $y=c\cosh(x/c)$ is then $c\cosh^2(x/c)=r_c$.

3 If $\lambda^2a^2=a^2+b^2$, $ds/d\theta=a(\lambda^2+\theta^2)^{\frac{1}{2}}$,
$\mathbf{t}=(\lambda^2+\theta^2)^{-\frac{1}{2}}[(\cos\theta-\theta\sin\theta)\mathbf{i}+(\sin\theta+\theta\cos\theta)\mathbf{j}+b\mathbf{k}]$.
$\kappa=a^{-1}(\lambda^2+\theta^2)^{-\frac{3}{2}}[(\lambda^2+\theta^2)(4+\theta^2)-\theta^2]$; $\mathbf{n}$ given by $\kappa^{-1}\,d\mathbf{t}/ds$, then $\mathbf{b}=\mathbf{t}\wedge\mathbf{n}$. Torsion given by $\tau=\mathbf{n}'\cdot\mathbf{b}$.

4 $ds/du = 3(2+u^2)$, $\mathbf{t} = (2+u^2)^{-1}(2\mathbf{i}+2u\mathbf{j}+u^2\mathbf{k})$.
$\kappa = \frac{2}{3}(2+u^2)^{-2}$, $\mathbf{n} = (-2u\mathbf{i}+(2-u^2)\mathbf{j}+2u\mathbf{k})(2+u^2)^{-1}$.
$\mathbf{b} = \mathbf{t} \wedge \mathbf{n} = -(2\mathbf{i}+2u\mathbf{j}-2\mathbf{k})(2+u^2)^{-1}$.
$\kappa = |\mathbf{b}'| = 4(2+u^2)^{-1}$.

5 $\kappa = 3/25a$, $\tau = 4/25a$.

6 $\dot{\mathbf{r}} = 3\mathbf{i}+6u\mathbf{j}+6u^2\mathbf{k}$, $\ddot{\mathbf{r}} = 6\mathbf{j}+12u\mathbf{k}$, $\dddot{\mathbf{r}} = 12\mathbf{k}$. (5.60) and (5.61) give $\kappa = \tau = \frac{2}{3}(2u^2+1)^{-2}$.

### Problems 5.4

1
$$\int_0^a dx \int_0^{a-x} dy\,(x+y+k)^{-1}$$
$$= \int_0^a dx\,[\ln(a+k) - \ln(x+k)]$$
$$= a + k\ln k - k\ln(a+k).$$

As $k \to 0$, $k \ln k \to 0$, so $\iint_\sigma d\sigma\,(x+y)^{-1} = a$.

2 $0 \leqslant \lambda < 1$, $0 \leqslant \xi < 2\pi$. Jacobian $= \lambda/ab$.
$$\int_0^1 d\lambda \int_0^{2\pi} d\xi\,\lambda/ab \cdot \lambda^3/a^2b\cos^2\xi\sin\xi = 4/15a^3b^2.$$

3 $a>0: u = +\sqrt{a}$, $-\infty < y < \infty$.
$a<0: v = +\sqrt{(-a)}$ and $v = -\sqrt{(-a)}$, $0 \leqslant u < \infty$.
Region of integration is rectangle bounded by
$u = +\sqrt{a_1}$, $u = +\sqrt{a_2}$, $v = -\sqrt{(-a_3)}$, $v = +\sqrt{(-a_3)}$.
Jacobian $= 4(u^2+v^2)$, $x^2+y^2 = (u^2+v^2)^2$.
$$\text{Integral} = \iint du\,dv \cdot 4 = 8\sqrt{(-a_3)}\,(\sqrt{a_1} - \sqrt{a_2}).$$

5 Angle $\alpha$ between $x$- and $y$-axes. Polar coordinates $(\rho, \phi)$ relative to $x$-axis: $x = \rho\sin(\alpha-\phi)/\sin\alpha$, $y = \rho\sin\phi/\sin\alpha$; $\partial(x,y)/\partial(\rho,\phi) = \rho/\sin\alpha$. Integral is
$$\int_0^\infty d\rho \int_0^\alpha d\phi\,(\rho/\sin\alpha)\exp(-\rho^2) = \alpha/2\sin\alpha.$$

7 At $\pm x$, arc length in $(y, z)$ plane is $2a\cos^{-1}(x/a)$. Total area $=$ $8a\int_{x=0}^{a} dx\cos^{-1}(x/a) = 8a^2$.

8 Surface parameters $(\rho, \phi)$ with $z = a\phi$: orthogonal system on surface with $h_1 = 1$, $h_2 = (\rho^2 + a^2)^{\frac{1}{2}}$. Integral of $\rho\cos\phi$ is

$$\int_0^{\frac{1}{2}\pi} \cos\phi \, d\phi \int_0^a d\rho\, \rho(\rho^2 + a^2)^{\frac{1}{2}} = \tfrac{1}{3}a^3(2\sqrt{2} - 1).$$

Area $= \pi a^2(\sqrt{2}\sinh^{-1}1)$.

9

$$\left|\frac{\partial \mathbf{r}}{\partial u} \wedge \frac{\partial \mathbf{r}}{\partial \phi}\right| = 2u^2[u^2 + \phi^2 + 4]^{\frac{1}{2}}.$$

## Problems 5.5

1

$$r^2 = c^2(\cosh^2\xi - \cos^2\eta) = c^2(\sinh^2\xi + \sin^2\eta).$$

$$\text{Volume} = 2\pi c^3 \int_0^a d(\sinh\xi) \int_0^b d(\sin\eta)(\sinh^2\xi + \sin^2\eta)$$

$$= \tfrac{2}{3}\pi c^3 \sinh a \sin b \,(\sinh^2 a + \sin^2 b).$$

Integral $= 2\pi c^3 \sinh a \sin b\,[\tfrac{1}{5}(\sinh^4 a - \sin^4 b) + \tfrac{1}{3}(\sinh^2 a + \sin^2 b)]$.

3 In positive octant, ranges are $0 \leqslant z \leqslant a$, $0 \leqslant x \leqslant z$, $0 \leqslant y \leqslant (a^2 - z^2)^{\frac{1}{2}}$. Total volume is

$$8\int_0^a dz\, z(a^2 - z^2)^{\frac{1}{2}} = \tfrac{8}{3}a^3.$$

5 Limit as $R \to \infty$ of

$$2\pi \int_0^R dr \frac{r^2}{(r^2 + a^2)^2} \int_0^\alpha d\theta \cos^2\theta$$

$$= \frac{\pi}{4a}[2\alpha + \sin 2\alpha]\tan^{-1}\frac{R}{a}.$$

As $R \to \infty$, $\tan^{-1}(R/a) \to \frac{1}{2}\pi$.

6 Jacobian $= 2|u^4 v^2 \phi|$. Volume $= 2\pi u_1^5/15$. $|\mathbf{r}| = uv(u^2 + 4\phi^2)^{\frac{1}{2}}$; integral is

$$\frac{1}{3}\left[u_1^2 \ln\left(\frac{u_1^2 + 16\pi^2}{u_1^2 + 4\pi^2}\right) + 24\pi^2 u_1 - 128\pi^3 \tan^{-1}\frac{u_1}{4\pi} + 16\pi^3 \tan^{-1}\frac{u_1}{2\pi}\right].$$

**Problems 6.1**

2 $\operatorname{div} \mathbf{v} = 2(x\,\mathrm{e}^z + y\,\mathrm{e}^z + z)$.

$$\iiint_\tau \operatorname{div} \mathbf{v}\, \mathrm{d}\tau = 2\int_0^a \mathrm{d}x \int_0^a \mathrm{d}y \int_0^a \mathrm{d}z\, (x\,\mathrm{e}^z + y\,\mathrm{e}^z + z)$$

$$= 2a^3(\mathrm{e}^a - 1) + a^4.$$

On $x = 0$, $y = 0$, $z = 0$, normal component $\mathbf{v}\cdot\mathbf{n} = 0$. On $x = a$, $\mathbf{v}\cdot\mathbf{n} = a^2\,\mathrm{e}^z$, so

$$\int_0^a \mathrm{d}y \int_0^a \mathrm{d}z\, v_n = a^3(\mathrm{e}^a - 1).$$

Other two terms likewise.

3 Use (6.23a) to give $\operatorname{div} \mathbf{v} = 2z = 2r\cos\theta$, also obtainable from (6.23b). Hemisphere integral

$$\int \mathrm{d}\tau\, \operatorname{div} \mathbf{v} = \int_0^a r^2\, \mathrm{d}r \int_0^{\frac{1}{2}\pi} \sin\theta\, \mathrm{d}\theta \int_0^{2\pi} \mathrm{d}\phi\, 2r\cos\theta$$

$$= \tfrac{1}{2}\pi a^4.$$

On hemispherical surface, $\mathbf{v}\cdot\mathbf{n} = v_r = 2r^2\sin^2\theta\cos\theta$, giving surface integral

$$2a^4 \int_0^{\frac{1}{2}\pi} \sin\theta\, \mathrm{d}\theta \int_0^{2\pi} \mathrm{d}\phi\, \sin^2\theta\cos\theta = \pi a^4.$$

On disc $\theta = \frac{1}{2}\pi$, $\mathbf{v}\cdot\mathbf{n} = -v_z = -r^2$, giving surface integral

$$-\int_0^a r\, \mathrm{d}r \int_{\phi=0}^{2\pi} \mathrm{d}\phi\, r^2 = -\tfrac{1}{2}\pi a^4.$$

Total surface integral $= \frac{1}{2}\pi a^4$, as required.

**Problems 6.2**

1 Rectangular components $\partial\psi/\partial x$, $\partial\psi/\partial y$, $\partial\psi/\partial z$, with $\psi = (2xy + z^2)(x^2 + y^2)^{-\frac{1}{2}}$. Cylindrical polars: (6.39) has $\partial\psi/\partial\rho = \sin 2\phi - \rho^{-2}z^2$, $\rho^{-1}\partial\psi/\partial\phi = 2\cos 2\phi$, $\partial\psi/\partial z = 2\rho^{-1}z$. Spherical polars: $\psi = r[\sin\theta\sin 2\phi + \cos\theta\cot\theta]$. Components given by (6.40).

2 By (6.40), grad $\psi = -2\mu r^{-3}\cos\theta\,\mathbf{t}_r - \mu r^{-3}\sin\theta\,\mathbf{t}_\theta$. Small variations $\delta r$, $\delta\theta$ on level surface $\psi =$ constant satisfy

$$\delta r\;\partial\psi/\partial r + \delta\theta\;\partial\psi/\partial\theta = 0$$

or

$$\delta r\;2\mu r^{-3}\cos\theta + r\,\delta\theta\;\mu r^{-3}\sin\theta = 0.$$

So a small displacement $\delta r\,\mathbf{t}_r + r\,\delta\theta\,\mathbf{t}_\theta$ on surface is orthogonal to grad $\psi$.

**Problems 6.3**

1
$$\begin{aligned}\text{curl }\mathbf{v} &= [8x^2y - 2z(x^2+y^2)]\mathbf{i} + 4xy(z-2y)\mathbf{j}\\ &= 2\rho^2(4\rho\sin\phi\cos^2\phi - z)\mathbf{i}\\ &\quad + 4\rho^2\sin\phi\cos\phi(z - 2\rho\sin\phi)\mathbf{j}.\end{aligned}$$

Substitute $\mathbf{i} = \mathbf{t}_\rho\cos\phi - \mathbf{t}_\phi\sin\phi$, $\mathbf{j} = \mathbf{t}_\rho\sin\phi + \mathbf{t}_\phi\cos\phi$.

2
$$\begin{aligned}\oint \mathbf{v}\cdot d\mathbf{s} &= \int_0^{\frac{1}{2}\pi} a\,d\theta\cdot e^{-a}\sin\theta + \int_0^{\frac{1}{2}\pi} a\,d\phi\\ &\quad - \int_0^{\frac{1}{2}\pi} a\,d\theta\,2^{-\frac{1}{2}}\,e^{-a}\sin\theta\\ &= a\,e^{-a}(1-2^{-\frac{1}{2}}) + \tfrac{1}{2}\pi a.\end{aligned}$$

$$(\text{curl }\mathbf{v})_r = (r^{-1}\sin 2\theta + \tfrac{1}{2}r^{-1}e^{-r}\sin\theta\sin\tfrac{1}{2}\theta)/\sin\theta.$$

$$\begin{aligned}&a^2\int_0^{\frac{1}{2}\pi}\sin\theta\,d\theta\int_0^{\frac{1}{2}\pi}d\phi\,(\text{curl }\mathbf{v})_r\\ &\quad = \int_0^{\frac{1}{2}\pi}d\theta\,[\tfrac{1}{2}\pi a\sin 2\theta + a\,e^{-a}\sin\theta(1-2^{-\frac{1}{2}})],\end{aligned}$$

equalling $\oint \mathbf{v}\cdot d\mathbf{s}$ on integration.

3
$$\begin{aligned}\text{curl }\mathbf{v} &= -(\rho\sin 2\phi + 2z)\mathbf{t}_\rho + 2\rho\cos\phi(1-\cos\phi)\mathbf{t}_\phi\\ &\quad + \rho^{-1}(3\rho^2 + z^2 + 2\rho z\sin\phi)\mathbf{t}_z.\end{aligned}$$

On $z = 0$, $\displaystyle\int_0^a \rho\,d\rho\int_0^{2\pi}d\phi\;\mathbf{t}_z\cdot\text{curl }\mathbf{v} = 2\pi\int_0^a \rho\,d\rho\;3\rho = 2\pi a^3.$

$$\begin{aligned}\mathbf{t}_r\cdot\text{curl }\mathbf{v} &= -(\rho\sin 2\phi + 2z)\sin\theta\\ &\quad + \rho^{-1}(3\rho^2 + z^2 + 2\rho z\sin\phi)\cos\theta\\ &= -r\sin\theta(\sin\theta\sin 2\phi + 2\cos\theta)\\ &\quad + r\cot\theta(1 + 2\sin^2\theta + \sin 2\theta\sin\phi).\end{aligned}$$

$$a^2 \int_0^{\frac{1}{2}\pi} \sin\theta \, d\theta \int_0^{2\pi} d\phi \, \mathbf{t}_r \cdot \text{curl}\, \mathbf{v}$$

$$= 2\pi a^3 \int_0^{\frac{1}{2}\pi} \sin\theta \, d\theta \, [-2 \sin^2\theta \cos\theta$$

$$+ \cot\theta \, (1 + 2\sin^2\theta)]$$

$$= 2\pi a^3, \text{ equal to integral over the disc.}$$

## Problems 6.4

3 For any constant $\mathbf{a}$, using (6.69) with $\text{div}\, \mathbf{a} = 0$,

$$\mathbf{a} \cdot \iint_\sigma d\boldsymbol{\sigma} \, \psi = \iiint_\tau d\tau \, \text{div}(\mathbf{a}\psi)$$

$$= \mathbf{a} \cdot \iiint d\tau \, \text{grad}\, \psi.$$

Using (6.71) with $\text{curl}\, \mathbf{a} = \mathbf{0}$,

$$\mathbf{a} \cdot \iint_\sigma d\boldsymbol{\sigma} \wedge \mathbf{v} = \iint d\boldsymbol{\sigma} \cdot (\mathbf{v} \wedge \mathbf{a})$$

$$= \iiint_\tau d\tau \, \text{div}(\mathbf{v} \wedge \mathbf{a})$$

$$= \mathbf{a} \cdot \iiint_\tau d\tau \, \text{curl}\, \mathbf{v}.$$

# INDEX